BASIC MATHEMATICS

a review

JAMES ROGERS
JAMES VanDYKE
JACK BARKER

Department of Mathematics,
Portland Community College,
Portland, Oregon

W. B. SAUNDERS COMPANY
Philadelphia • London • Toronto

W. B. Saunders Company: West Washington Square
Philadelphia, PA 19105

1 St. Anne's Road
Eastbourne, East Sussex BN21 3UN, England

1 Goldthorne Avenue
Toronto, Ontario M8Z 5T9, Canada

Library of Congress Cataloging in Publication Data

Rogers, James V.
    Basic mathematics.

    1. Mathematics—1961-   I. Van Dyke, James,
joint author. II. Barker, Jack, joint author.
III. Title.
QA39.2.R65   510   77-77107
ISBN 0-7216-7633-2

Basic Mathematics: A Review   ISBN 0-7216-7633-2

Last digit is the print number: 9   8   7   6   5   4   3

PREFACE

APPLICATION:

A review text for students who have had an exposure to arithmetic and algebra. This exposure probably took place in high school a few years prior to their entry into college. College placement points out the need for a review before the student can enter a vocational program or a more advanced mathematics course.

OBJECTIVE:

To provide a write-in text for students who need to review the skills of basic mathematics as a prerequisite for business mathematics, advanced mathematics or technical mathematics.

VOCABULARY:

A *review text* is one that allows a student to quickly recall and practice a skill as opposed to one that offers an in-depth presentation of the topic. *Modularized material* allows for an instructor to choose sections of material for use in a classroom situation. *Individualized units* are keyed to unit tests which can be used to determine which objectives need additional attention.

THE HOW AND WHY OF IT:

The material is divided into *twelve* major units with objectives explicitly stated. The first three chapters are devoted to arithmetic. Chapter Four covers denominate numbers and measurement. Chapters Five through Eleven review algebra, including a treatment of ratio and percent. The review of percent from the algebraic viewpoint has proved quite successful in classroom and mathematics laboratory use. The final chapter is an introduction to trigonometry.

Each section begins with an application drawn from the world of experience or related to a section in the text. These applications demonstrate the usefulness of the skill and serve to motivate the student's interest in the topic.

Each skill expected of the student is stated in the form of one or more behavioral objectives.

The technical vocabulary to be used is defined, described, or used in context. The text is written with a minimum of rigor, and the use of formal mathematical vocabulary is limited as much as possible. For example, there are no objectives stated in terms of the associative, commutative, or distributive laws.

An explanation of how the process is performed and why it works is included. Emphasis is placed on the "how," since it is a review text. The student is "talked" through the process and most sections end with a simple rule for quick recall of the procedure.

EXAMPLES:

Several examples are worked out in detail so students can follow the process step by step. This allows them to become comfortable with the process before trying it on their own.

EXERCISES:

There are over 2700 exercises in the text, of which approximately 15 percent are applications. The exercises are usually divided into three categories. In most sections the first ten problems can be used as oral exercises in class so that the instructor can check on how well the skill has been learned. These are followed by a set of more difficult problems to be used by the student to gain proficiency in using the skill. The final group consists of applications of the skill drawn from experiences available in today's world. These applications are placed in each section so the student can study them as they relate to the skill being practiced. This avoids a section devoted entirely to applications with no apparent linkage to the preceding skills. The applications, when placed in this manner, serve to promote the transfer of learning to the area in which the student will be applying the mathematical skill.

TIMETABLE:

The text can be used in a variety of classroom situations, depending upon the needs of the students. Four such possibilities follow:

1. A three quarter or two semester course for those students reviewing the basic skills prior to enrolling in a technical math-

ematics program. In such a course all of the topics would normally be covered.

2. A two quarter or one semester course for vocational students. Such a course would cover Chapters One through Six and Eight. This omits the more advanced topics of algebra and the unit on trigonometry.

3. A one quarter or one semester course for those students preparing for business mathematics or business courses. This course would cover Chapters One through Four and Eight, along with some selected objectives in algebra.

4. A one quarter or one semester course as a review prior to beginning a college algebra program. This course would cover Chapters Five through Seven and Nine through Twelve.

The authors are appreciative of their wives, Elinore Rogers, Carol Van Dyke, and Mary Barker, without whose unfailing patience the work would never have reached completion. Special thanks are due to John Snyder and William Karjane of W. B. Saunders Co. for their suggestions and support during the development of this text.

Acknowledgment is given for the excellent contributions to the text by the following reviewers: Jean Newton, St. Petersburg Junior College; Richard DeTar, Sacramento City College; Albert Liberi, Westchester Community College; Robert Riner, Ohio State University.

Thanks are also due to Claire Simpson and Bina Bate for fine typing under deadlines.

JAMES ROGERS
JAMES VAN DYKE
JACK BARKER

CONTENTS

CHAPTER 1

WHOLE NUMBERS

1.1   WHOLE NUMBERS: PLACE VALUE AND WORD NAMES

The purchasing agent for the Upjohn Corporation received a telephone bid of twenty-three thousand eighty-one dollars as the price of a new printing press. What is the numeral form of the bid?   APPLICATION

1) Write numeral form from word names.   OBJECTIVES
2) Write word names from numeral form.

The *digits* are 0, 1, 2, 3, 4, 5, 6, 7, 8, and 9. These are the symbols   VOCABULARY
(numerals) that name the whole numbers from zero through nine. Numbers larger than nine are written in *numeral form* (492) by placing the digits (4, 9, and 2) in a certain order according to the standard *place value*.

<table>
<tr><td>hundred's
place</td><td>ten's
place</td><td>one's
place</td></tr>
</table>

Words, spoken or written, that represent numbers are called *word names* (four hundred ninety-two).

In our number system (called the Hindu-Arabic system), the digits   THE HOW AND
are the only symbols used besides commas. This system is also called a   WHY OF IT
base ten (decimal) system. From right to left, the first three place value names are one, ten, and hundred.

hundred	ten	one

FIGURE 1.1

For instance, in 573,   3 has place value one   (1)
7 has place value ten   (10)
5 has place value hundred (100)

Continuing to the left, the digits are grouped in threes. The first five groups are named (from right to left) unit, thousand, million, billion, and trillion. The group on the extreme left may contain one, two, or three digits, while all other groups *must* contain three digits. Within each group the names are the same.

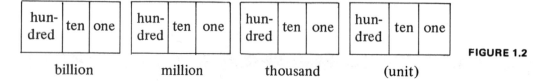

hun-dred	ten	one	hun-dred	ten	one	hun-dred	ten	one	hun-dred	ten	one
billion			million			thousand			(unit)		

FIGURE 1.2

The place value of any position is one, ten, or hundred, followed by the group name.

> Any whole number can be written by writing the word name for the number represented by the digits in a group of three followed by the group name. The "units" group name is usually not read or written in naming a number.

So 34,573 is written

	3	4
	thousand	

5	7	3
	(unit)	

thirty-four thousand, five hundred seventy three

> The numeral form of a number can be written from the word name by writing the digits in each group of three and replacing the group names with commas.

EXAMPLES

a) Consider the number whose numeral form is 17,698,453.

Place value notation:	17,	698,	453
Word name of number in each group:	seventeen,	six hundred ninety-eight,	four hundred fifty-three
Group name:	million,	thousand,	(unit)
Word name:	Seventeen million, six hundred ninety-eight thousand, four hundred fifty-three.		

b) Reversing the process, we have:

Word name: Two million, two hundred thirty-seven thousand, five hundred sixty-four.

group
name

group
name

(The group name, unit, is understood.)

Numeral form: 2,   2 3 7,   5 6 4

million   thousand   (unit)

Write (or read) the word names of these numbers.

1) 2502   2) 2520   3) 2052   4) 25,200   5) 252,000

Write (or read) the numeral form of these numbers.

6) three hundred six   7) three hundred sixty

8) three thousand six   9) three thousand sixty

10) three hundred sixty thousand

Write the word names of these numbers.

11) 542   12) 540

13) 504   14) 5042

15) 500,042   16) 50,050,500

17) 502,520,052   18) 5,713,522,117

Write the numeral form for these numbers.

19) Two hundred forty-three.

20) Two hundred forty-three thousand, seven.

21) Two hundred three thousand, forty-seven.

22) Four million, two thousand, seven hundred.

23) Four hundred six million, two hundred forty-two thousand, seven hundred thirteen.

24) Three hundred fifteen million, five hundred seventy-two.

25) Six million, six hundred six.

26) Fifty-four billion, fifty-five million, fifty-six thousand, fifty-seven.

27) In the application at the beginning of this section, what is the numeral form of the bid?

28) Michelle is paid 15¢ per mile when she uses her car for company business. She must report the mileage she has driven her auto in both numeral and word form. If she drives 785 miles during the month of June, what is the word name she should write for the mileage?

29) The world population during 1976 exceeded four billion. Write the number four billion in numeral form.

1.2 ADDITION AND SUBTRACTION OF WHOLE NUMBERS

APPLICATION

Firebird Auto is offering a $366 discount on all new cars priced above $5000. What is the cost of a car priced at $6785?

OBJECTIVES

3) Find the sum of two or more whole numbers.
4) Find the difference of two whole numbers.

Recall that a natural number is any number from the group 1, 2, 3, **VOCABULARY** 4, 5, . . . Notice that 0 is a whole number but not a natural number.

Words used with addition and subtraction problems are:

$$\left.\begin{array}{r}16 \\ 23 \\ \underline{7}\end{array}\right\} \text{addends} \qquad \begin{array}{r}35 \leftarrow \text{minuend} \\ \underline{18} \leftarrow \text{subtrahend} \\ 17 \leftarrow \text{difference}\end{array}$$

$$46 \leftarrow \text{sum}$$

To perform addition and subtraction, there are certain facts that **THE HOW AND** must be memorized. The facts are commonly listed in an "addition table," **WHY OF IT** which shows the sums of all possible combinations of two single-digit numbers. It is assumed from this point on that you know these facts.

> To add two or more whole numbers, list the numbers in columns so that all digits which have the same place value are lined up. Find the sum of the digits in each column, starting with the units digits and working to the left. If any sums are greater than 9, record the units digit and carry the number named by the remaining digit(s) to the next column.

To subtract 2 from 8 asks us to find the number that, when added to 2, gives a sum of 8. That is, $8 - 2$ wants to know $2 + ? = 8$. Since $2 + 6 = 8$, then $8 - 2 = 6$.

What is $47 - 15$? Since $15 + 32 = 47$, this tells us that $47 - 15 = 32$. There is a second subtraction problem associated with $15 + 32 = 47$. Do you see that this tells us that $47 - 32 = 15$?

> To subtract whole numbers, we must always subtract digits with the same place value (in the same column). When digits with the same place value (in the same column) cannot be subtracted, we must regroup by borrowing from the place value column(s) to the left of the column in which we are working.

EXAMPLES

a) **Add:**   Think:

121	
1682	ones column $2 + 1 + 9 = 12$, carry the "1" to the tens column
4491	tens column $1 + 8 + 9 + 2 = 20$, carry the "2" to the hundreds column
$\underline{7629}$	
13802	hundreds column $2 + 6 + 4 + 6 = 18$, carry the "1" to the thousands column
	thousands column $1 + 1 + 4 + 7 = 13$; since there are no digits in ten thousands column, no carry is necessary; record the answer.

b) 48 + 232 + 4 + 2834 = ?

We write the problem vertically, making sure to line up the place values correctly.

```
  111
   48
  232
    4
 2834
 3118
```

c) Subtract: Think:

```
  752
  295
```

We must do two regroupings. We cannot subtract 5 ones from 2 ones, and we cannot subtract 9 tens from 5 tens. Therefore, we borrow 1 ten (10 ones) from the tens to regroup with the ones (for a total of 12 ones) and then borrow 1 hundred (which is 10 tens) from the hundreds to regroup with the tens (for a total of 14 tens).

```
 6 14
   4̸ 12
 7̸ 5̸ 2̸
 2  9  5
 4  5  7
```

A check on the answer can be made by seeing whether the sum of 295 and 457 is 752.

```
Check:    295
          457
          752
```

EXERCISES   Add.

1) 37 + 6   2) 28 + 5   3) 16 + 8   4) 124 + 7

5) 130 + 29

Subtract.

6) 39 − 7   7) 37 − 9   8) 93 − 7   9) 79 − 23

10) 73 − 14

Add.

11) 93
     42
     71

12) 75
     94
     83
     20
     37

13) 94
     87
     20
     43
     91
     89

14) 419
     638
     215
     482

15) 654
     382
     197
     846
     321

16) 491
     309
     380
     976
     621

17) 1684
     2873
     4091
     5004

18) 4629
     1083
     2194
     4083
     6219

19) 20000
     60904
      9130
        72

20) $48 + 271 + 4 + 1956$

21) $28004 + 78 + 987 + 1435$

22) $13 + 8271 + 4082 + 16752$

Subtract each of the following.

23) 2184
      998

24) 6802
     4765

25) 7002
     3654

26) 2704
      809

27) 40000
     18214

28) 7985
     2006

29) 46709
     30510

30) $4125 - 687$

31) $2008 - 193$

32) $6800 - 2002$

33) In the application at the beginning of this section, what is the cost of the car?

34) If there are 58275 tickets available for a certain baseball game, and all but 2863 are sold, how many tickets have been sold?

35) On Monday, Mr. John drove his car 182 miles, on Tuesday 364 miles, on Wednesday 235 miles, on Thursday 317 miles and on Friday 293 miles. What was the total mileage driven in the five days?

36) If Car *A* costs $597 more than Car *B*, what is the price of Car *A* if the price of Car *B* is $2987?

1.3 MULTIPLICATION AND DIVISION OF WHOLE NUMBERS

APPLICATION   If the Mighty Good Dog Food Company puts a dozen tins of dog food in a case, how many cases and how many extra tins will there be in a lot of 1069 tins?

OBJECTIVES
5) Find the product of two whole numbers.
6) Divide two whole numbers.

VOCABULARY   Words used with multiplication and division problems are:

$$(3)(17) = 51$$

factors   product

$$\left.\begin{array}{r} 291 \\ 3 \end{array}\right\} \text{factors}$$
$$\overline{873} \leftarrow \text{product}$$

$$\begin{array}{r} 29 \leftarrow \text{quotient} \\ 16\,\overline{)464} \end{array}$$

divisor   dividend

partial quotient   remainder

$$\begin{array}{r} 31 \quad \text{R 3} \\ 16\,\overline{)499} \end{array}$$

divisor   dividend

Multiplication is often thought of as a short method for addition. $6 \cdot 8$ or 6×8 or $(6)(8)$ asks for the sum of six eights. You can now find the sum, $8 + 8 + 8 + 8 + 8 + 8 = 48$, or $6 \cdot 8 = 48$.

To perform multiplication without repeated addition, there are certain facts that should be memorized. These facts are commonly listed in a "multiplication table," which shows the products of all possible combinations of two single-digit numbers. It is assumed from this point on, that you know these facts.

Using the multiplication facts along with the concept of place value, the product can be found as follows:

$$
\begin{array}{rl}
629 & \\
\underline{46} & \\
54 & \text{6 times 9} \\
120 & \text{6 times 20} \\
3600 & \text{6 times 600} \\
360 & \text{40 times 9} \\
800 & \text{40 times 20} \\
\underline{24000} & \text{40 times 600} \\
28934 &
\end{array}
$$

As you get better at this method of multiplying, you might want to take some shortcuts. The usual shortcut is to find the sum (mentally) of all the numbers multiplied by 6 and the sum of all the numbers multiplied by 40. This is shown below:

$$
\begin{array}{r}
13 \\
1\cancel{8} \\
629 \\
\underline{46} \\
3774 \\
\underline{25160} \\
28934
\end{array}
$$

$(24)(?) = 288$ is the same as $288 \div 24 = ?$ From this idea, division is referred to as the inverse of multiplication or as a process of undoing multiplication.

Recall that multiplication can be interpreted as summing up a number of groups. Division then is the process of discovering how many groups were summed.

The most common difficulty is that the maximum number of ones, tens, hundreds or whatever must be guessed correctly at each step of the way.

This process is shown in the following example:

$$
\begin{array}{r}
745 \\
23\overline{)17135} \\
16100 \\
\overline{1035} \\
920 \\
\overline{115} \\
115 \\
\overline{0}
\end{array}
$$

Think: One group of 1000 twenty-threes is 23000, which is more than 17135; so there are no groups of 1000 twenty-threes. Try 7 one hundred groups of twenty-three or 700 × 23 = 16100. Subtract this from 17135. Since the difference is less than 100 × 23 = 2300, 7 is the maximum number of 100 groups. Now try 4 ten groups of twenty-three or 40 × 23 = 920. Subtract 920 from 1035. Since the difference is less than 10 × 23 = 230, 4 is the maximum number of ten groups. Continuing, try 5 groups of 23 or 5 × 23 = 115. Subtract 115 from 115. Since the difference is 0, the division is complete. Checking, (745) (23) = 17135, so the answer is correct.

Not all division problems involving whole numbers have a divisor (factor) that is summed a whole number of times. In

$$
\begin{array}{r}
3 \\
12\overline{)41} \\
36 \\
\overline{5}
\end{array}
$$

we see that 41 contains 3 twelves and 5 toward the next group of twelve. The answer is now written as 3 remainder 5.

The word "remainder" is usually abbreviated "R," and then the result is

$$41 \div 12 = 3 \text{ R } 5$$

A check can be made by taking (12) (3) and adding the remainder: (12) (3) + 5 = 41.

EXAMPLES

a)   0·37 = 0

b)   38·74

$$
\begin{array}{r}
1 \\
\cancel{3} \\
74 \\
38 \\
\overline{592} \\
2220 \\
\overline{2812}
\end{array}
$$

c)   601

    7)4207

      4200

          7

          0

          7

          7

          0

(Note that a "0" must be placed above the tens digit if the place values of the 6 and 1 are to be correct.)

Check: (7) (601) = 4207

d)   30 R 47   Check: 55

    55)1697   30

      1650   00

        47   165

                        1650

                          47

                      1697

EXERCISES

Multiply.

1)   48
     1

2)   14
     2

3)   122
      4

4)   18
     3

5)   64
     0

Divide.

6) 7)70

7) 7)147

8) 7)714

9) 7)78

10) 7)359

Multiply.

11)   86
     9

12)   25
     16

13)   47
     39

14)   469
      5

15)   723
     26

16)   837
     49

17) (400)(93)

18)   3218
       9

19) 756
    359

20) 4009
    908

21) 6125
    2234

Divide.

22) $6\overline{)372}$

23) $7\overline{)147}$

24) $8\overline{)256}$

25) $33\overline{)1273}$

26) $68\overline{)2614}$

27) $20\overline{)697}$

28) $47\overline{)14476}$

29) $18\overline{)2880}$

30) $44080 \div 76$

31) $408\overline{)126075}$

32) $507\overline{)97051}$

33) $903\overline{)3613816}$

34) In the application at the beginning of this section, how many cases and how many extra tins are in the lot?

35) The Good Food Grocery Store ordered 234 cases of pears. If each case cost 7 dollars, what was the total cost of the pears?

36) Ellie ordered three hundred twenty-two 8-packs of Hot'n Cola for sale at her store. How many bottles will there be?

37) If a contractor can build a house in 27 days, how many houses can be completed in 297 days?

38) The estate of Norm Barker totaled $27,835. It is to be shared equally by his five nephews. How much money will each nephew receive?

39) If it takes 33 bales of hay to make a ton, how many tons and how many extra bales are there in a field that yields 3691 bales?

1.4 EXPONENTS

A Congressional committee proposes to increase the national debt by 13×10^9 dollars. Write this amount in numeral form. **APPLICATION**

7) Identify the exponent when a number is written in exponential form. **OBJECTIVES**
8) Identify the base when a number is written in exponential form.
9) Find the value of expressions written in exponential form.

3^4 is an exponential form of a number, where 4 is the *exponent* and 3 is the *base*. The exponent shows how many times the base is used as a factor. The *power* or *value* is the result of the multiplication. **VOCABULARY**

A *natural* or *counting* number is a number used in the ordinary process of counting (1,2,3,4,5, and so on). *Whole numbers* include both counting numbers and the number zero (0,1,2,3,4, . . .).

Factors are numbers used as multipliers. In the expression 6×5 or $6 \cdot 5$, 6 and 5 are factors.

The exponential form of writing a number is often used to write certain multiplication problems in a shorter form. **THE HOW AND WHY OF IT**

Any whole number can be used as an exponent.

> When a whole number greater than one is used as an exponent, it indicates how many times the base is used as a factor.

For example, $3^4 = 3 \cdot 3 \cdot 3 \cdot 3$. The value of 3^4 is 81, since $3 \cdot 3 \cdot 3 \cdot 3 = 9 \cdot 3 \cdot 3 = 27 \cdot 3 = 81$, and 81 is sometimes referred to as the fourth power of three.

$$\text{base} \longrightarrow 3^{\overset{\displaystyle \text{exponent}}{\downarrow}4} \quad = 81 \longleftarrow \text{value}$$

If 1 is used as the exponent, the number named is equal to the base.

That is, $7^1 = 7$.
If 0 is used as the exponent, the base cannot be 0.

By definition, any natural number with 0 for the exponent is 1.

For example, $2^0 = 1$, $11^0 = 1$, and $6^0 = 1$.

EXAMPLES

a) 2^3 The exponent is 3.
 The base is 2. 2^3 means $2 \cdot 2 \cdot 2$
 The value is 8. $2^3 = 8$

b) 8^1 The exponent is 1.
 The base is 8. 8^1 means 8
 The value is 8. $8^1 = 8$

c) 3^5 The exponent is 5.
 The base is 3. 3^5 means $3 \cdot 3 \cdot 3 \cdot 3 \cdot 3$
 The value is 243. $3^5 = 243$

d) 7^0 The exponent is 0.
 The base is 7. 7^0 means 1
 The value is 1. $7^0 = 1$

EXERCISES   Identify the exponent of each expression.

1) 2^4 2) 5^6 3) 10^4

4) 2^0 5) 1^6 6) 9^1

Identify the base of each expression.

7) 6^2 8) 0^7 9) 7^5

10) 10^5 11) 13^0 12) 4^{21}

Find the value of each expression.

13) 3^2

14) 4^3

15) 1^9

16) 5^1

17) 10^4

18) 10^6

19) 10^8

20) 99^0

21) In the application at the beginning of this section, what is the numeral form of the increase in the national debt?

22) A western state had a fiscal surplus of 16×10^7 dollars for a given year. Write this number of dollars in both numeral and word form.

23) The distance from earth to the nearest star (Alpha Centauri) is approximately 255×10^{11} miles. Write this distance in numeral form.

1.5 ORDER OF OPERATIONS (PROBLEMS WITH TWO OR MORE OPERATIONS)

APPLICATION

The cash price for a used car is $2795. If Mr. Macy pays $500 down and makes 30 payments of $99 each, the full price he pays is given by $P = 500 + 99(30)$. Is this more or less than the cash price? How much is the difference?

OBJECTIVE

10) Perform any combination of operations (addition, subtraction, multiplication, and/or division) on whole numbers, in the conventional order.

VOCABULARY   Parentheses ( ) and brackets [ ] are used in mathematics as grouping symbols. These symbols indicate that the operations inside are to be performed first. Thus,

$$(24 \div 4) \div 2 = 6 \div 2 = 3$$

$$\text{and } 24 \div (4 \div 2) = 24 \div 2 = 12$$

THE HOW AND WHY OF IT   If we are asked to work the problem $2 \cdot 3 + 4$, it is possible to get two answers. If we do the multiplication first and then the addition, we get $2 \cdot 3 + 4 = 6 + 4 = 10$. If we do the addition first and then the multiplication, we have $2 \cdot 3 + 4 = 2 \cdot 7 = 14$.

If we are asked to work the problem $16 \div 2 + 6$, again it is possible to get two answers. That is, we could work it as follows: $16 \div 2 + 6 = 8 + 6 = 14$ or $16 \div 2 + 6 = 16 \div 8 = 2$.

The order in which the operations are performed is important. Because of this, an agreement has been made. It can be stated as follows:

> In any problem with two or more operations and no grouping symbols, start from the left, work to the right doing *only* the multiplications and divisions as you come to them, and then go back to the left and work to the right doing the additions and subtractions. If there are grouping symbols, do what is inside the symbols first, following the above convention, and then proceed as before.

Neither multiplication nor division takes preference over the other. They are performed in the order in which they appear. The same is true with regard to addition and subtraction.

EXAMPLES

a) $7 \cdot 9 + 6 \cdot 2 = 63 + 12 = 75$

b) $25 - 6 \div 3 + 8 \cdot 4 = 25 - 2 + 32 = 23 + 32 = 55$

c) $5 \cdot 9 + 9 - 6 (7 + 1) = 5 \cdot 9 + 9 - 6 \cdot 8 = 45 + 9 - 48 = 54 - 48 = 6$

d) $45 \div 9 - 2 + 28 \div 14 + 10 = 5 - 2 + 2 + 10 = 3 + 2 + 10 = 5 + 10 = 15$

e) $(4 + 7 + 9) \div (8 - 3) = (11 + 9) \div 5 = 20 \div 5 = 4$

Perform the indicated operations.

1) $5 \cdot 6 + 2$

2) $5 + 6 \cdot 2$

3) $20 \div 4 - 2$

4) $20 - 4 \div 2$

5) $20 \div 4 + 1$

6) $20 \div (4 + 1)$

7) $12 \div 6 \cdot 2$

8) $12 \cdot 6 \div 2$

9) $16 \div 2 + 3 \cdot 2$

10) $16 \div (2 + 3 \cdot 2)$

11) $4 \cdot 7 + 12$

12) $15 + 9 \cdot 2$

13) $2 \cdot 9 \div 3$

14) $19 - 3 \cdot 2$

15) $14 \cdot 2 + 5 \cdot 3$

16) $12 \cdot 3 - 5 \cdot 4$

17) $12(4 + 2) - 24 \div 12$

18) $12 \div 3 \cdot 2$

19) $24 - 9 \cdot 2 + 6 \cdot 3$

20) $(5 \cdot 2 + 3) \cdot 2 - 10$

21) $8 \div 2 - 2 + 9 \cdot 2 - 8$

22) $120 \div (5 \cdot 5 - 5)$

23)   In the application at the beginning of this section which method is the most expensive? How much is the difference?

24)   On March 1 the balance in a checking account was $432. During the month a check was written for $63, three checks for $18 and a check for $89. There was one deposit of $22 and a service charge of $1. The balance at the end of the month can be calculated by finding the value of
$$432 - 63 - 3(18) - 89 + 22 - 1.$$
What was the balance at the end of the month?

25)   Charles Mitchell is planning to carpet two rooms in his home. One has dimensions 15 feet by 15 feet, the other 25 feet by 17 feet. The number of square feet of carpet he needs can be calculated by finding the value of
$$15 \cdot 15 + 25 \cdot 17.$$
How many square feet of carpet does he need?

26)   Greg wants to calculate the total resistance in a power supply circuit in a television set. The set has two resistors in parallel. If one resistor is 2400 ohms and the second is 1200 ohms, the total resistance can be calculated by finding the value of
$$R_t = 2400 \div (2400 \div 1200 + 1)$$
What is the total resistance in the circuit?

27)   To check that $x = 14$ is a solution to the equation
$$125 - 6x + 4 = 45,$$
we substitute 14 for x and find the value of
$$125 - 6(14) + 4.$$
Verify that the above expression equals 45.

28) Substitute $x = 24$ in the following equation and check that it is a solution (makes the equation true).

$$\frac{x}{8} + \frac{x}{12} + 4x = 101$$

that is, $x \div 8 + x \div 12 + 4x = 101$.

1.6 DIVISORS AND FACTORS

A television station has 130 minutes of programming to fill. In how many ways can this time be scheduled if all programs are to be the same length and a whole number of minutes? (For example, ten programs each thirteen minutes in length, $10 \cdot 13 = 130$.) APPLICATION

11) List all factors (divisors) of a given whole number. OBJECTIVES
12) Write a whole number as the product of two factors in as many ways as possible.

Remember that 6 is a factor of 24 since $(6)(4) = 24$. 6 is also called a VOCABULARY
divisor of 24 since $24 \div 6 = 4$. Another way of saying this is that 24 is *divisible* by 6.

When asked to list all the factors of 250, you are asked to find all THE HOW AND
whole numbers that are divisors of 250. Since $250 \div 10 = 25$, 10 is a WHY OF IT
factor (divisor) of 250. Rather than continue at random, the following process will assure us of obtaining all such factors. It will also express 250 as a product of two factors in as many ways as possible. First, write down the natural numbers from 1 up to a natural number whose square (second power) is larger than 250.

1	6	11	16	Since $16 \cdot 16 = 256$, we stop here.
2	7	12		
3	8	13		
4	9	14		
5	10	15		

If any of these numbers divide 250, that number and the quotient are factors of 250. If 16 or any other number larger than 16 is a factor of 250, it will show up as:

(a number less than 16) (that number) = 250.

For example, 25 is a factor, and it shows up when you discover that 10 is a factor. Of course, if a number does not divide 250, it is not a factor and we can cross out that number.

$$
\begin{array}{llll}
1 \cdot 250 & \cancel{6} & \cancel{11} & \cancel{16} \\
2 \cdot 125 & \cancel{7} & \cancel{12} & \\
\cancel{3} & \cancel{8} & \cancel{13} & \\
\cancel{4} & \cancel{9} & \cancel{14} & \\
5 \cdot 50 & 10 \cdot 25 & \cancel{15} &
\end{array}
$$

We now have a complete listing of all factors of 250, namely, 1, 2, 5, 10, 25, 50, 125, and 250. We also have all possible ways to express 250 as a product of two factors: 1·250, 2·125, 5·50, and 10·25.

There are other ways that you can express 250 as a product involving more than two factors. One of these methods will be of interest to us in another section.

> To find all the factors of a given number, write down the natural numbers from 1 up to the number whose square is just larger than the given number. If any of these numbers divide the given number, it and the quotient are factors of the given number.

EXAMPLES

a) The factors of 68 can be found as follows:

$$
\begin{array}{lll}
1 \cdot 68 & 4 \cdot 17 & \cancel{7} \\
2 \cdot 34 & \cancel{5} & \cancel{8} \quad \text{Since } 9 \cdot 9 = 81, \text{ we stop here.} \\
\cancel{3} & \cancel{6} & \cancel{9}
\end{array}
$$

The list of factors is 1, 2, 4, 17, 34, 68. The list of products is 1·68, 2·34, and 4·17.

b) The factors of 180 are:

$$
\begin{array}{lll}
1 \cdot 180 & 6 \cdot 30 & \cancel{11} \\
2 \cdot 90 & \cancel{7} & 12 \cdot 15 \\
3 \cdot 60 & \cancel{8} & \cancel{13} \\
4 \cdot 45 & 9 \cdot 20 & \cancel{14} \quad \text{Since } 14 \cdot 14 = 196, \text{ we stop here.} \\
5 \cdot 36 & 10 \cdot 18 &
\end{array}
$$

The list of factors is 1, 2, 3, 4, 5, 6, 9, 10, 12, 15, 18, 20, 30, 36, 45, 60, 90, 180. The list of products is 1·180, 2·90, 3·60, 4·45, 5·36, 6·30, 9·20, 10·18, 12·15.

c)   The factors of 29 are:

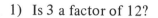

The list of factors is 1, 29. The list of products is 1·29.

1) Is 3 a factor of 12?

2) Is 4 a factor of 12?

3) Is 5 a factor of 12?

4) Is 6 a factor of 12?

5) Is 7 a factor of 7?

6) Is 22 a factor of 11?

7) List all the ways to write 15 as the product of two numbers.

8) List all the ways to write 16 as the product of two numbers.

9) List all the ways to write 17 as the product of two numbers.

10) List all the ways to write 18 as the product of two numbers.

List all the factors of each of the following whole numbers.

11) 50   12) 36   13) 40

14) 80   15) 96   16) 100

17) 101   18) 220   19) 111

20) 245

Write each of the following whole numbers as the product of two factors in as many ways as possible.

21) 100   22) 80   23) 96

24) 48   25) 128   26) 131

27) 847   28) 500   29) 720

30) 1311

31) In the application at the beginning of this section, how many ways can the programming take place?

1.7 PRIMES AND COMPOSITES

Primes can be used to find the lowest common denominator of frac-   APPLICATION
tions $\left(\dfrac{5}{56} + \dfrac{27}{64} \right)$ and to reduce some fractions $\left(\dfrac{456}{792} \right)$.

OBJECTIVE   13) Identify a given whole number as prime or composite.

VOCABULARY   A *prime number* is a whole number greater than one with exactly two different factors (divisors). These two factors are the number 1 and the whole number itself. Whole numbers greater than 1 with more than two different factors are called *composite numbers.*

THE HOW AND WHY OF IT   Primes and composites are one useful way to classify whole numbers. The whole numbers zero (0) and one (1) are neither prime nor composite. They are in a class by themselves.

Two (2) is the first prime number ($2 = 2 \cdot 1$), since 2 and 1 are the only factors of 2.

Four (4) is a composite number ($4 = 4 \cdot 1$ and $4 = 2 \cdot 2$), since it has more than two factors. Four has the three factors 1, 2, and 4. It is possible to study each whole number and list its factors or divisors to see if it has exactly two or more than two factors.

> To tell whether a number is prime or composite, make a table of possible divisors and try each one. If the number has exactly two divisors, it is prime.

To determine if 51 is prime or composite, make a table of possible divisors and try each one.

$$1 \cdot 51 \qquad \cancel{4} \qquad \cancel{7}$$
$$\cancel{2} \qquad \cancel{5} \qquad \cancel{8}$$
$$3 \cdot 17 \qquad \cancel{6}$$

The factors of 51 are 1, 3, 17, and 51. Since it has four factors, it is a composite number.

EXAMPLES

a)   31 is a prime number since it has exactly two factors (1 and 31).

b)   101 is a prime number since it has exactly two factors (1 and 101).

c)   91 is composite since it has more than two factors (1, 7, 13, and 91).

d)   Is 323 prime or composite? By division 2, 3, 5, 7, 11, and 13 are not factors. Note that we need only test prime divisors, since composite numbers include prime factors. Next try 17. Since $323 \div 17 = 19$, 323 has more than two factors (1, 17, 19, 323) and is therefore composite.

e)   Is 331 prime or composite? By division 2, 3, 5, 7, 11, 13, 17, and 19 are not factors. Since 19^2 is larger than 331, it is fruitless to try larger primes; therefore, 331 is prime.

Identify each of these numbers as prime or composite:

1) 12   2) 13

3) 14   4) 15

5) 16   6) 17

7) 27   8) 37

9) 47   10) 57

11) 97   12) 197

13) 297   14) 397

15) 497   16) 597

17) 697   18) 797

19) 897   20) 997

21) 1097   22) 1197

1.8 PRIME FACTORIZATION

APPLICATION   Prime factorization will be used to reduce fractions, to find the least common multiple of numbers, and to find the lowest common denominator of fractions.

OBJECTIVE   14) Write the prime factorization of a given whole number.

VOCABULARY   The *prime factorization* (prime factored form) of a number is the number written as a product of primes. A number is said to be *completely factored* when it is in prime factored form $(18 = 2 \cdot 3 \cdot 3)$.

THE HOW AND   To express a number as a product of primes, try to divide the
WHY OF IT   number by each prime in consecutive order starting with 2. When a prime divides the number, the quotient is then divided by the same prime if possible. Continue the process until the quotient is one. The indicated product of all the primes that are divisors is the prime factorization of the whole number.

For instance:

$$
\begin{array}{r|l}
2 & 48 \\
2 & 24 \\
2 & 12 \\
2 & 6 \\
3 & 3 \\
& 1
\end{array}
$$

$48 = 2 \cdot 2 \cdot 2 \cdot 2 \cdot 3 = 2^4 \cdot 3$

If a whole number or any of the resulting quotients is a large prime, you will recognize this when you have tried each prime whose square is smaller than the number and found no divisor. Then divide by the larger prime to complete the process. For instance:

$$
\begin{array}{r|l}
3 & 1179 \\
3 & 393 \\
131 & 131 \\
& 1
\end{array}
$$

131 is not divisible by 2, 3, 5, 7, or 11; also, $13 \cdot 13 = 169$. Therefore, 131 is prime.

$1179 = 3 \cdot 3 \cdot 131 = 3^2 \cdot 131$

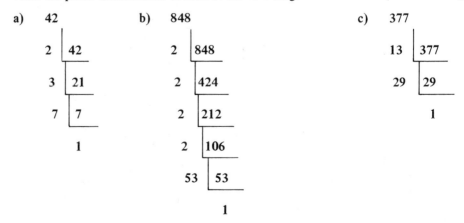

EXAMPLES

Write the prime factorization of each of the following.

a) 42

$$
\begin{array}{r|l}
2 & 42 \\
3 & 21 \\
7 & 7 \\
& 1
\end{array}
$$

b) 848

$$
\begin{array}{r|l}
2 & 848 \\
2 & 424 \\
2 & 212 \\
2 & 106 \\
53 & 53 \\
& 1
\end{array}
$$

c) 377

$$
\begin{array}{r|l}
13 & 377 \\
29 & 29 \\
& 1
\end{array}
$$

$42 = 2 \cdot 3 \cdot 7$

$848 = 2 \cdot 2 \cdot 2 \cdot 2 \cdot 53 = 2^4 \cdot 53$

$377 = 13 \cdot 29$

EXERCISES   Write the prime factorization of each of the following whole numbers.

1) 8   2) 9   3) 10   4) 11

5) 12   6) 15   7) 18   8) 21

9) 22   10) 27   11) 30   12) 36

13) 48   14) 50   15) 64   16) 51

17) 71   18) 91   19) 97   20) 100

21) 120   22) 150   23) 198   24) 325

25) 414   26) 768   27) 563   28) 888

29) 512   30) 657   31) 818   32) 891

33) 1428   34) 2000   35) 3672

1.9 LEAST COMMON MULTIPLE

Find the least common multiple of the denominators of the following fractions:
$$\frac{5}{6}, \frac{2}{9}, \frac{7}{12}, \text{ and } \frac{5}{18}.$$
The least common multiple can be used in adding fractions.

APPLICATION

15) Find the least common multiple of two or more whole numbers.

OBJECTIVE

A multiple of a number is the product of that number and a natural
number.

The *least common multiple* of two or more whole numbers is the
smallest natural number that is a multiple of each of the given numbers.
LCM is an abbreviation of Least Common Multiple. The LCM of 4 and 6 is
12.

VOCABULARY

Find the LCM of 30 and 42. This can be done by listing the multiples
of each and finding the smallest one which is in each group of multiples.

**THE HOW AND
WHY OF IT**

Multiples of 30 = 30, 60, 90, 120, 150, 180, $\boxed{210}$, 240, 270, . . .

Multiples of 42 = 42, 84, 126, 168, $\boxed{210}$, 252, 294, . . .

The LCM of 30 and 42 is 210. Finding the LCM in this manner has a
drawback. You might have to list hundreds of multiples before finding a
common multiple. For this reason, we look at a second method for finding the LCM.

> Write the prime factorizations of the numbers in exponential form. The LCM is the product of the highest power of each prime factor found in the numbers.

Consider the LCM of 40 and 48. First write the prime factorization of each number in exponential form.

$$48 = 2 \cdot 2 \cdot 2 \cdot 2 \cdot 3 = 2^4 \cdot 3^1$$
$$40 = 2 \cdot 2 \cdot 2 \cdot 5 = 2^3 \cdot 5^1$$

3^1 is the only power of 3, therefore the highest.
5^1 is the only power of 5, therefore the highest.
Of 2^3 and 2^4, 2^4 is the highest power. Therefore, LCM =
$$2^4 \cdot 3^1 \cdot 5^1 = 240.$$

EXAMPLES

a)   What is the LCM of 12 and 20?

$12 = 2 \cdot 2 \cdot 3 = 2^2 \cdot 3^1$
$20 = 2 \cdot 2 \cdot 5 = 2^2 \cdot 5^1$
LCM $= 2^2 \cdot 3^1 \cdot 5^1 = 60$

b)   Find the LCM of 18, 24, and 30.

$18 = 2 \cdot 3 \cdot 3 \quad = 2^1 \cdot 3^2$
$24 = 2 \cdot 2 \cdot 2 \cdot 3 = 2^3 \cdot 3^1$
$30 = 2 \cdot 3 \cdot 5 \quad = 2^1 \cdot 3^1 \cdot 5^1$
LCM $= 2^3 \cdot 3^2 \cdot 5 = 8 \cdot 9 \cdot 5 = 360$

c)   What is the LCM of 360 and 1200?

$360 = 2^3 \cdot 3^2 \cdot 5^1$
$1200 = 2^4 \cdot 3^1 \cdot 5^2 = 3600$
LCM $= 2^4 \cdot 3^2 \cdot 5^2 = 3600$

EXERCISES   Find the least common multiple (LCM) of each of the following groups of numbers.

1) 3, 4   2) 3, 5   3) 3, 6

4) 3, 7

5) 4, 5

6) 4, 6

7) 4, 7

8) 4, 8

9) 3, 4, 6

10) 4, 6, 8

11) 6, 8

12) 10, 15

13) 12, 10, 15

14) 20, 18

15) 100, 75

16) 48, 60

17) 36, 60

18) 42, 28

19) 55, 33, 15

20) 120, 96

21) 80, 48, 72

22) 250, 75, 150

23) 4, 10, 15, 12

24) 32, 72, 60

25) 90, 70, 21

26) 36, 240

27) 17, 51, 68

28) 91, 14, 35, 49   29) 38, 57, 114, 171   30) 144, 128, 300, 180

31) In the application at the beginning of this section, what is the least common multiple of the denominators?

1. (Obj. 11) List all the factors of 56. _____

2. (Obj. 15) What is the LCM of 8, 12, and 36? _____

3. (Obj. 4) Subtract. $\begin{array}{r} 2009 \\ \underline{432} \end{array}$ _____

4. (Obj. 6) Divide. $24\overline{)8904}$

5. (Obj. 3) Add. $\begin{array}{r} 67 \\ 21 \\ 34 \\ 52 \\ 87 \\ \underline{69} \end{array}$ _____

6. (Obj. 5) Multiply. $\begin{array}{r} 723 \\ \underline{37} \end{array}$ _____

7. (Obj. 9) Find the value of 2^5. _____

8. (Obj. 3) Add. $\begin{array}{r} 1437 \\ 215 \\ 9213 \\ 14 \\ 30 \\ \underline{8021} \end{array}$ _____

9. (Obj. 10) Perform the indicated operations.

$$45 - 8 \cdot 2 \div 4 + 12$$ _____

10. (Obj. 6) Divide. $48\overline{)62149}$ _____

11. (Obj. 14) Write the prime factorization of 624. _____

12. (Obj. 10) Perform the indicated operations.

$$3 \cdot 8 + 14$$ _____

13. (Obj. 3) Add. $\begin{array}{r} 45 \\ 37 \\ \underline{21} \end{array}$ _____

14. (Obj. 9) Find the value of 5^3. _____

15. (Obj. 13) Is 103 a prime or a composite number? _____

16. (Obj. 4) Subtract.   2372
                           897
                           _____   _____

17. (Obj. 10) Perform the indicated operations.

$$36 - 12 \div 4$$ _____

18. (Obj. 4) Subtract.   6023
                           2914
                           _____   _____

19. (Obj. 6) Divide.   7)364   _____

20. (Obj. 14) Write the prime factorization of 180.   _____

21. (Obj. 5) Multiply.   2093
                             59
                           _____   _____

22. (Obj. 6) The distance around an automobile wheel
              is 45 inches. How many revolutions will it
              make in one mile? (1 mile = 63,360 inches)   _____

23. (Obj. 6) An automobile traveled 792 miles and used
              36 gallons of gasoline. What were the miles
              per gallon for the trip?   _____

24. (Obj. 5) If the speed of sound is 1100 feet per sec-
              ond, how far away is a skyrocket if the
              sound is heard 9 seconds after the explo-
              sion is observed?   _____

25. (Obj. 10) On a construction project 40 tons of fill
              gravel are to be used. If Mr. Rock has
              hauled four loads that weighed 12,974;
              13,642; 11,921; and 14,672 pounds, how
              many more pounds are left to be hauled?
              (1 ton = 2000 pounds)   _____

CHAPTER 2

FRACTIONS AND MIXED NUMBERS

2.1 FRACTIONS AND MIXED NUMBERS

If $\frac{1}{2}$ inch represents 10 feet on a scale drawing, how many feet does $\quad$ APPLICATION $2\frac{1}{2}$ inches represent?

16) Write a fraction to describe parts of units (unit regions or unit $\quad$ OBJECTIVES groups).

17) Change improper fractions to mixed numbers.

18) Change mixed numbers to improper fractions.

A *fraction* is a name of a number $\left(\text{such as } \frac{6}{7}\right)$ written in the form $\quad$ VOCABULARY $\frac{a}{b}$. The upper numeral (6) is the *numerator* and is a whole number. The lower numeral (7) is the *denominator* and is a natural number. That is,

$$\frac{\text{numerator}}{\text{denominator}}.$$

If the numerator is smaller than the denominator $\left(\text{such as } \frac{4}{9}\right)$, the fraction is called a *proper fraction* and names a number less than one. If the numerator is not smaller than the denominator $\left(\text{such as } \frac{8}{3}\right)$, the fraction is called an *improper fraction.*

A *mixed numeral,* or, more commonly, a *mixed number,* is the sum of a whole number and a fraction $\left(3 + \frac{1}{2}\right)$ with the plus sign omitted $\left(3\frac{1}{2}\right)$.

A *unit region* is a square, triangle, circle, rectangle, or any other figure (or group of such figures) used to illustrate a whole (or the number

1). *Congruent regions* (or parts) are figures that are the same size and the same shape.

THE HOW AND WHY OF IT

When using a unit region to illustrate the fraction $\frac{a}{b}$, the letter b represents the number of congruent parts into which each unit region is divided. The letter a represents the number of these parts under discussion.

The following unit region is subdivided into 7 congruent parts, and 6 of the parts are shaded. The fraction $\frac{6}{7}$ represents the shaded part of Figure 2.1. The denominator (7) tells us the number of congruent parts in the unit. The numerator (6) tells us the number of these parts under discussion. The fraction $\frac{1}{7}$ represents the unshaded part of the unit region.

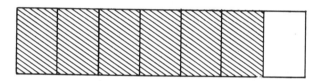

FIGURE 2.1

The fractions $\frac{a}{a}$ and $\frac{0}{a}$ have special significance. $\frac{a}{a} = 1$ and $\frac{0}{a} = 0$.

To change an improper fraction to a mixed number, divide the numerator by the denominator. If there is a remainder, write the whole number and then write the fraction, $\frac{\text{remainder}}{\text{divisor}}$.

$$\frac{7}{3} = 7 \div 3 \quad \text{or} \quad 3\overline{)7} \;\; \begin{array}{r} 2 \\ \underline{6} \\ 1 \end{array}$$

So, $\frac{7}{3} = 2\frac{1}{3}$

To change a mixed number to an improper fraction, multiply the denominator times the whole number, add the numerator, and put the sum over the denominator.

So, $1\frac{3}{7} = \frac{7 \cdot 1 + 3}{7} = \frac{7 + 3}{7} = \frac{10}{7}$

EXAMPLES

These are examples of proper fractions:

a) $\dfrac{3}{5}$

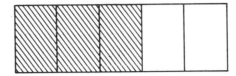

b) $\dfrac{0}{3}$

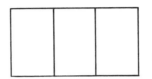

(no shaded parts)

Change each improper fraction to a mixed number.

c) $\dfrac{8}{7} = 7\overline{)\,8\,} = 1\dfrac{1}{7}$

d) $\dfrac{11}{3} = 3\overline{)\,11\,} = 3\dfrac{2}{3}$

Change each mixed number to an improper fraction.

e) $2\dfrac{4}{5} = \dfrac{5\cdot 2 + 4}{5} = \dfrac{10 + 4}{5} = \dfrac{14}{5}$

f) $3\dfrac{1}{10} = \dfrac{10\cdot 3 + 1}{10} = \dfrac{30 + 1}{10} = \dfrac{31}{10}$

g) $7 = 7\dfrac{0}{1} = \dfrac{1\cdot 7 + 0}{1} = \dfrac{7}{1}$

Write the fraction which describes the shaded part of each figure.

EXERCISES

1)

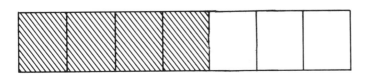

2)

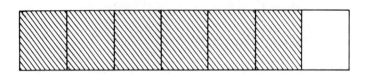

3)

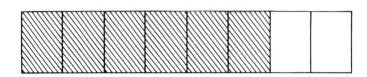

4)

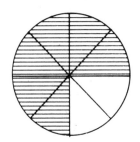

5)

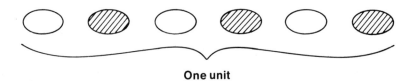

One unit

Change each improper fraction to a mixed number.

6) $\frac{17}{8}$ 7) $\frac{49}{12}$ 8) $\frac{43}{4}$ 9) $\frac{87}{10}$

10) $\frac{132}{11}$ 11) $\frac{133}{15}$ 12) $\frac{195}{3}$ 13) $\frac{32}{23}$

14) $\frac{75}{35}$ 15) $\frac{79}{6}$ 16) $\frac{200}{3}$

Change each mixed number to an improper fraction.

17) $1\frac{1}{2}$ 18) $1\frac{2}{3}$ 19) $3\frac{3}{4}$ 20) $3\frac{2}{7}$

21) $2\frac{0}{3}$ 22) $5\frac{3}{8}$ 23) 5 24) $10\frac{7}{10}$

25) $60\frac{5}{6}$ 26) $16\frac{2}{3}$ 27) $3\frac{17}{100}$

28) In the application at the beginning of this section, how many feet does $2\frac{1}{2}$ inches represent?

29) Hal's construction company must place section barriers, each $\frac{1}{8}$ mile long, between the two sides of a freeway. How many such sections will be needed for $24\frac{3}{8}$ miles of freeway?

30) A scale is marked (as usual) with a whole number at each pound. What whole number mark will be closest to a measurement of $\frac{50}{16}$ lb?

31) A ruler is marked (as usual) with a whole number at each centimeter. What whole number mark will be closest to a measurement of $\frac{87}{10}$ cm?

32) The fraction $\dfrac{373}{10}$ falls between what two whole numbers?

33) How many congruent parts must be shaded in a figure of 18 unit regions to illustrate the mixed number $17\dfrac{3}{8}$?

2.2 MULTIPLICATION OF FRACTIONS

APPLICATION

Lois spends $\dfrac{1}{2}$ of the family income on rent, utilities, and food. She pays $\dfrac{3}{7}$ of this amount for rent. What fraction of the family income goes for rent?

OBJECTIVE

19) Find the product of two or more numbers written as fractions.

VOCABULARY

A *product* is the answer to a multiplication problem.

THE HOW AND WHY OF IT

What is $\dfrac{1}{2}$ of $\dfrac{1}{3}$? (This is more commonly written as $\dfrac{1}{2} \cdot \dfrac{1}{3}$ = ?) Refer to Figure 2.2 in which the unit region is divided into three congruent parts (thirds); the $\dfrac{1}{3}$ which we want to multiply by $\dfrac{1}{2}$ is shaded. Since we are interested in one half of this one third, let us divide each of the thirds in the unit region into two congruent parts. As a result of doing this, we see in Figure 2.3 that the unit region is now divided into six congruent parts.

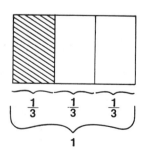

FIGURE 2.2

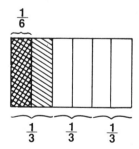

FIGURE 2.3

Since we want one half of the shaded one third, we can see that $\frac{1}{2}$ of $\frac{1}{3}$ must be $\frac{1}{6}$. That is, $\frac{1}{2} \cdot \frac{1}{3} = \frac{1}{6}$. We could also write $\left(\frac{1}{2}\right)\left(\frac{1}{3}\right) = \frac{1}{6}$.

Note that we get $\frac{1}{6}$ if we multiply the numerators and multiply the denominators and make these products the numerator and denominator, respectively.

The product of two or more numbers written as fractions is the product of the numerators over the product of the denominators.

EXAMPLES

a) $\quad \frac{3}{4} \cdot \frac{5}{8} = \frac{3 \cdot 5}{4 \cdot 8} = \frac{15}{32}$

b) $\quad \frac{5}{6} \cdot \frac{7}{9} = \frac{5 \cdot 7}{6 \cdot 9} = \frac{35}{54}$

c) $\quad \frac{1}{2} \cdot \frac{3}{4} \cdot \frac{5}{8} = \frac{1 \cdot 3 \cdot 5}{2 \cdot 4 \cdot 8} = \frac{15}{64}$

d) $\quad \frac{5}{2} \cdot 3 = \frac{5}{2} \cdot \frac{3}{1} = \frac{15}{2}$

e) $\quad 5 \cdot \frac{3}{16} = \frac{5}{1} \cdot \frac{3}{16} = \frac{15}{16}$

Multiply.

1) $\frac{1}{3} \cdot \frac{8}{9}$

2) $\frac{2}{5} \cdot \frac{7}{9}$

3) $\frac{5}{8} \cdot \frac{7}{3}$

4) $\frac{1}{2} \cdot \frac{1}{8}$

5) $\frac{5}{3} \cdot \frac{7}{8}$

6) $\frac{10}{11} \cdot \frac{2}{3}$

7) $\frac{4}{5} \cdot \frac{4}{5}$

8) $\frac{1}{2} \cdot \frac{1}{3} \cdot \frac{1}{4}$

9) $\frac{2}{3} \cdot \frac{1}{5} \cdot \frac{2}{3}$

10) $\frac{5}{4} \cdot \frac{3}{2} \cdot \frac{1}{4}$

11) $\frac{1}{2} \cdot \frac{3}{4} \cdot \frac{5}{11}$

12) $\frac{15}{2} \cdot \frac{9}{4}$

13) $\frac{1}{2} \cdot \frac{11}{3} \cdot \frac{5}{6}$

14) $\frac{2}{3} \cdot \frac{10}{7} \cdot \frac{8}{9}$

15) $\frac{15}{16} \cdot \frac{9}{2} \cdot \frac{3}{4}$

16) $\frac{4}{5} \cdot \frac{8}{1}$

17) $\frac{3}{7} \cdot 4$

18) $\left(\frac{8}{9}\right) \frac{10}{11}$

19) $\frac{9}{10} \left(\frac{13}{16}\right)$

20) $\frac{1}{23} \cdot \frac{1}{19}$

21) $\frac{1}{23} \cdot 19$

22) $\frac{15}{2} \cdot \frac{21}{2}$

23) $25 \cdot \frac{9}{17}$

24) $\left(\frac{15}{19}\right) \left(\frac{17}{16}\right)$

25) $\frac{3}{5} \cdot \frac{1}{2} \cdot \frac{7}{4}$

26) $\frac{2}{3} \cdot 7 \cdot \frac{2}{11}$

27) In the application at the beginning of this section, what fraction of the family income goes for rent?

28) An article was priced to sell for $6. During a sale the price was reduced $\frac{1}{3}$. By how many dollars was the price reduced?

29) In a certain school, $\frac{3}{8}$ of the student body take a foreign language. Of those who take a foreign language, $\frac{1}{2}$ take French. What part of the student body takes French?

2.3 BUILDING FRACTIONS

The main application of this procedure is in adding and subtracting fractions.   APPLICATION

20) Write a fraction with a required denominator that is equivalent   OBJECTIVE
to a given fraction.

Equivalent fractions are fractions that are different names for the   VOCABULARY
same number. $\frac{3}{6}$ and $\frac{4}{8}$ are equivalent since both represent one half of a unit region.

THE HOW AND WHY OF IT

The process of writing equivalent fractions described below is sometimes referred to as "building" fractions. It may be thought of as the opposite of the process called "reducing" fractions, which we will study later in this chapter. It is based on the fact that multiplying a number by 1 does not change its value (multiplication property of one). We may build new fractions from any given fraction.

$$\frac{3}{5} = \frac{3}{5} \cdot 1 = \frac{3}{5} \cdot \frac{2}{2} = \frac{6}{10}$$

$$\frac{3}{5} = \frac{3}{5} \cdot 1 = \frac{3}{5} \cdot \frac{3}{3} = \frac{9}{15}$$

$$\frac{3}{5} = \frac{3}{5} \cdot 1 = \frac{3}{5} \cdot \frac{4}{4} = \frac{12}{20}$$

and so on. We may write

$$\frac{3}{5} = \frac{6}{10} = \frac{9}{15} = \frac{12}{20} = \frac{15}{25} = \frac{18}{30} = \frac{21}{35} = \cdots$$

Multiplying any fraction by $\frac{2}{2}, \frac{3}{3}, \frac{4}{4}$ (or any form of the number one) builds the fraction to an equivalent form but does not change its value. A common problem is illustrated by $\frac{3}{5} = \frac{?}{60}$, where the denominator 60 is a requirement. The fraction $\frac{12}{12}$ is a useful form here for the number 1, because we can write $\frac{3}{5} \cdot 1 = \frac{3}{5} \cdot \frac{12}{12} = \frac{36}{60}$. So $\frac{36}{60}$ is the required equivalent fraction. We chose $\frac{12}{12}$ because $60 \div 5 = 12$, and therefore $5 \cdot 12 = 60$.

> To write a fraction with a required denominator that is equivalent to a given fraction, divide the required denominator by the given denominator. Multiply the result by the numerator of the given fraction. This product is the numerator of the new fraction.
>
> In general, $\frac{a}{b} \cdot 1 = \frac{a}{b} \cdot \frac{c}{c} = \frac{ac}{bc}$ provided b is not 0 and c is not 0.

EXAMPLES

a) $\frac{3}{5} = \frac{?}{80}$ $\frac{3}{5} \cdot \frac{16}{16} = \frac{48}{80}$ (Note that $80 \div 5 = 16$.)

b) $\frac{2}{7} = \frac{?}{56}$ $\frac{2}{7} \cdot \frac{8}{8} = \frac{16}{56}$

c) $\frac{13}{10} = \frac{?}{120}$ $\frac{13}{10} \cdot \frac{12}{12} = \frac{156}{120}$

Find the missing numerators.   **EXERCISES**

1) $\frac{1}{2} = \frac{}{10}$

2) $\frac{3}{2} = \frac{}{10}$

3) $\frac{7}{2} = \frac{}{10}$

4) $\frac{7}{2} = \frac{}{18}$

5) $\frac{5}{8} = \frac{}{48}$

6) $\frac{3}{4} = \frac{}{100}$

7) $\frac{7}{4} = \frac{}{16}$

8) $\frac{}{12} = \frac{2}{3}$

9) $\frac{3}{11} = \frac{}{66}$

10) $\frac{23}{6} = \frac{}{12}$

11) $\frac{9}{5} = \frac{}{100}$

12) $\frac{}{300} = \frac{7}{15}$

13) $\frac{3}{4} = \frac{}{68}$

14) $\frac{6}{9} = \frac{}{108}$

15) $\frac{}{90} = \frac{11}{15}$

16) $\frac{}{126} = \frac{19}{42}$

17) $\frac{16}{7} = \frac{}{147}$

18) $\frac{7}{10} = \frac{}{1000}$

19) $\dfrac{7}{8} = \dfrac{}{1000}$ 20) $\dfrac{1}{5} = \dfrac{}{1000}$

21) Write a fraction equivalent to $\dfrac{3}{4}$ that has a denominator of 28.

22) What fraction with denominator 132 is equivalent to $\dfrac{2}{3}$?

2.4   LISTING FRACTIONS IN ORDER OF VALUE

APPLICATION   A container of a chemical was weighed by two people. Mary recorded the weight as $3\dfrac{1}{8}$ lb. George read the weight as $3\dfrac{3}{16}$ lb. Whose measurement was heavier?

OBJECTIVE   21) Arrange a given group of fractions in order of value from smallest to largest.

VOCABULARY   When two or more fractions have the same denominator, we say that they have a *common denominator*.

THE HOW AND   It is clear from unit region illustrations (Fig. 2.4) that $\dfrac{4}{5}$ is larger than
WHY OF IT $\dfrac{2}{5}$.

FIGURE 2.4

Whenever two fractions have the same denominator, the one with the larger numerator has the larger value.

If fractions to be compared do not have the same denominator, then one or both may be "built" up. These equivalent fractions then have a common denominator and can be compared more easily. The preferred common denominator of a group of fractions is the least common multiple of all the denominators.

EXAMPLES

a) Which has larger value: $\frac{6}{11}$ or $\frac{8}{11}$? $\frac{8}{11}$ is larger because 8 is larger than 6.

b) Which has larger value: $\frac{6}{11}$ or $\frac{1}{2}$? It is easier to determine the answer if we build both fractions to a denominator of 22. (The LCM of 2 and 11 is 22.) $\frac{6}{11} = \frac{12}{22}$ and $\frac{1}{2} = \frac{11}{22}$, so $\frac{6}{11}$ is larger.

c) Arrange the fractions $\frac{2}{3}, \frac{3}{8}$, and $\frac{3}{4}$ in order of increasing value. (24 is the LCM of 3, 8, and 4.) $\frac{2}{3} = \frac{16}{24}, \frac{3}{8} = \frac{9}{24}$, and $\frac{3}{4} = \frac{18}{24}$, so in order of increasing value the fractions are $\frac{3}{8}, \frac{2}{3}, \frac{3}{4}$.

Arrange these fractions in order of increasing value.

EXERCISES

1) $\frac{3}{4}, \frac{1}{4}$

2) $\frac{13}{17}, \frac{18}{17}$

3) $\frac{1}{2}, \frac{5}{6}$

4) $\frac{5}{6}, \frac{2}{3}$

5) $\frac{3}{5}, \frac{1}{2}$

6) $\frac{2}{7}, \frac{7}{2}$

7) $\frac{2}{3}, \frac{2}{5}$

8) $\frac{5}{2}, \frac{3}{2}$

9) $\frac{1}{2}, \frac{2}{3}, \frac{1}{6}$

10) $1\frac{2}{5}, 1\frac{3}{10}$

Which fraction has the larger value?

11) $\frac{2}{3}, \frac{4}{5}$

12) $\frac{8}{9}, \frac{9}{10}$

13) $1\frac{3}{4}, 1\frac{7}{10}$

14) $\frac{5}{3}, \frac{3}{2}$

15) $\frac{7}{6}, \frac{11}{6}$

16) $\frac{6}{7}, \frac{6}{11}$

Arrange these fractions in order of increasing value.

17) $\frac{2}{5}, \frac{2}{3}, \frac{4}{7}$

18) $\frac{4}{5}, \frac{2}{3}, \frac{3}{4}$

19) $\frac{5}{8}, \frac{7}{10}, \frac{3}{4}$

20) $\frac{5}{7}, \frac{7}{9}, \frac{9}{11}$

21) In the application at the beginning of this section, whose measurement was heavier?

22) Two rulers are marked in inches, but on one the spaces are divided into tenths and on the other they are divided into sixteenths. Both are used to measure a line on a scale drawing. The nearest mark on one ruler is $5\frac{7}{10}$ inches, and the nearest mark on the other is $5\frac{11}{16}$ inches. Which is the larger (longer) measurement?

23) The night nurse at Malcolm X Community Hospital found bottles containing codeine tablets out of the usual order. The bottles contained tablets having the following strengths of codeine: $\frac{1}{8}, \frac{3}{32}, \frac{5}{16}, \frac{3}{8}, \frac{9}{16}, \frac{1}{2}$, and $\frac{1}{4}$ grain, respectively. Arrange the bottles in order of the strength of codeine from the smallest to the largest.

24) Joe, an apprentice helper, was given the task of sorting a bin of bolts according to their diameters. The bolts had the following diameters: $\frac{11}{16}, \frac{7}{8}, 1\frac{1}{16}, \frac{3}{4}, 1\frac{1}{8}$, and $1\frac{3}{32}$ inches. How should he list the diameters from smallest to largest?

2.5 REDUCING FRACTIONS

On a sheet metal layout the diameter of a hole is given as $\frac{28}{32}$ inch. **APPLICATION** Write this diameter as a reduced fraction.

22) Reduce a given fraction to lowest terms.　　　　　　**OBJECTIVE**

VOCABULARY

Reducing a fraction is writing a simpler fraction that is equivalent to the given fraction. A simpler fraction is one whose numerator and denominator are smaller. Reducing a fraction to *lowest terms* is writing an equivalent fraction that has the smallest possible numerator and denominator. For instance, $\frac{6}{12} = \frac{1}{2}$.

THE HOW AND WHY OF IT

Recall from the unit on building fractions that we were able to change $\frac{1}{2}$ to its equivalent fraction $\frac{2}{4}$ by using the multiplication property of one. We now want to reverse that process. We know that $\frac{1}{2} = \frac{2}{4}$ since $\frac{1}{2} = \frac{1}{2} \cdot 1 = \frac{1}{2} \cdot \frac{2}{2} = \frac{2}{4}$. Reversing the process, we do the following:

$$\frac{2}{4} = \frac{1 \cdot 2}{2 \cdot 2} = \frac{1}{2} \cdot \frac{2}{2} = \frac{1}{2} \cdot 1 = \frac{1}{2}$$

We could say that we use the multiplication property of 1 to "unmultiply."

> To reduce a fraction, we eliminate the common factors in the numerator and denominator. If *all* common factors (except the number 1) have been eliminated, the fraction is reduced to lowest terms. This rule is stated in two parts:
>
> $$\frac{ac}{bc} = \frac{a}{b} \cdot \frac{c}{c} = \frac{a}{b} \cdot 1 = \frac{a}{b}$$
>
> $$\frac{a}{ab} = \frac{a \cdot 1}{a \cdot b} = \frac{1}{b}$$

If the common factors cannot be discovered easily, they can always be found by writing the numerator and denominator in prime factored form.

EXAMPLES

a) $\quad \dfrac{28}{21} = \dfrac{4 \cdot 7}{3 \cdot 7} = \dfrac{4}{3}$

b) $\quad \dfrac{16}{24} = \dfrac{8 \cdot 2}{12 \cdot 2} = \dfrac{8}{12} = \dfrac{4 \cdot 2}{6 \cdot 2} = \dfrac{4}{6} = \dfrac{2 \cdot 2}{3 \cdot 2} = \dfrac{2}{3}$ ◀

(completely reduced or reduced to lowest terms)

or $\dfrac{16}{24} = \dfrac{2 \cdot 8}{3 \cdot 8} = \dfrac{2}{3}$ ◀————— reduced to lowest terms)

c) $\frac{16}{8} = \frac{8 \cdot 2}{8 \cdot 1} = \frac{2}{1} = 2$

d) $\frac{126}{144} = \frac{2 \cdot 3 \cdot 3 \cdot 7}{2 \cdot 2 \cdot 2 \cdot 2 \cdot 3 \cdot 3}$ **Prime factor, since the numbers are large**

$= \frac{7}{2 \cdot 2 \cdot 2}$ **Eliminate the common factors**

$= \frac{7}{8}$ **Multiply**

e) $\frac{8}{9} = \frac{2 \cdot 2 \cdot 2}{3 \cdot 3} = \frac{8}{9}$ **No common factors other than 1**

Reduce to lowest terms.   **EXERCISES**

1) $\frac{12}{14}$ 2) $\frac{14}{24}$ 3) $\frac{24}{36}$ 4) $\frac{4}{10}$

5) $\frac{6}{9}$ 6) $\frac{8}{12}$ 7) $\frac{10}{25}$ 8) $\frac{18}{30}$

9) $\frac{20}{22}$ 10) $\frac{21}{24}$ 11) $\frac{30}{75}$ 12) $\frac{45}{75}$

13) $\frac{65}{75}$ 14) $\frac{8}{40}$ 15) $\frac{10}{40}$ 16) $\frac{15}{40}$

17) $\frac{35}{40}$ 18) $\frac{21}{36}$ 19) $\frac{48}{30}$ 20) $\frac{16}{4}$

21) $\dfrac{45}{32}$ 22) $\dfrac{64}{72}$ 23) $\dfrac{90}{126}$ 24) $\dfrac{36}{100}$

25) $\dfrac{121}{132}$ 26) $\dfrac{98}{210}$ 27) $\dfrac{268}{402}$ 28) $\dfrac{97}{101}$

29) $\dfrac{98}{102}$ 30) $\dfrac{153}{255}$ 31) $\dfrac{200}{330}$ 32) $\dfrac{546}{910}$

33) $\dfrac{630}{1050}$ 34) $\dfrac{504}{1764}$ 35) $\dfrac{294}{1617}$

36) In the application at the beginning of this section, what reduced fraction represents the diameter of the hole?

37) The number of board feet in a piece of lumber that measures 2 inches by 5 inches by 10 inches is given by

$$\text{ft. b. m.} = \frac{(2)\,(5)}{12} \cdot 10 = \frac{100}{12}$$

Reduce the ft. b. m. to lowest terms.

38) The volume of a small box with dimensions $\frac{2}{3}$ inch by $\frac{3}{4}$ inch by $\frac{5}{6}$ inch is given by

$$V = \left(\frac{2}{3}\right)\left(\frac{3}{4}\right)\left(\frac{5}{6}\right) = \frac{30}{72}$$

Reduce the volume to lowest terms.

2.6 MULTIPLICATION OF FRACTIONS AND MIXED NUMBERS

Bobby drives his automobile at an average rate of 40 miles per hour for $2\frac{1}{2}$ hours. How many miles does he travel? **APPLICATION**

23) Find the product of numbers written as fractions or mixed numbers and reduce the product to lowest terms. **OBJECTIVES**

24) Find the reciprocal of a given natural number, fraction, or mixed number.

If two fractions have a product of 1, either one is called the *reciprocal* of the other. For example, $\frac{2}{3}$ is the reciprocal of $\frac{3}{2}$. **VOCABULARY**

To multiply mixed numbers, change each mixed number to its equivalent improper fraction and multiply. **THE HOW AND WHY OF IT**

The product of two or more fractions can often be found more quickly in reduced form (lowest terms) by reducing *before* performing the multiplication. Check each of the numerators and denominators to determine if there are common factors in any numerator *and* denominator. If there is a common factor in a numerator and a denominator, factor each of these whole numbers so that this common factor is seen. The multiplication is indicated by writing the product as a single fraction whose numerator and denominator are in factored form. For instance,

$$\frac{12}{35} \cdot \frac{25}{18} = \frac{2 \cdot 6 \cdot 5 \cdot 5}{7 \cdot 5 \cdot 3 \cdot 6}$$

The common factors in the numerator and denominator can then be eliminated, and the indicated multiplication can be performed. The answer will be reduced completely. So,

$$\frac{12}{35} \cdot \frac{25}{18} = \frac{2 \cdot 5}{7 \cdot 3} = \frac{10}{21}$$

If the common factors cannot be seen easily, use the prime factorization of each numerator and denominator. (See Example f.)

> The reciprocal of any fraction may be found by interchanging the numerator and denominator.

This is often called "inverting" the fraction. For instance, the reciprocal of $\frac{3}{7}$ is $\frac{7}{3}$. We check by verifying that the product is 1:

$$\frac{3}{7} \cdot \frac{7}{3} = \frac{21}{21} = 1$$

EXAMPLES

a) $\left(1\frac{1}{2}\right) \cdot \frac{3}{4} = \frac{3}{2} \cdot \frac{3}{4} = \frac{9}{8}$

(Note that the 8 and 12 have a common factor of 4.)

b) $\frac{8}{15} \cdot \frac{5}{12} = \frac{4 \cdot 2}{5 \cdot 3} \cdot \frac{5 \cdot 1}{4 \cdot 3} = \frac{4 \cdot 2 \cdot 5 \cdot 1}{5 \cdot 3 \cdot 4 \cdot 3} = \frac{2 \cdot 1}{3 \cdot 3} = \frac{2}{9}$

(5 and 15 have a common factor of 5.)

c) $\left(3\frac{3}{4}\right)\left(2\frac{2}{5}\right) = \frac{15}{4} \cdot \frac{12}{5} = \frac{3 \cdot 5}{4} \cdot \frac{4 \cdot 3}{5} = \frac{3 \cdot 5 \cdot 4 \cdot 3}{4 \cdot 5} = \frac{3 \cdot 3}{1} = 9$

d) The reciprocal of $\frac{7}{10}$ is $\frac{10}{7}$. Check: $\frac{7}{10} \cdot \frac{10}{7} = \frac{70}{70} = 1$

e) The reciprocal of $1\frac{4}{5}$ is ? First change $1\frac{4}{5}$ to an improper fraction:

$1\frac{4}{5} = \frac{9}{5}$. The reciprocal of $1\frac{4}{5}$ is $\frac{5}{9}$. Check: $\frac{9}{5} \cdot \frac{5}{9} = 1$.

f) $\frac{126}{144} = \frac{2 \cdot 3 \cdot 3 \cdot 7}{2 \cdot 2 \cdot 2 \cdot 2 \cdot 3 \cdot 3} = \frac{7}{2 \cdot 2 \cdot 2} = \frac{7}{8}$

Multiply and reduce each answer to lowest terms.   **EXERCISES**

1) $\dfrac{1}{2} \cdot \dfrac{4}{3}$

2) $\dfrac{3}{4} \cdot \dfrac{4}{7}$

3) $\dfrac{2}{3} \cdot \dfrac{5}{6}$

4) $\dfrac{3}{2} \cdot \dfrac{8}{7}$

5) $\dfrac{5}{6} \cdot \dfrac{12}{11}$

6) $\dfrac{1}{8} \cdot \dfrac{12}{7}$

7) $\dfrac{1}{2} \cdot \dfrac{2}{3} \cdot \dfrac{3}{4}$

8) $\dfrac{3}{7} \cdot \dfrac{14}{9}$

9) $\dfrac{4}{5} \cdot \dfrac{7}{8}$

10) $\dfrac{7}{8} \cdot \dfrac{8}{21}$

11) $\dfrac{5}{18} \cdot \dfrac{2}{3} \cdot \dfrac{27}{35}$

12) $\dfrac{15}{8} \left(1\dfrac{3}{5}\right)$

13) $\left(7\dfrac{1}{2}\right) \left(4\dfrac{3}{4}\right)$

14) $\left(\dfrac{5}{8}\right) \left(\dfrac{16}{7}\right) \left(2\dfrac{3}{4}\right)$

15) $\dfrac{4}{6} \cdot \dfrac{9}{30}$

16) $\dfrac{2}{3} \cdot \dfrac{4}{5} \cdot \dfrac{6}{7}$

17) $5 \cdot \dfrac{1}{4} \cdot \dfrac{8}{15}$

18) $1\dfrac{4}{5} \cdot \dfrac{7}{8} \cdot \dfrac{8}{21}$

19) $\dfrac{5}{6} \cdot 1\dfrac{1}{5}$

20) $\left(4\dfrac{1}{2}\right) \left(\dfrac{2}{9}\right) \left(\dfrac{7}{8}\right)$

21) $\dfrac{3}{8} \cdot \dfrac{4}{7} \cdot \dfrac{8}{3} \cdot \dfrac{14}{24}$

22) $\dfrac{3}{4} \cdot \dfrac{4}{5} \cdot \dfrac{10}{21} \cdot \dfrac{7}{2}$ 23) $\dfrac{12}{20} \cdot \dfrac{35}{24}$ 24) $\left(\dfrac{7}{2}\right)\left(3\dfrac{1}{2}\right)\left(\dfrac{4}{7}\right)$

25) $\left(2\dfrac{1}{4}\right)\left(2\dfrac{2}{3}\right)\left(5\dfrac{1}{6}\right)$ 26) $\left(15\dfrac{2}{3}\right)\left(7\dfrac{5}{8}\right)$ 27) $\dfrac{56}{65} \cdot \dfrac{39}{48} \cdot \dfrac{18}{25}$

28) What is the reciprocal of $\dfrac{5}{8}$?

29) What is the reciprocal of $2\dfrac{2}{3}$?

30) What is the reciprocal of 3?

31) What is the reciprocal of $\dfrac{1}{8}$?

32) In the application at the beginning of this section, how many miles did Bobby travel?

33) If Tammy can make $2\dfrac{2}{3}$ drawings in an hour, how many can she make in $6\dfrac{3}{4}$ hours?

34) If a steel bar weighs $\dfrac{3}{4}$ pound per foot, what is the weight of a bar that is $20\dfrac{1}{2}$ feet in length?

35) The owner of a sporting goods store usually sells 1200 tennis racquets a year at a price of $18 each. He decides to raise the price per racquet by $\dfrac{1}{8}$, which will in turn decrease his sales by $\dfrac{1}{6}$. What will

be his net income under the new system? Is it more or less than under the old pricing system?

36) Each student in the television repair class is allowed $6\frac{3}{4}$ in of wire solder for his work. How many inches of wire must be ordered for a class of 32 students?

37) The water pressure during a bad fire is reduced to $\frac{5}{9}$ of its original pressure at the hydrant. What is the reduced pressure if the original pressure was $73\frac{2}{3}$ lb/in² ?

2.7 DIVISION OF FRACTIONS AND MIXED NUMBERS

If each turn of a wood screw sinks it $\frac{3}{16}$ of an inch, how many turns are needed to sink it $1\frac{1}{4}$ inches?   APPLICATION

25) Find the quotient of two numbers written as fractions, mixed numbers, or whole numbers.   OBJECTIVE

Recall that the reciprocal of $\frac{3}{5}$ is $\frac{5}{3}$ because $\frac{3}{5} \cdot \frac{5}{3} = 1$.   VOCABULARY

Consider $\frac{3}{4} \div \frac{1}{3} = ?$ This can be thought of as $\frac{1}{3} \cdot (?) = \frac{3}{4}$, since multiplication and division are inverse operations. Replace the "?" with $\frac{3}{1} \cdot \frac{3}{4}$ and you have the right answer.   THE HOW AND WHY OF IT

$$\frac{1}{3}\left(\frac{3}{1}\cdot\frac{3}{4}\right)=\frac{3}{4}$$

$$1\cdot\frac{3}{4}=\frac{3}{4} \qquad \left(\text{Note: } \frac{1}{3}\cdot\frac{3}{1}=1\right)$$

$$\frac{3}{4}=\frac{3}{4}$$

So, $\frac{3}{4}\div\frac{1}{3}=\frac{3}{4}\cdot\frac{3}{1}=\frac{9}{4}$; $\frac{3}{4}$ is multiplied by the reciprocal of the divisor $\frac{1}{3}$.

> In general, to divide two fractions, change the problem to multiplication by using the reciprocal of the divisor (that is, invert the divisor and multiply). The pattern is:
>
> $$\frac{d}{e}\div\frac{f}{g}=\frac{d}{e}\cdot\frac{g}{f}=\frac{dg}{ef}$$
>
> To divide whole numbers and/or mixed numbers:
> 1) write them as improper fractions, and
> 2) divide, following the procedures for fractions.

EXAMPLES

a) $\quad\dfrac{8}{3}\div\dfrac{4}{5}=\dfrac{8}{3}\cdot\dfrac{5}{4}=\dfrac{4\cdot2\cdot5}{3\cdot4}=\dfrac{10}{3}=3\dfrac{1}{3}$

b) $\quad3\dfrac{1}{3}\div6\dfrac{7}{8}=\dfrac{10}{3}\div\dfrac{55}{8}=\dfrac{10}{3}\cdot\dfrac{8}{55}=\dfrac{5\cdot2\cdot8}{3\cdot5\cdot11}=\dfrac{16}{33}$

c) $\quad12\dfrac{6}{10}\div21\dfrac{3}{5}=\dfrac{126}{10}\div\dfrac{108}{5}=\dfrac{126}{10}\cdot\dfrac{5}{108}=\dfrac{2\cdot3\cdot3\cdot7\cdot5}{2\cdot5\cdot2\cdot2\cdot3\cdot3\cdot3}=\dfrac{7}{12}$

d) $\quad4\div3\dfrac{1}{3}=\dfrac{4}{1}\div\dfrac{10}{3}=\dfrac{4}{1}\cdot\dfrac{3}{10}=\dfrac{2\cdot2\cdot3}{1\cdot2\cdot5}=\dfrac{6}{5}=1\dfrac{1}{5}$

EXERCISES

Divide. Reduce answers to lowest terms and express as mixed numbers where possible.

1) $\dfrac{3}{7}\div\dfrac{1}{2}$

2) $\dfrac{1}{2}\div\dfrac{1}{3}$

3) $\dfrac{1}{3}\div\dfrac{1}{2}$

4) $\dfrac{2}{3}\div\dfrac{5}{7}$

5) $\dfrac{2}{5}\div\dfrac{5}{7}$

6) $\dfrac{1}{10}\div3$

7) $7 \div \frac{1}{5}$

8) $\frac{3}{2} \div \frac{1}{3}$

9) $\frac{5}{7} \div \frac{5}{6}$

10) $\frac{1}{4} \div \frac{3}{4}$

11) $\frac{8}{9} \div \frac{2}{9}$

12) $\frac{8}{9} \div \frac{8}{3}$

13) $\frac{8}{9} \div \frac{2}{3}$

14) $\frac{5}{8} \div 1\frac{3}{4}$

15) $3\frac{7}{8} \div 7\frac{3}{4}$

16) $\frac{2}{5} \div \frac{3}{5}$

17) $12\frac{2}{9} \div \frac{2}{3}$

18) $\frac{7}{8} \div 3\frac{3}{4}$

19) $\frac{30}{49} \div \frac{20}{21}$

20) $9 \div 6$

21) $8 \div 12$

22) $\frac{4}{15} \div \frac{5}{8}$

23) $26 \div 4\frac{5}{16}$

24) $2\frac{4}{9} \div 3$

25) $\frac{51}{10} \div \frac{14}{15}$

26) $5\frac{11}{15} \div 2\frac{4}{5}$

27) $2\frac{11}{15} \div 5\frac{4}{5}$

28) $\frac{75}{132} \div \frac{245}{308}$

29) In the application at the beginning of this section, how many turns are needed?

30) Carol has $\frac{7}{8}$ lb of cheese. If an omelet recipe calls for $\frac{1}{4}$ lb of cheese, how many omelets can she make?

31) A machinist takes $74\frac{1}{2}$ minutes to machine 5 pins. How long will it take to machine one pin?

32) A welder is to drill 13 equally spaced holes on the center line of a flat metal bar. The two end holes are to be 1 inch from each end of the bar. What will be the distance between the centers of the holes if the bar is 52 inches long?

33) As part of her job at a pet store, Becky feeds each gerbil $\frac{1}{8}$ cup of seeds each day. If the seeds come in packages of $\frac{3}{4}$ cup, how many gerbils can be fed from one package?

34) Norma has a ribbon that is $\frac{3}{4}$ of a yard long. From this ribbon, she wants to cut 6 pieces of equal length. How long will each piece be?

35) How many boards of length $3\frac{1}{3}$ ft can Eddie cut from a board of length $13\frac{1}{3}$ ft?

2.8 ADDITION OF FRACTIONS

What is the distance around (perimeter) this triangle?   APPLICATION

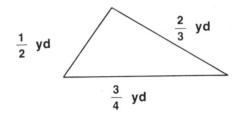

26) Find the sum of two or more numbers written as like fractions.   OBJECTIVES
27) Find the sum of two or more numbers written as unlike fractions.

Like fractions are fractions with the same (or common) denominators. For instance, $\frac{1}{4}$ and $\frac{3}{4}$ are like fractions.   VOCABULARY

The sum $\frac{1}{5} + \frac{2}{5}$ is ? Remember that the denominators tell the number   THE HOW AND
of congruent parts in the unit region. The numerator tells us how many of   WHY OF IT
these parts we have. By adding the numerators we find the total number
of these parts, and the common denominator keeps track of the correct
size of the parts. This is illustrated in Figure 2.5, using rectangular unit
regions.

FIGURE 2.5

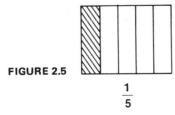
$$\frac{1}{5}$$

$+$

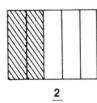
$$\frac{2}{5}$$

$=$

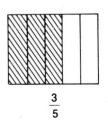
$$\frac{3}{5}$$

> To find the sum of two or more numbers written as like fractions, add the numerators and write the sum over the common denominator.
>
> $$\frac{a}{c} + \frac{b}{c} = \frac{a+b}{c}$$
>
> To add unlike fractions, write an equivalent fraction for each, so that they have a common denominator. Add these fractions as before.

EXAMPLES

a) $\dfrac{3}{8} + \dfrac{4}{8} = \dfrac{7}{8}$

b) $\dfrac{1}{3} + \dfrac{2}{3} + \dfrac{1}{3} = \dfrac{4}{3}$

c) $\dfrac{3}{8} + \dfrac{1}{4} = ?$ The LCM of 8 and 4 is 8. Now we have $\dfrac{3}{8} = \dfrac{3}{8}$ and $\dfrac{1}{4} = \dfrac{1}{4} \cdot \dfrac{2}{2} = \dfrac{2}{8}$, so $\dfrac{3}{8} + \dfrac{1}{4} = \dfrac{3}{8} + \dfrac{2}{8} = \dfrac{5}{8}$.

d) $\dfrac{5}{12} + \dfrac{2}{9} = ?$ The LCM of 12 and 9 is 36. Now we have $\dfrac{5}{12} = \dfrac{5}{12} \cdot \dfrac{3}{3} = \dfrac{15}{36}$ and $\dfrac{2}{9} = \dfrac{2}{9} \cdot \dfrac{4}{4} = \dfrac{8}{36}$, so $\dfrac{5}{12} + \dfrac{2}{9} = \dfrac{15}{36} + \dfrac{8}{36} = \dfrac{23}{36}$.

e) $\dfrac{3}{4} + \dfrac{7}{10} = ?$ The LCM of 4 and 10 is 20. Now we have $\dfrac{3}{4} + \dfrac{7}{10} = \dfrac{15}{20} + \dfrac{14}{20} = \dfrac{29}{20}$.

EXERCISES

Add. Write answers as proper or improper fractions reduced to lowest terms.

1) $\dfrac{1}{2} + \dfrac{1}{2}$

2) $\dfrac{1}{4} + \dfrac{2}{4}$

3) $\dfrac{3}{10} + \dfrac{4}{10} + \dfrac{1}{10}$

4) $\dfrac{5}{12} + \dfrac{4}{12}$

5) $\dfrac{8}{11} + \dfrac{2}{11}$

6) $\dfrac{2}{15} + \dfrac{4}{15} + \dfrac{3}{15}$

7) $\dfrac{1}{2}+\dfrac{1}{4}$

8) $\dfrac{2}{3}+\dfrac{1}{6}$

9) $\dfrac{1}{2}+\dfrac{1}{3}+\dfrac{1}{6}$

10) $\dfrac{3}{5}+\dfrac{1}{4}$

11) $\dfrac{2}{7}+\dfrac{3}{14}$

12) $\dfrac{1}{2}+\dfrac{3}{4}+\dfrac{3}{16}$

13) $\dfrac{3}{15}+\dfrac{7}{12}$

14) $\dfrac{5}{9}+\dfrac{5}{12}$

15) $\dfrac{3}{35}+\dfrac{8}{21}$

16) $\dfrac{7}{16}+\dfrac{3}{20}+\dfrac{1}{5}$

17) $\dfrac{5}{12}+\dfrac{9}{16}+\dfrac{7}{24}$

18) $\dfrac{3}{10}+\dfrac{7}{20}+\dfrac{11}{30}$

19) $\dfrac{9}{14}+\dfrac{5}{21}$

20) $\dfrac{1}{10}+\dfrac{2}{5}+\dfrac{5}{6}+\dfrac{1}{15}$

21) $\dfrac{5}{12}+\dfrac{3}{16}+\dfrac{7}{20}$

22) $\dfrac{5}{18}+\dfrac{4}{21}$

23) $\dfrac{8}{65}+\dfrac{7}{39}$

24) $\dfrac{19}{54}+\dfrac{7}{48}$

25) $\dfrac{21}{25}+\dfrac{11}{50}+\dfrac{7}{75}$

26) $\dfrac{1}{12}+\dfrac{1}{36}+\dfrac{1}{54}$

27) $\dfrac{23}{96}+\dfrac{14}{64}+\dfrac{7}{80}$

28) $\dfrac{117}{240}+\dfrac{41}{270}$

29) In the application at the beginning of this section, what is the perimeter of the triangle?

30) On the New York Stock Exchange, Vern's stock rose $\frac{1}{8}$ of a point in the morning and an additional $\frac{3}{16}$ of a point during the afternoon. What was the total rise for the day?

31) A Boise Cascade lumber mill produces dry lumber that measures $\frac{11}{16}$ inch thick. What was the original thickness if the shrinkage from green size is $\frac{3}{32}$ inch?

32) Three lamps are connected in series. The total resistance is the sum of the individual resistances. What is the total resistance in ohms if the individual resistances are $\frac{1}{4}$ ohm, $\frac{5}{8}$ ohm, and $\frac{5}{6}$ ohm?

2.9 ADDITION OF MIXED NUMBERS

APPLICATION

Michelle worked $4\frac{1}{2}$ hours on both Monday and Wednesday, $6\frac{3}{4}$ hours on Tuesday, and 8 hours on Saturday. How many hours did she work during the week?

OBJECTIVE

28) Find the sum of two or more numbers written as whole numbers and/or mixed numbers.

VOCABULARY

Recall that a *mixed number* is the sum of a whole number and a fraction.

THE HOW AND WHY OF IT

$3\frac{1}{4} + 5\frac{1}{6} = ?$ This can also be written $\left(3 + \frac{1}{4}\right) + \left(5 + \frac{1}{6}\right)$. By using the commutative and associative laws of addition, the following statement is true:

$$\left(3 + \frac{1}{4}\right) + \left(5 + \frac{1}{6}\right) = (3 + 5) + \left(\frac{1}{4} + \frac{1}{6}\right) = 8 + \left(\frac{3}{12} + \frac{2}{12}\right) = 8\frac{5}{12}$$

If we write the problem in vertical form, this grouping takes place naturally:

$$3\frac{1}{4} = 3\frac{3}{12}$$ Add the whole numbers

$$5\frac{1}{6} = 5\frac{2}{12}$$ and add the fractions as above.

$$8\frac{5}{12}$$

When we are adding mixed numbers, it is possible that the fractions have a sum that is greater than 1. In this case, the resulting improper fraction can be changed to a mixed number and added to the whole number part. For example:

$$11\frac{7}{10} = 11\frac{21}{30}$$

$$23\frac{8}{15} = 23\frac{16}{30}$$

$$34\frac{37}{30} = 34 + 1\frac{7}{30} = 35\frac{7}{30}$$

To find the sum of numbers written as whole numbers or mixed numbers, add the whole number parts and the fraction parts. The sum is usually written as a mixed number whose fraction part is a reduced proper fraction.

EXAMPLES

a) $$5\frac{5}{12} = 5\frac{5}{12}$$

$$1\frac{3}{4} = 1\frac{9}{12}$$

$$6\frac{14}{12} = 6 + 1\frac{2}{12} = 7\frac{2}{12} = 7\frac{1}{6}$$

b) $$25\frac{7}{8} = 25\frac{63}{72}$$

$$13\frac{5}{9} = 13\frac{40}{72}$$

$$7\frac{1}{6} = 7\frac{12}{72}$$

$$45\frac{115}{72} = 45 + 1\frac{43}{72} = 46\frac{43}{72}$$

c) $8 = 8\frac{0}{3}$

$5\frac{2}{3} = 5\frac{2}{3}$

$\overline{\quad\quad\quad 13\frac{2}{3}}$

EXERCISES   Add. State all answers as mixed numbers where possible.

1) $7\frac{3}{8}$

$\underline{4\frac{2}{8}}$

2) $15\frac{3}{5}$

$\underline{17\frac{1}{5}}$

3) $8\frac{5}{12}$

$\underline{6\frac{1}{12}}$

4) $5\frac{1}{4}$

$2\frac{3}{4}$

$\underline{7\frac{1}{4}}$

5) $6\frac{2}{7}$

$\underline{4\frac{3}{7}}$

6) $12\frac{2}{3}$

$\underline{1\frac{1}{3}}$

7) $3\frac{4}{9}$

$2\frac{1}{9}$

$\underline{1\frac{1}{9}}$

8) $8\frac{2}{3}$

$\underline{1\frac{2}{3}}$

9) $5\frac{1}{2}$

$\underline{2\frac{1}{4}}$

10) $7\dfrac{1}{6}$

$2\dfrac{1}{3}$
___

11) $14\dfrac{7}{10}$

$11\dfrac{3}{5}$
___

12) $9\dfrac{5}{24}$

$101\dfrac{7}{12}$
___

13) $213\dfrac{5}{18}$

$506\dfrac{7}{12}$
___

14) $213\dfrac{5}{6}$

$347\dfrac{3}{10}$
___

15) $142\dfrac{7}{11} + 93$

16) $47\dfrac{1}{5} + 23\dfrac{2}{3} + 15\dfrac{1}{2}$

17) $47\dfrac{3}{8}$

23

$42\dfrac{5}{12}$
___

18) $7\dfrac{7}{20}$

$\dfrac{3}{4}$

$8\dfrac{9}{40}$
___

19) $82\dfrac{41}{45}$

$97\dfrac{25}{27}$
___

20) $2\dfrac{2}{5} + 7\dfrac{1}{6} + 1\dfrac{4}{15} + 3\dfrac{1}{10}$

21) $6\dfrac{4}{11} + 7\dfrac{5}{6}$

22) $24\dfrac{8}{9} + 423\dfrac{5}{6}$

23) $39\dfrac{2}{3} + 49\dfrac{3}{4} + 99\dfrac{2}{9}$

24) In the application at the beginning of this section, how many hours did Michelle work?

25) On successive days, Wendy picked $21\frac{3}{4}$, $31\frac{5}{8}$, and $27\frac{3}{16}$ pounds of beans. How many pounds of beans did she pick during the three days?

26) John rode his bicycle $2\frac{3}{8}$ miles Tuesday, $7\frac{3}{4}$ miles Wednesday, $10\frac{2}{3}$ miles Thursday, and $4\frac{5}{6}$ miles Friday. What was his total mileage during the four days?

27) A brass rod was cut into lengths of $3\frac{1}{2}$, $6\frac{1}{4}$, $4\frac{5}{8}$, and $6\frac{3}{4}$ inches. How long was the original piece of stock if $\frac{1}{16}$ inch is allowed for each cut?

2.10 SUBTRACTION OF FRACTIONS

APPLICATION   Dan has a bolt that is $\frac{3}{8}$ of an inch in length. He found that it is $\frac{3}{16}$ of an inch too long. What is the length of the bolt Dan needs?

OBJECTIVE   29) Find the difference of two numbers written as fractions.

VOCABULARY   No new vocabulary.

$\frac{2}{3} - \frac{1}{3} = ?$ Looking at the unit region representation in Figure 2.6, we can see that all that is necessary is to subtract the numerators and retain the common denominator.

THE HOW AND WHY OF IT

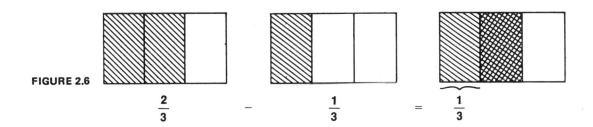

FIGURE 2.6

$$\frac{2}{3} \qquad - \qquad \frac{1}{3} \qquad = \qquad \frac{1}{3}$$

To subtract two numbers written as fractions, rewrite the fractions so that they have a common denominator. Subtract the numerators and write this difference over the common denominator.

$$\frac{3}{4} - \frac{1}{3} = \frac{9}{12} - \frac{4}{12} = \frac{5}{12}$$

EXAMPLES

a) $\frac{11}{20} - \frac{5}{20} = \frac{6}{20} = \frac{3}{10}$

b) $\frac{7}{8} - \frac{2}{3} = \frac{21}{24} - \frac{16}{24} = \frac{5}{24}$

c) $\frac{7}{15} - \frac{1}{4} = \frac{28}{60} - \frac{15}{60} = \frac{13}{60}$

Subtract. Reduce answers to lowest terms.

EXERCISES

1) $\frac{5}{7} - \frac{3}{7}$

2) $\frac{12}{15} - \frac{4}{15}$

3) $\frac{7}{2} - \frac{1}{2}$

4) $\frac{15}{16} - \frac{7}{16}$

5) $\frac{3}{14} - \frac{1}{14}$

6) $\frac{9}{10} - \frac{7}{10}$

7) $\dfrac{7}{4} - \dfrac{3}{4}$

8) $\dfrac{5}{16} - \dfrac{1}{8}$

9) $\dfrac{13}{20} - \dfrac{3}{10}$

10) $\dfrac{3}{4} - \dfrac{5}{12}$

11) $\dfrac{5}{6} - \dfrac{3}{5}$

12) $\dfrac{17}{30} - \dfrac{7}{30}$

13) $\dfrac{11}{15} - \dfrac{2}{5}$

14) $\dfrac{2}{3} - \dfrac{3}{8}$

15) $\dfrac{3}{4} - \dfrac{5}{16}$

16) $\dfrac{11}{12} - \dfrac{11}{15}$

17) $\dfrac{9}{50} - \dfrac{3}{40}$

18) $\dfrac{13}{12} - \dfrac{17}{18}$

19) $\dfrac{33}{35} - \dfrac{17}{28}$

20) $\dfrac{7}{10} - \dfrac{9}{20}$

21) $\dfrac{3}{10} - \dfrac{1}{15}$

22) $\dfrac{7}{16} - \dfrac{1}{6}$

23) $\dfrac{7}{24} - \dfrac{5}{18}$

24) $\dfrac{23}{48} - \dfrac{21}{80}$

25) $\dfrac{5}{9} - \dfrac{2}{75}$

26) $\dfrac{22}{33} - \dfrac{9}{44}$

27) Find the difference of $\dfrac{1}{2}$ and $\dfrac{2}{5}$.

28) In the application at the beginning of this section, what is the length of the bolt Dan needs?

29) A cake recipe calls for $\frac{3}{4}$ cup of milk. Carol has $\frac{1}{3}$ cup of milk. How much milk will she need to borrow from her neighbor in order to make the cake?

30) Mary is given $\frac{7}{8}$ lb of dried fruit. She uses $\frac{2}{3}$ lb to make a fruit cake. How much dried fruit does she have left?

31) A loan is to be paid as follows: $\frac{1}{5}$ the first year, $\frac{1}{4}$ the second year, $\frac{1}{3}$ the third year, and the remainder the fourth year. How much of the loan is paid the fourth year?

2.11 SUBTRACTION OF MIXED NUMBERS

Frank poured $8\frac{3}{10}$ yards of cement for a fountain. Another fountain took $5\frac{3}{8}$ yards. How much more cement was needed for the larger fountain than for the smaller one? **APPLICATION**

30) Find the difference of two numbers written as mixed numbers. **OBJECTIVE**

No new vocabulary. **VOCABULARY**

In general, the pattern for subtraction of mixed numbers is similar to the pattern for addition of mixed numbers. **THE HOW AND WHY OF IT**

To subtract two mixed numbers where the fractional part being subtracted is the smaller of the two fractions, subtract the fractional parts and then subtract the whole number parts.

To subtract two mixed numbers where the fractional part being subtracted is the larger of the two fractions, rename the smaller fraction by borrowing one (1) from the whole number and adding it to the fraction; then subtract.

To rename $32\frac{1}{3}$ with a larger fractional part, borrow one (1) from 32 and add it to the fraction.

$$32\frac{1}{3} = 31 + 1\frac{1}{3} = 31 + \frac{4}{3} = 31\frac{4}{3}$$

Any mixed number can be given a new name in this way.

EXAMPLES

a)
$$47\frac{5}{8} = 47\frac{35}{56}$$
$$\underline{36\frac{3}{7} = 36\frac{24}{56}}$$
$$11\frac{11}{56}$$

b)
$$28\frac{9}{16} = 28\frac{27}{48}$$
$$\underline{13\frac{1}{3} = 13\frac{16}{48}}$$
$$15\frac{11}{48}$$

c)
$$15\frac{3}{10} = 14 + 1\frac{3}{10} = 14\frac{13}{10}$$
$$\underline{7\frac{7}{10} =} \qquad \underline{7\frac{7}{10} =} \quad \underline{7\frac{7}{10}}$$
$$7\frac{6}{10} = 7\frac{3}{5}$$

d)
$$32\frac{1}{3} = 32\frac{4}{12} = 31 + 1\frac{4}{12} = 31\frac{16}{12}$$
$$\underline{27\frac{3}{4} = 27\frac{9}{12} =} \quad \underline{27\frac{9}{12} = 27\frac{9}{12}}$$
$$4\frac{7}{12}$$

In the next two examples, one of the numbers is a whole number. Any whole number can be written as a mixed number by adding on a fraction whose numerator is zero and whose denominator is a natural number (remember that the value of such a fraction is zero). In these examples, the natural number is chosen so the denominators will be the same.

e) $4\frac{2}{3} = 4\frac{2}{3}$

$\underline{3 = 3\frac{0}{3}}$

$1\frac{2}{3}$

f) $11 = 11\frac{0}{9} = 10 + 1\frac{0}{9} = 10\frac{9}{9}$

$\underline{2\frac{2}{9} = 2\frac{2}{9} = 2\frac{2}{9} = 2\frac{2}{9}}$

$8\frac{7}{9}$

Subtract. Reduce answers to lowest terms.

1) $3\frac{3}{5}$

$\underline{1\frac{2}{5}}$

2) $7\frac{8}{9}$

$\underline{2\frac{4}{9}}$

3) $17\frac{5}{6}$

$\underline{6\frac{1}{6}}$

4) $1\frac{7}{12}$

$\underline{\frac{3}{12}}$

5) $6\frac{28}{29}$

$\underline{5\frac{19}{29}}$

6) $9\frac{3}{5}$

$\underline{2\frac{3}{5}}$

7) $23\frac{5}{8}$

$\underline{15}$

8) 13

$\underline{2\frac{1}{2}}$

9) $12\frac{5}{8}$

$\underline{3\frac{1}{4}}$

10) $32\frac{5}{9}$

$25\frac{1}{3}$

11) $13\frac{6}{7}$

$8\frac{4}{7}$

12) $27\frac{5}{9}$

$23\frac{2}{9}$

13) $212\frac{37}{80}$

$109\frac{21}{80}$

14) $145\frac{2}{3}$

$27\frac{1}{2}$

15) $6\frac{5}{6}$

$3\frac{3}{10}$

16) $4\frac{7}{9}$

$1\frac{5}{12}$

17) $28\frac{3}{10}$

$14\frac{7}{10}$

18) $16\frac{1}{8}$

$3\frac{5}{8}$

19) $212\frac{1}{9}$

57

20) $36\frac{7}{16}$

$19\frac{11}{16}$

21) 45

$16\frac{2}{3}$

22) $81\frac{3}{14}$

$16\frac{5}{7}$

23) $15\frac{5}{12}$

$13\frac{17}{18}$

24) $9\frac{2}{11}$

$6\frac{23}{44}$

25) $14\frac{3}{14} - 5$

26) $14 - 5\frac{3}{14}$

27) $6\frac{2}{3} - 1\frac{5}{6}$

28) $1\frac{5}{12} - \frac{5}{8}$

29) $3\frac{4}{15} - 2\frac{7}{10}$

30) In the application at the beginning of this section, how much more cement was needed?

31) Cecil has a piece of lumber that measures $10\frac{7}{12}$ ft, and it is to be used in a spot that calls for a length of $8\frac{3}{4}$ ft. How much must he cut off of the board?

32) Dick harvested $30\frac{3}{4}$ tons of wheat. He sold $12\frac{3}{10}$ tons to the Cartwright Flour Mill. How many tons of wheat does he have left?

33) Larry starts with a bar of steel of length $21\frac{3}{4}$ inches and cuts off pieces of length $3\frac{1}{8}$ inches, $2\frac{1}{4}$ inches, and $5\frac{1}{2}$ inches. How much of the bar is left, allowing for a waste of $\frac{3}{32}$ inch for each cut?

34) The actual weight of a box of Whammy O's cereal must be within $\frac{1}{4}$ oz of the weight shown on the box. What are the maximum and minimum weights allowable if the weight on the box is $10\frac{1}{2}$ oz?

35) The outlet voltage on a 110 volt line that has a voltage drop of $6\frac{7}{8}$ volts from the panel bus to the outlet is given by $V = 110 - 6\frac{7}{8}$. What is the voltage at the outlet?

2.12 ORDER OF OPERATIONS

APPLICATION

The inside height of a china hutch is to be $66\frac{1}{4}$ inches. Five shelves are to be inserted, each one $\frac{3}{4}$ inch in thickness and spaced equally. The distance between the shelves is given by

$$D = (66\frac{1}{4} - 5 \cdot \frac{3}{4}) \div 5$$

Calculate the distance.

OBJECTIVE

31) Perform any combination of operations (addition, subtraction, multiplication, and/or division) on numbers, written in fraction form, in the conventional order.

VOCABULARY

No new vocabulary.

THE HOW AND WHY OF IT

The conventional order of operations for fractions is the same as for whole numbers.

EXAMPLES

a) $\frac{5}{6} - \frac{1}{2} \cdot \frac{2}{3} = \frac{5}{6} - \frac{1}{3}$ (multiplication performed first)

$= \frac{3}{6}$ (subtraction performed next)

$= \frac{1}{2}$ (reduce to lowest terms)

b) $\frac{1}{2} \div \frac{2}{3} \cdot \frac{1}{4} = \frac{3}{4} \cdot \frac{1}{4}$ (division performed first)

$= \frac{3}{16}$ (multiplication performed next)

Perform the indicated operations.

1) $\frac{1}{3} \cdot \frac{4}{5} + \frac{4}{15}$

2) $\frac{3}{10} + \frac{4}{5} \cdot \frac{1}{2}$

3) $\frac{7}{8} - \frac{1}{2} \cdot \frac{3}{4}$

4) $\frac{3}{4} \div 2 + \frac{1}{8}$

5) $\frac{1}{3} \cdot \frac{3}{7} \div \frac{1}{4}$

6) $\frac{1}{3} \div \frac{2}{3} \cdot \frac{4}{7}$

7) $\frac{7}{10} - \frac{3}{10} + \frac{1}{10}$

8) $\frac{5}{3} \cdot \frac{2}{3} - \frac{7}{9}$

9) $\frac{1}{4} + \frac{3}{8} \div \frac{1}{2}$

10) $\frac{1}{4} \div \frac{3}{8} + \frac{1}{2}$

11) $\frac{3}{4} \div \frac{1}{3} \cdot \frac{1}{6}$

12) $\frac{3}{4} \div \left(\frac{1}{3} \cdot \frac{1}{6} \right)$

13) $\frac{1}{3} \cdot \frac{1}{6} \div \frac{3}{4}$

14) $\frac{4}{5} - \frac{2}{5} \cdot \frac{1}{2}$

15) $\frac{4}{5} - \frac{2}{5} + \frac{1}{2}$

16) $\frac{7}{9} - \frac{1}{3} + \frac{1}{2} \div \frac{2}{5} \cdot \frac{4}{5}$

17) $\left(\frac{7}{9} - \frac{1}{3} + \frac{1}{2} \right) \div \frac{2}{5} \cdot \frac{4}{5}$

18) $\frac{17}{9} - \left(\frac{1}{3} + \frac{1}{2} \div \frac{5}{2} \right) \cdot \frac{5}{4}$

19) In the application at the beginning of this section, what is the distance between the shelves?

20) The resistance of a series-parallel electric circuit is given by

$$\Omega = 4\frac{1}{3} + 3\frac{1}{6} + 1 \div \left(\frac{1}{2} + \frac{1}{4} + \frac{2}{3}\right) \text{ ohms}$$

Find the resistance of the circuit.

21) To check that $x = \frac{2}{3}$ is a solution to the equation $6 - 5x + 3(x + 1) = 7\frac{2}{3}$, we substitute $\frac{2}{3}$ for x and find the value of

$$6 - 5\left(\frac{2}{3}\right) + 3\left(\frac{2}{3} + 1\right)$$

Verify that the above expression equals $7\frac{2}{3}$.

22) Substitute $x = 1\frac{1}{6}$ in the following equation and check that it is a solution (makes the equation true).

$$3x - 2\frac{1}{4} = 1\frac{1}{4}$$

1. (Obj. 28) Add. $4\frac{3}{5}$

   $1\frac{2}{3}$

   $7\frac{1}{2}$

   $4\frac{5}{6}$

   _____

2. (Obj. 31) Perform the indicated operations.

   $\frac{2}{3} \cdot \frac{9}{10} - \frac{4}{5} \div \frac{8}{3}$ _____

3. (Obj. 25) Divide. $\frac{2}{3} \div \frac{5}{8}$ _____

4. (Obj. 23) Multiply. $3\frac{1}{2} \cdot 4\frac{2}{3} \cdot 1\frac{5}{7}$ _____

5. (Obj. 27) Add. $\frac{1}{2} + \frac{1}{4} + \frac{1}{3}$ _____

6. (Obj. 25) Divide. $3\frac{1}{2} \div 3\frac{1}{4}$ _____

7. (Obj. 30) Subtract. $15\frac{5}{7}$

   $5\frac{5}{9}$

   _____

8. (Obj. 20) Find the missing numerator. $\frac{2}{9} = \frac{?}{108}$ _____

9. (Obj. 30) Subtract. $17 - 4\frac{5}{6}$ _____

10. (Obj. 24) What is the reciprocal of $3\frac{2}{5}$?   _____

11. (Obj. 29) Subtract. $\frac{7}{8} - \frac{2}{3}$ _____

12. (Obj. 23) Multiply. $\dfrac{3}{4} \cdot \dfrac{2}{7} \cdot \dfrac{14}{9}$ _____

13. (Obj. 31) Perform the indicated operations.

$$\dfrac{1}{2} + \dfrac{1}{3} \cdot \dfrac{5}{6}$$ _____

14. (Obj. 19) Multiply. $\dfrac{4}{5} \cdot \dfrac{9}{7}$ _____

15. (Obj. 21) List the following fractions in order from smallest to largest.

$$\dfrac{3}{10}, \ \dfrac{3}{8}, \ \dfrac{2}{5}$$ _____

16. (Obj. 30) Subtract.

$$11\dfrac{2}{3}$$

$$\underline{5\dfrac{7}{8}}$$ _____

17. (Obj. 28) Add. $3\dfrac{5}{8} + 4\dfrac{1}{3} + \dfrac{5}{6}$ _____

18. (Obj. 25) Divide. $\dfrac{7}{8} \div 21$ _____

19. (Obj. 26) Add. $\dfrac{27}{96} + \dfrac{13}{96}$ _____

20. (Obj. 21) List the following fractions in order from smallest to largest.

$$\dfrac{5}{9}, \dfrac{4}{7}, \dfrac{2}{3}, \dfrac{10}{21}$$ _____

21. (Obj. 22) Reduce to lowest terms. $\dfrac{42}{63}$ _____

22. (Obj. 28) Four poured brass castings have weights of $38\dfrac{3}{4}$, $37\dfrac{1}{2}$, $42\dfrac{7}{8}$, and $27\dfrac{5}{16}$ lb. What is the total weight of the metal in the four castings? _____

23. (Obj. 30) A metal bar weighing $48\frac{5}{8}$ lb is machined and then reweighed. If the bar then weighs $42\frac{3}{4}$ lb, how much metal was removed? _____

24. (Obj. 23) The distance around a circle is about $3\frac{1}{7}$ times its diameter. What is the approximate distance around a circle whose diameter is $4\frac{1}{2}$ inches? _____

25. (Obj. 25) How many pieces may be cut from a 2 X 4 that is 16 ft long if each piece is to be $1\frac{1}{3}$ ft in length? Disregard any waste from the saw cut. _____

CHAPTER 3

DECIMALS

3.1 DECIMAL NUMBERS: NUMERAL FORM AND WORD NAMES

An accountant must write a company check for $73.07. How should she write the amount in words on the check?

32) Write numeral forms from word names.
33) Write word names from numeral forms.

Recall that the symbols 0, 1, 2, 3, 4, 5, 6, 7, 8, and 9 are called digits. These digits, and a period called a *decimal point*, are used to write other names for fractions and mixed numbers, ($\frac{1}{4}$ = .25). These other names are called *decimal numerals*, or more commonly *decimal numbers* or *decimals*. This is an extension of the place value system for whole numbers, since place value in decimals is used in a similar manner.

The number of digits to the right of the decimal point is sometimes referred to as the number of *decimal places*. For example, 3.47 is said to have two decimal places.

Decimals are written by using a standard place value in the same manner as the whole numbers. The place value in the case of decimals is:

1) the same as whole numbers for digits to the left of the decimal point, and

2) a fraction whose denominator is a power of 10 for digits to the right of the decimal point.

The digits to the right of the decimal point have place values of $\frac{1}{10^1}$, $\frac{1}{10^2}$, $\frac{1}{10^3}$, . . . in that order from left to right. When the denominator of a fraction is a power of ten, it can be written as a decimal. Any fraction equivalent to such a fraction can also be written as a decimal. This includes all whole numbers, since any whole number can be written with a denominator of 1 (or 10^0).

Using the ones place as the central position (the place value 10^0), the place value of a decimal looks like the following:

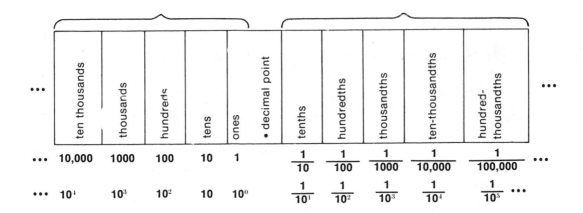

Note that the decimal point separates the whole number part from the fraction part.

When reading and writing decimals with digits only to the right of the decimal point, read the whole number formed by the digits (ignoring the decimal point) followed by the place value of the digit farthest to the right. When reading and writing numerals with digits to the left *and* to the right of the decimal point, the digits to the left are read as a whole number, the word "and" is used to indicate the decimal point, and then the digits to the right are read according to the directions just given. These decimal numerals are like mixed numbers, since each has a whole number part and a fraction part.

EXAMPLES

a)   The word name for .58 is "fifty-eight hundredths." This can also be written $\frac{58}{100}$.

b)   The word name for .003 is "three thousandths." This can also be written $\frac{3}{1000}$.

c)   The word name for 2.24 is "two and twenty-four hundredths." This can also be written $2\frac{24}{100}$.

d)   The numeral form for "fifteen ten-thousandths" is .0015.

e)   The numeral form for "four hundred five and four hundred five thousandths" is 405.405.

f)   A single wire strand in an insulating cable is fifty ten-thousandths inch in diameter. In numeral form this is written .0050 inch.

Write (or read) the word name of each of the following.   **EXERCISES**

1) 2.5   2) .016

3) 36.05   4) 16.16

5) .0018

Write (or read) the numeral form of each of the following:

6) Four tenths

7) Two and two hundredths

8) Thirteen and fifteen thousandths

9) Six hundredths

10) One hundred and one hundredth

Write the word name of each of the following:

11) .4   12) 0.219

13) 3.26   14) 48.0004

15) .542   16) .504

17) 5.04   18) 50.4

19) 500.4   20) 50.04

21) 5.004   22) 18.0205

23) 45.   24) 384.0

25) .345   26) 300.045

Write the numeral form for each of the following.

27) Fifteen hundredths

28) Five hundred and five thousandths

29) Five thousand and five hundredths

30) Five hundred-thousandths

31) Five thousand

32) One thousand and five thousandths

33) Two hundred and thirty-one thousandths

34) Two hundred thirty-one thousandths

35) In the application at the beginning of this section, what words did she use to write the amount of the check?

36) Pete Rios is typing up the bid specification for a new press from an audio tape. How does he record the following specifications in numeral form? The minimum sheet size is to be five and twenty-five hundredths by eight and thirty hundredths. The maximum sheet size is to be ten and six hundredths by twenty-one and seventy-five hundredths.

3.2 LISTING DECIMALS IN ORDER OF VALUE

APPLICATION   An Insapik camera has a fixed exposure time of .042 second. A Trupix camera has an exposure time of .0325 second. Which camera has the shortest exposure time?

OBJECTIVE   34) Arrange a given group of decimals in order of value from smallest to largest. The decimals .2, .35, and .471 are in order from smallest to largest.

VOCABULARY   A group of numbers is said to be *ordered* when they are arranged in order of value from smallest to largest.

THE HOW AND
WHY OF IT   Fractions can be ordered when they have a common denominator by ordering the numerators. This idea can be extended to decimals when they have the same number of places to the right of the decimal point. This assures us of common denominators when we write them as fractions. Thus, $.26 = \frac{26}{100}$ and $.37 = \frac{37}{100}$ have a common denominator when written in fraction form; therefore we know that .26 is less than .37. If the

decimals do not contain the same number of places to the right of the decimal point, zeros can be inserted on the right-hand end. So, .3 and .15 have a common denominator when written in fraction form as .3 = .30 = $\frac{30}{100}$ and .15 = $\frac{15}{100}$. Therefore, .15 is less than .3.

Decimals can be listed in order by the following procedure:

Step 1) Make sure that all numbers have the same number of decimal places by inserting zeros on the right when necessary.

Step 2) Remove the decimal points.

Step 3) Write the numbers in order from smallest to largest.

Step 4) Replace the decimal points and remove the extra zeros.

EXAMPLES

a) .62, .637, .6159, .621. First write all numbers with the same number of decimal places to the right of the decimal point.

.6200, .6370, .6159, .6210

Now order the whole numbers formed by removing the decimal points.

6159, 6200, 6210, 6370

Replace the decimal points and remove the extra zeros. The numbers are now ordered.

.6159, .62, .621, .637

b) 1.357, 1.361, 1.3534, 1.358

1.3570, 1.3610, 1.3534, 1.3580

13534, 13570, 13580, 13610

1.3534, 1.357, 1.358, 1.361

List the decimals in each group in order of value from smallest to largest. **EXERCISES**

1) .8, .81, .801

2) .06, .05, .07

3)   1.3, 1.29, 1.31

4)   17.0, 16.9, 17.5

5)   .0051, .0049, .00491

6)   .11, .09, .13

7)   .113, .115, .112, .119

8)   100.6, 99.9, 99.08

9)   4.35, 3.96, 4.87, 2.49

10)   1.67, 1.65, 1.6, 1.7

11)   .565, .556, .566, .555

12)   .9, .899, .86, .91, .903

13)   .0031, .00305, .003, .00312

14)   .1163, .116, .1159, .117

15) 17.05, 17.16, 17.0506, 17.057

16) .0729, .073001, .072, .073, .073015

17) .30009, .301, .3008, .30101

18) .888, .88799, .8881, .88759

19) 1.999, 2., 1., 1.5, 2.006

20) 8.36, 8.2975, 8.3599, 8.3401

21) In the application at the beginning of this section, which camera has the shortest exposure time?

22) Greg and Larry each measured the diameter of a coin and got the following measurements respectively: .916 centimeter and .921 centimeter. Which measurement is the smallest?

23) The Davis Meat Co. bid 98.375 cents per pound to provide meat to the Ajax Grocery. Circle K Meats put in a bid of 98.35 cents, and J & K Meats made a bid of 98.3801 cents. Which is the best bid for the Ajax Grocery?

24) A vial contains 2.3059 grams of charcoal. Dan weighs the charcoal and determines that it weighs 2.31 grams. Is Dan's weighing too light or too heavy?

3.3 APPROXIMATION BY ROUNDING

APPLICATION   The displacement of a V-8 diesel is about 4941.3829 cc. What is the displacement to the nearest hundredth?

OBJECTIVE   35) Round off a given decimal to a specified place value.

VOCABULARY   Decimals can be considered to be exact or approximate, depending upon whether they name the number under discussion *(exact decimal)* or whether they are an approximation of a number to a specified place value *(approximate decimal).*

The exact decimal 15.3 is said to be *rounded off* to the nearest whole number when expressed as 15. In this case, 15 is an approximate decimal. ≈ is a symbol for "approximately equal to," so 15.3 ≈ 15.

In practical applications, most decimals are the result of reading measurements, and therefore are approximate decimals.

THE HOW AND WHY OF IT   Examine the following portion of a measuring device:

If we are asked to give an approximation of the number indicated by the arrow to the nearest tenth, the answer is 2.6, because the arrow is over half way between the 2.5 and the 2.6. Rounded to the nearest hundredth, the answer is 2.56, because the arrow is less than half way between 2.56 and 2.57.

To round off, all we need to do is see if the number we have is less than half way between the two approximate measurements. If it is less than half way, choose the smaller approximation. If it is not less than half way, choose the larger approximation.

To round 6.3265 to the nearest hundredth, draw an arrow under the hundredths place to identify the round off position.

6.3265
↑

We must choose between 6.32 and 6.33. Since the first digit to the right of the round off position is 6, the number is more than half way to 6.33. So,

$$6.3265 \approx 6.33$$

To round off a decimal number to a given place value:

Step 1) Draw an arrow under the given place value.

Step 2) Change all digits following the given place value to zeros.

Step 3) Select two approximations. The first is the number found in step 2. The second is found by adding one (1) to the digit in the place value indicated by the arrow in the first approximation.

Step 4) Choose the approximation by using

the larger approximation: ⎰ if the digit in the place value after the arrow in step 1 is 5, 6, 7, 8, or 9

the smaller approximation: ⎰ if the digit in the place value after the arrow in step 1 is 0, 1, 2, 3, or 4

EXAMPLES

a)   .3582 rounded to the nearest hundredth is .36.
         ↑
     .36 is chosen instead of .35 because the 8 indicates that the number is more than half way between .35 and .36. So, .3582 ≈ .36.

b)   3582 rounded to the nearest thousand is 4000, because the digit to the right of
     ↑
     the round off position is 5.

c)   16.349 rounded to the nearest tenth is 16.3, because the digit to the right of the
          ↑
     round off position is 4.

d)   249.7 rounded to the nearest unit is 250. The choice is between 249 and 250;
         ↑
     the 7 tells us to choose the larger approximation.

e)   3.996 rounded to the nearest hundredth is 4.00. The choice is between 3.99 and
         ↑
     3.99 + .01 = 4.00. The 6 tells us to choose the larger approximation.

EXERCISES   Round each of the following decimals to the nearest tenth, hundredth, and thousandth.

1) 2.6532   2) .8359   3) 12.3015

4) 1.3347   5) 9.9892   6) 10.0752

7) 53.3125   8) 9.7765   9) 2.1789

10) 3.0007   11) 0.7934   12) 14.5553

13) 0.7891   14) 1.1139   15) .9999

Round each of the following to the nearest hundred.

16) 6,921   17) 75,671   18) 3,971

19) 3700.89   20) 3049.5

Round each of the following to the nearest ten.

21) 26,775   22) 379.15   23) 396.51

24) 14.95   25) 7,453

Round each of the following to the nearest thousand.

26) 23,786   27) 4,978   28) 965.0348

29) 43,215   30) 999,501

Round each of the following to the nearest whole number.

31) 36.58   32) 100.495   33) 15.50

34) 5.399   35) 78.905

36) In the application at the beginning of this section, what is the displacement to the nearest hundredth?

37) A chemist made an analysis of the ingredients in a box of Wheaties. He finds 10.063 grams of non-nutritive fiber. Round this weight to the nearest tenth of a gram.

38) A micrometer is used to measure the thickness of a metal bar. The operator reads the measurement as .4978 inch. What is the thickness of the bar to the nearest hundredth of an inch?

3.4 ADDITION AND SUBTRACTION OF DECIMALS

APPLICATION   A train has, in part, 2 carloads of wheat which weigh 41.93 tons each. There are 3 carloads of corn which weigh 38.09 tons each. How much weight do the five carloads add to the train's load? How much more does a carload of wheat weigh than a carload of corn?

OBJECTIVES   36) Find the sum of two or more numbers written as decimals.
37) Find the difference of two numbers written as decimals.

VOCABULARY   The vocabulary for the addition and subtraction of decimals is the same as for whole numbers. Recall that the number of digits after the decimal point is called the number of decimal places.

THE HOW AND WHY OF IT   The addition facts commonly listed in the "addition table" are used in the addition of decimals in much the same way they are used in the addition of whole numbers.

$$6.3 + 2.5 = ?$$

We make use of the expanded notation of our place value system to explain the process.

6.3 = 6 ones + 3 tenths
2.5 = 2 ones + 5 tenths
8 ones + 8 tenths = 8.8

The vertical form gives us a natural grouping of the ones and tenths.

> Decimals can be added as if they were whole numbers by writing them in vertical columns and by keeping the digits arranged in columns having the same place value.

Thus, 2.8 + 13.4 + 6.22 can be written 2.80 + 13.40 + 6.22 (with the same number of decimal places as you use when ordering decimals) and then,

2.80
13.40
6.22 (see example a)

Consider 6.59 − 2.34. Written vertically in expanded form, we have:

6.59 = 6 ones + 5 tenths + 9 hundredths
2.34 = 2 ones + 3 tenths + 4 hundredths
----
         4 ones + 2 tenths + 5 hundredths

So, 6.59 − 2.34 is 4.25. Note that we subtracted ones from ones, tenths from tenths, and hundredths from hundredths.

> Subtraction of decimals is based on place value in the same manner as subtraction of whole numbers.

When necessary, we can regroup or "borrow" as with whole numbers. 6.271 − 3.845 = ? In expanded form we have:

6.271 = 6 ones + 2 tenths + 7 hundredths + 1 thousandth
3.845 = 3 ones + 8 tenths + 4 hundredths + 5 thousandths

Since we cannot subtract 5 thousandths from 1 thousandth, or 8 tenths from 2 tenths, we regroup. "Borrow" one of the hundredths from the seven hundredths and combine with the 1 thousandth (.01 = .010, and .010 + .001 = .011 or 11 thousandths). Also "borrow" one of the ones from the 6 ones and combine with the 2 tenths (1 = 1.0, and 1.0 + .2 = 1.2 or 12 tenths). Then we have:

6.271 = 5 ones + 12 tenths + 6 hundredths + 11 thousandths
3.845 = 3 ones + 8 tenths + 4 hundredths + 5 thousandths
----
         2 ones + 4 tenths + 2 hundredths + 6 thousandths

So, 6.271 − 3.845 = 2.426. The examples show the common shortcuts.

EXAMPLES

a)
```
   11
   2.80
  13.40
   6.22
  22.42
```

Think:

hundredths column:	0 + 0 + 2 = 2, no carry
tenths column:	8 + 4 + 2 = 14, carry the "1" to the ones column and write the "4" in the tenths column (14 tenths = 1.4)
ones column:	1 + 2 + 3 + 6 = 12, carry the "1" to the tens column and write the "2" in the ones column
tens column:	1 + 1 = 2, no carry

b)   1.05 + .723 + 72.6 + 8 = ? Write each decimal with the same number (three) of decimal places: 1.050 + .723 + 72.600 + 8.000 or

$$\begin{array}{r} 1.050 \\ .723 \\ 72.600 \\ \underline{8.000} \\ 82.373 \end{array}$$

c)   Subtract: Think:

5.831   We must regroup since we cannot subtract 7 thousandths from 1
.287   thousandth. Therefore, we borrow 1 hundredth from the hundredths
          to regroup with the thousandths.

$$\begin{array}{r} 2\ 11 \\ 5.8\cancel{3}\ \cancel{1} \\ \underline{.28\ 7} \end{array}$$

$$\begin{array}{r} 12 \\ 7\ \cancel{2}\ 11 \\ 5.\cancel{8}\ \cancel{3}\ \cancel{1} \\ \underline{.2\ 8\ 7} \\ 5.5\ 4\ 4 \end{array}$$ Now we cannot subtract 8 hundredths from 2 hundredths, so we borrow one tenth from the tenths to regroup with the hundredths.

Check: $$\begin{array}{r} .287 \\ \underline{5.544} \\ 5.831 \end{array}$$

d)   Subtract:

6
2.94

$$\begin{array}{r} 5\ 10 \\ \cancel{6}.\cancel{0}\ 0 \\ \underline{2.9\ 4} \end{array}$$ Think:

The "6" can be rewritten as 6.00 so both numbers will have the
$$\begin{array}{r} 9 \\ 5\ \cancel{10}\ 10 \\ \cancel{6}.\cancel{0}\ \cancel{0} \\ \underline{2.9\ 4} \\ 3.0\ 6 \end{array}$$ same number of decimal places. We borrow 1 one and add it to the tenths place (10 tenths) and then borrow 1 tenth and add it to the hundredths place. We can now subtract in each column.

Check: $$\begin{array}{r} 2.94 \\ \underline{3.06} \\ 6.00 \end{array}$$

Add.

1) .6
    .5

2) .03
    .27

3) 6.11
    .98

4) 7.23
    2.54

5) 5.42
    3.5

Subtract.

6) .8
    .5

7) .75
    .09

8) 16.31
    3.25

9) 4.63
    1.22

10) 7.7
    5.23

Add.

11) 5.302
     9.01
     .9479
     14.85

12) 5.19
    9.957
    4.3

13) 87.5521
     572.63
     98.007
     113.98

14) 213.6
    48.089
    117.35
    92.91

15) 1.904 + 3.33 + 7.9 + 16.63

16) 12.2 + 3.72 + 5.14 Round sum to the nearest tenth.

17) .0075 + .012 + .0009 + .003 Round sum to the nearest thousandth.

Subtract.

18) 3.457
   2.509

19) .303
   .178

20) 12.1
   9.34

21) .845
   .076

22) 9.006
   3.257

23) 10
   3.582

24) 3.18 − .083

25) 162.375 − 87.496. Round difference to nearest hundredth.

26) 38.06 − 14.5

27) 3 − 2.651. Round difference to nearest tenth.

28) Find the difference of .529 and .057.

29) In the application at the beginning of this section, how much weight was added to the train and how much more does wheat weigh per carload?

30) During a chemistry experiment, Ross had 10.3 cc of a solution. He used 7.6 cc. How much of the solution did he have left?

31) Jack went shopping with $25.50 in cash. He bought a record for $4.89 and a sweater for $11.88. On the way home he stopped for $4.75 worth of gas. How much cash did he have left?

32) Russ was hitchhiking across the city. The first car to offer him a ride took him 3.8 miles. The second took him 8.5 miles, and the third car went 7.4 miles before he got out. How many miles did he ride altogether?

33) On a short vacation trip, Paul stopped for gas four times. The first time he got 8.6 gallons. At the second station he got 14.9 gallons, and at the third he got 15.4 gallons. At the last stop he got 13.5 gallons. How much gas did he buy on the trip?

34) Heather wrote five checks in the amounts of $34.85, $17.37, $67.48, $21.10, and $8.10. She has $152.29 in her checking account. Does she have enough money deposited to cover the five checks?

3.5 MULTIPLICATION OF DECIMALS

Gasoline costs 69.9 cents a gallon. How much will 56.3 gallons cost?   APPLICATION

38) Find the product of two numbers written as decimals.   OBJECTIVE

VOCABULARY   No new vocabulary. (See multiplication of whole numbers.)

THE HOW AND   The "multiplication table" for whole numbers is used in the multipli-
WHY OF IT   cation of decimals in the same way it is used to multiply whole numbers.
In fact, the multiplication procedure for decimals is, with one exception,
the same as for whole numbers. The one exception is in the placement of
the decimal point in the product. To determine the placement of the
decimal in (.3) (.8), we write the factors as fractions:

$$\frac{3}{10} \times \frac{8}{10} = \frac{24}{100}$$

In decimal form this becomes: .3 × .8 = .24

> To find the product of two decimals, first multiply the numbers
> as if they were whole numbers. To determine the placement of the
> decimal point, we count the number of decimal places (to the right)
> in each of the factors. The sum of these two counts is the number of
> decimal places the product must have.

This is true because in fraction form the denominators are multiples
of ten, and the product of these is the same or a larger multiple of ten.
Thus, the product 7.2 × .13 requires three decimal places. Since 72 × 13
= 936, the product is

$$7.2 \times .13 = .936 \quad \text{or} \qquad \begin{array}{r} 7.2 \\ .13 \\ \hline 216 \\ 72 \\ \hline .936 \end{array}$$

It may happen that the product of two decimals does not contain
enough digits for the necessary number of decimal places. For instance,
for .2 × .3 the product of 2 × 3, which is 6, is a *one* digit numeral and we
need *two* decimal places in the product. In this case, we insert a zero
before the six:

$$.2 \times .3 = .06$$

This is true because

$$\frac{2}{10} \times \frac{3}{10} = \frac{6}{100} = .06$$

EXAMPLES

a) .7 × 6 = 4.2

b) 7 × .6 = 4.2

c) .7 × .6 = .42

d) 11 × .33 = 3.63

e) .53   f)   2.31   g)   .21   Note that we must insert a zero be-
    15   3.4   .14   fore the "2" because the product
   265   924   84   must have four decimal places.
    53   693   21
  7.95   7.854   .0294

Multiply.

1) .3
   .6
	2) 3(3.24)	3) 1.05
		      7

4) 5(2.16)	5) 2.6
	    .003
		6) 7.51
		    .03

7) 8.65
   .001
	8) .33
	   .06
		9) .5(2.3)

10) .6(1.09)	11) 1.4
	    2.1
		12) 8.97
		    .61

13) 34.6
    6.5
	14) .272
	    3.1
		15) 9.04
		    .47

16) 7.41
    .206
	17) 8.2
	    .68
		18) 7.213
		     88

19) 3.32
.04

20) 6.005
75

21) .059
6.4

22) 98
.079

23) .345
4.8

24) (5.31) (5.9)

25) (.011) (.032)

26) (2.52) (1.37)

27) (6.02) (3.59)

28) (9.08) (.078)

29) (.0649) (2.5)

30) (18) (5.09) (9.3)

31) (3.99) (9.4) (2.1)

32) Find the product of 4.07 and 7.17. Round to the nearest thousandth.

33) In the application at the beginning of this section, how much did the gasoline cost?

34) Joe earns $2.37 an hour. How much did he earn during the week if he worked 30.25 hours? (Round to the nearest hundredth of a dollar.)

35) If gasoline costs 65.9 cents per gallon, how much must Dale pay for 12.7 gallons?

36) Melanie bought six records on sale. The advertised sale price was two records for $3.87. How much did she pay for them?

37) Fred and Freda deposit $300 in a savings account that pays 6% interest compounded annually. The formula for finding the value of their savings to the nearest cent at the end of three years is

$$A = 300 \times (1.06)^3$$

Find the value of their savings.

3.6 DIVISION OF DECIMALS

A compact car averages 26.3 miles per gallon for a long trip. How many gallons would be needed for a trip of 1,480 miles? State the number of gallons to the nearest tenth of a gallon.

APPLICATION

39) Find the quotient (or approximate quotient) of two numbers written as decimals.

OBJECTIVE

No new vocabulary. (See division of whole numbers.)

VOCABULARY

Division of a decimal by a counting number is the same as division of whole numbers except for the placement of the decimal point. The decimal point is written above the decimal point of the dividend.

THE HOW AND WHY OF IT

Division is the inverse of multiplication; therefore, the quotient of a decimal divided by a counting number whose prime factors are 2 and/or 5 must have the same number of decimal places as the dividend. The decimal point can be placed correctly by writing it directly above the one in the dividend. After that, the division can be done as if the numbers were whole numbers. In the dividend, zeros may be inserted after the last digit following the decimal point, if necessary, to continue the division process. It should be noted that the quotients where the divisors are counting numbers with only prime factors of two and/or five are always exact decimals.

Consider:

```
        .0019
20).0380
     20
    180
    180
      0
```

Check: .0019
         20
       ‾‾‾‾‾‾
       0.0380

Think: Write the decimal point for the quotient above the decimal point in the dividend. Zero divided by twenty is zero. Write "0" above the "0." There are no groups of twenty in 3, so write "0" above the "3." Next, divide thirty-eight by twenty. There is one group of twenty in thirty-eight, so write "1" above the "8." Multiply and subtract as with whole numbers. The difference is 180. 180 divided by 20 is nine. Write "9" above the "0." Multiply and subtract. Since the remainder is zero, the division is complete. Note that the quotient and the dividend both have four decimal places.

When a decimal is divided by 7, or any other whole number that has a prime factor other than 2 or 5, the division process may not have a remainder of zero at any step.

```
      .33
7)2.34
  2 1
   24
   21
    3
```

At this step we can insert zeros after 2.34 (for example, 2.34000) as before.

```
      .33428
7)2.34000
  2 1
   24
   21
   30
   28
   20
   14
   60
   56
    4
```

It appears that we might go on inserting zeros and continue endlessly. This, indeed, is what would happen. In practical applications, we stop the division process one place value beyond the accuracy required by the situation, and then round off. Therefore,

$2.34 \div 7 \approx .33$ to the nearest hundredth and

$2.34 \div 7 \approx .3343$ to the nearest ten-thousandth.

Division problems in which the divisor also contains a decimal point are easiest to handle by changing to an equivalent problem in which the divisor is a counting number. This can be done by multiplying both the divisor and the dividend by a power of 10. Since every indicated division can be treated as a fraction, this is the same as multiplying by some form of 1 which results in an equivalent fraction. For instance,

$$2.338 \div .7 = \frac{2.338}{.7} \cdot \frac{10}{10} = \frac{23.38}{7}$$

$$7.8 \div .014 = \frac{7.8}{.014} \cdot \frac{10^3}{10^3} = \frac{7800}{14}$$

$$4.865 \div .23 = 486.5 \div 23$$

To divide two numbers written as decimals, first write (if necessary) an equivalent division problem with a counting number for the divisor. This can be done by moving the decimal point to the right the same number of places in both the dividend and the divisor to make the divisor a counting number. Then divide as before, rounding off to the desired place value.

EXAMPLES

a)
```
     .23
  8)1.88
    1 6
     28
     24
      4
```
Here the remainder is not zero, so the division is not complete. We may insert a zero (1.880) without changing the value of the dividend, and continue dividing.

```
     .235
  8)1.880
    1 6
     28
     24
      40
      40
       0
```
Note that the quotient (.235) and the rewritten dividend (1.880) both have three decimal places.

Check: .235
           8
         1.880

b) 486.5 ÷ 23   Round to the nearest hundredth.

$$\begin{array}{r}21.152\\23\overline{)486.500}\\46\\\hline26\\23\\\hline35\\23\\\hline120\\115\\\hline50\\46\\\hline4\end{array}$$

It is necessary to insert two zeros in order to round off to hundredths.

Hence, $486.5 \div 23 \approx 21.15$.

c) Find the quotient correct to the nearest thousandth.

$.072\overline{).47891}$

Multiply dividend and divisor by 1000. (As the arrows show, move the decimal point in each number three places to the right.)

$$\begin{array}{r}6.6515\\72\overline{)478.9100}\\432\\\hline46\;9\\43\;2\\\hline3\;71\\3\;60\\\hline110\\72\\\hline380\\360\\\hline20\end{array}$$

Hence, $.47891 \div .072 \approx 6.652$.

EXERCISES   Divide.

1) $6\overline{).24}$ 2) $6\overline{)2.4}$ 3) $.6\overline{)24}$

4) $.6\overline{).24}$ 5) $.06\overline{)2.4}$ 6) $.3\overline{)6.96}$

7) $.5\overline{)1.2}$ 8) $.032 \div 4$ 9) $25\overline{)10}$

10) $.25\overline{)1.5}$ 11) $25\overline{)12.3}$ 12) $4\overline{)2.95}$

13) $80\overline{)104.8}$ 14) $20\overline{)7}$ 15) $200\overline{)3.6}$

16) $16.64 \div 32$ 17) $6.5 \div 125$ 18) $64\overline{)6211.84}$

19) $32\overline{)201.824}$

Divide. Round off to the nearest hundredth.

20) $3\overline{)13.7}$ 21) $9\overline{)2.202}$

22) $7.4\overline{)17.7}$ 23) $.41\overline{).716}$

Divide. Round off to the nearest thousandth.

24) $1.09\overline{)6.48}$ 25) $2.2\overline{)34.22}$

26) $24.9\overline{)60.363}$ 27) $131\overline{)29}$

Divide. Round off to the nearest tenth.

28) $.029\overline{).226}$ 29) $3.46\overline{)56.99}$

30) $518\overline{)3582}$ 31) $72.9\overline{)486.5}$

32) In the application at the beginning of this section, how many gallons of gas will be needed?

33) Vern bought a pair of red socks. The socks were on sale at 3 pairs for $2.60. How much did he pay for the socks?

34) Ross drove 214.6 miles on 11.3 gallons of gas. What was his mileage (miles per gallon)? Round to the nearest mile.

35) Pat rides his motor bike 237.8 miles in 5 hours. What is his average speed? (Round to the nearest mile per hour.)

36) If soup is on sale at eight cans for $1.20, what is the price of one can?

37) Eighty alumni of Tech U. donated $3560.72 to the University. What was the average donation?

3.7 FRACTIONS TO DECIMALS

Chef Carl is in charge of food preparation at a banquet for several clubs. He has already decided to bake 25 loaves of bread when he is told that one club, about $\frac{1}{8}$ of the total, cannot come. What decimal (equal to $\frac{7}{8}$) should he multiply by 25 to find how many loaves of bread will be needed now?

APPLICATION

40) Change a fraction or mixed number to either an exact decimal or to an approximate decimal to the nearest tenth, hundredth, or thousandth.

OBJECTIVE

No new vocabulary.

VOCABULARY

Recall that every fraction can be interpreted as the indicated division of two whole numbers, $\frac{a}{b} = a \div b$.

THE HOW AND WHY OF IT

To change a fraction to a decimal, divide the numerator by the denominator.

To change a mixed number to a decimal, change the fraction part to a decimal and add to the whole number part.

To find an approximate decimal name for a fraction, divide the numerator by the denominator and round off the result to the desired decimal place.

To change a mixed number to an approximate decimal, find the decimal approximation of the fraction part and add it to the whole number part.

EXAMPLES

a) $\frac{7}{20}$

$$\begin{array}{r} .35 \\ 20\overline{)7.00} \\ \underline{6\ 0} \\ 1\ 00 \\ \underline{1\ 00} \\ 0 \end{array}$$

$\frac{7}{20} = .35$

b) $\frac{9}{16}$

$$\begin{array}{r} .5625 \\ 16\overline{)9.0000} \\ \underline{8\ 0} \\ 1\ 00 \\ \underline{96} \\ 40 \\ \underline{32} \\ 80 \\ \underline{80} \\ 0 \end{array}$$

$\frac{9}{16} = .5625$

c) $\frac{7}{12} = 7 \div 12.$

If we divide 7 by 12 we can find the desired approximate decimal.

$$\begin{array}{r} .583 \\ 12\overline{)7.000} \\ \underline{6\ 0} \\ 1\ 00 \\ \underline{96} \\ 40 \\ \underline{36} \\ 4 \end{array}$$

$\frac{7}{12}$ to the nearest hundredth is .58

d) $21\frac{3}{7}$

$$
\begin{array}{r}
.428 \\
7\overline{)3.000} \\
2\,8 \\
\hline
20 \\
14 \\
\hline
60 \\
56 \\
\hline
4
\end{array}
$$

Therefore $\frac{3}{7} \approx .43$ to the nearest hundredth.

$21\frac{3}{7} = 21 + \frac{3}{7} \approx 21 + .43 = 21.43$

Write each of the following as a decimal.

EXERCISES

1) $\frac{3}{5}$ 2) $\frac{4}{5}$ 3) $\frac{3}{4}$ 4) $\frac{1}{4}$

5) $\frac{1}{2}$ 6) $\frac{1}{8}$ 7) $\frac{11}{20}$ 8) $\frac{1}{20}$

9) $\frac{3}{8}$ 10) $\frac{14}{20}$ 11) $\frac{11}{40}$ 12) $\frac{3}{16}$

13) $\frac{99}{125}$ 14) $\frac{7}{8}$ 15) $3\frac{5}{8}$ 16) $\frac{77}{200}$

17) $\frac{59}{800}$ 18) $\frac{7}{625}$ 19) $\frac{837}{250}$ 20) $\frac{393}{400}$

Find the approximate decimal to the nearest tenth for each of the following.

21) $\dfrac{1}{3}$ 22) $\dfrac{5}{6}$ 23) $\dfrac{78}{81}$ 24) $\dfrac{28}{33}$

Find the approximate decimal to the nearest hundredth for each of the following.

25) $\dfrac{2}{7}$ 26) $\dfrac{18}{23}$ 27) $5\dfrac{8}{27}$ 28) $\dfrac{45}{67}$

Find the approximate decimal to the nearest thousandth for each of the following.

29) $\dfrac{15}{43}$ 30) $\dfrac{7}{9}$ 31) $\dfrac{7}{12}$ 32) $\dfrac{9}{11}$

33) In the application at the beginning of this section what decimal should Chef Carl use?

34) Mary was paid $3\dfrac{5}{7}$ dollars per hour. How much is this per hour to the nearest cent?

35) Stock prices are listed in dollars and fractions of dollars. If a stock is selling for 6\frac{7}{8}$, write the price in decimal form.

36) In playing the game of craps, the mathematical probability that the person throwing the dice will win is $\frac{244}{495}$, and the probability he will lose is $\frac{251}{495}$. Express these fractions as approximate decimals to the nearest hundredth.

37) The chances of filling an inside straight in a poker game are 4 out of 47 ($\frac{4}{47}$). Express the chances as a decimal to the nearest hundredth.

3.8 REVIEW: ORDER OF OPERATIONS

A mechanic is assigned a service call that requires him to be out of town for four days. He is paid $.15 per mile and $25.00 a day for food and lodging while on service calls. At the beginning of his trip the odometer reading of his car is 24872.3 miles and when he returns it is 25264.8 miles. How much expense money was he paid?

APPLICATION

41) Perform any combination of operations (addition, subtraction, multiplication, and/or division) on numbers, written as decimals, in the conventional order.

OBJECTIVE

No new vocabulary.

VOCABULARY

THE HOW AND   The conventional order of operations for decimals is the same as for
WHY OF IT   whole numbers.

EXAMPLES

a) $.45 - (.32)(.75) = .45 - .24$ (multiplication performed first)
    $= .21$ (subtraction performed next)

b) $.75 \div (.03)(1.6) = (25)(1.6)$ (division performed first)
    $= 40$ (multiplication performed next)

EXERCISES   Perform the indicated operations.

1) $(.3)(2) + .4$ 2) $6.2 - 3(.2)$

3) $.6 \div .2 + 4(.2)$ 4) $(.2)(.5) + .3(.7)$

5) $.30 \div .3 - .2$ 6) $(.6)(.2) \div 2$

7) $.2(.4 - .2)$ 8) $.1(.65 - .34)$

9) $.5 + .3(.2 - .1)$ 10) $10 - 3(.8)$

11) $(1.05)(.75) + .18$ 12) $.08 + 2.37 - 1.6 + .98$

13) $(6.75)(1.3) - 2.61 + (3.2)(.5)$ 14) $4.97 \div (.07)(3.1)$

15) $1.56 - .216 \div .18$ 16) $.0033 \div .88 + 1.075 - .0976$

17) $(25.6)(.08) \div .02$ 18) $.0709 + .096 - .052 - .002$

19) $4.067 - (3.7)(.33) + 1.108$ 20) $(7.86)(5.06) - 3.4 + 7.09$

21) $5.03 - 2.19 + .75$ 22) $6.157 - (3.05)(1.6) + .09$

23) $(2.06)(.13) - (1.17)(.08)$ 24) $36.3 - 8.95 + 6 - 3.1$

25) $4.33 - 2.75 \div 5$ 26) $3.62 \div .02 + (8.6)(.51) - 82.6$

27) In the application at the beginning of this section, how much is he paid for expenses?

28) A welder is paid $7.32 per hour and time and one-half for any hours over 40 in any one week. What is his gross pay if he worked the following hours in one week: 8, 9.5, 8.75, 10, 9.25, and 4.5?

29) A house is valued at $48,500. The yearly rate for fire insurance is $3.42 per thousand. What is the cost of a five-year policy if the charge is 4.2 times the annual rate?

30) Ten years ago a land developer purchased a piece of land that contained 57.04 acres for $85,560. He now plans to divide it into lots of .62 acre each. If he sells each lot for $8000, what is the difference between the original purchase price and the total selling price?

1. (Obj. 38) Multiply.   2.3
          4.6   _____

2. (Obj. 37) Subtract.   7.3862
          7.2986   _____

3. (Obj. 37) Subtract.   .1572 − .0213   _____

4. (Obj. 38) Multiply.   (4.2) (8.3) (.614)   _____

5. (Obj. 34) List the following in order of value from the smallest to largest.

   .762, .0763, .7062, .0702   _____

6. (Obj. 37) Subtract.   15 − 2.14   _____

7. (Obj. 39) Divide. Round answer to nearest tenth.

   $2.3\,\overline{)16.219}$ _____

8. (Obj. 36) Add.   2.34 + .582 + 15.21 + 31 + 2.6   _____

9. (Obj. 40) Write $\frac{3}{7}$ as an approximate decimal rounded to the nearest thousandth.   _____

10. (Obj. 39) Divide. $7\,\overline{).861}$ _____

11. (Obj. 41) Perform the indicated operations.

    $15.36 \div .04 − (.27)(1.8)$ _____

12. (Obj. 39) Divide. Round the answer to the nearest hundredth.

    $.25\,\overline{).2731}$ _____

13. (Obj. 34) List the following in order of value from smallest to largest.

    .0043, .00425, .004, .00431   _____

14. (Obj. 39) Divide. $.21\,\overline{).6342}$ _____

15. (Obj. 41) Perform the indicated operations.

$$7.82 - (1.8)(2.05) + 6.214$$

_____

16. (Obj. 36) Add.   13.264
                      .8132
                    7.1431
                      .021

_____

17. (Obj. 40) Write $\frac{5}{8}$ as a decimal.

_____

18. (Obj. 36) Add.   .612
                  2.145
                  3
                  4.1

_____

19. (Obj. 38) Multiply.   6.234
                      .025

_____

20. (Obj. 41) Perform the indicated operations.

$$(2.3)(1.4) - .62$$

_____

21. (Obj. 36) Find the perimeter (distance around) of a five-sided building lot that has the following measurements: 150.6 m, 81.34 m, 79.6 m, 72.93 m, and 74.9 m.

_____

22. (Obj. 39) A copper wire has a diameter of .125 inch. How many turns of the wire can be wound on a metal core that is 4.5 inches long?

_____

23. (Obj. 38) Norm's pay rate is $4.24 per hour. If he works 7.5 hours per day, what is his salary per day? How much would he make in a five-day week?

_____

24. (Obj. 37) A piece of metal rod 3.375 inches long is cut from a rod that is 29 inches long. Allowing .0625 inch for waste, what is the length of the remaining piece?

_____

25. (Obj. 41) What is the total height of a pile that has 25 sheets of metal that are each .045 inch thick and 42 sheets that are each .025 inch thick?

_____

CHAPTER 4

MEASUREMENT

4.1 ENGLISH DENOMINATE NUMBERS

The Corner Grocery sold 20 lb 6 oz of hamburger on Wednesday, 13 lb 8 oz on Thursday, and 17 lb 10 oz on Friday. How much hamburger was sold during the three days?

42) Express an English denominate number in an equivalent English measure when the relationship between the two units of measure is known.

43) Express in specified units the sum or difference of English denominate numbers.

A *denominate number* is a symbol formed by using a number along with a unit of measure to indicate "how many" or "how much of a measurable quantity." 5 miles is a denominate number. *English denominate numbers* are denominate numbers whose unit of measure has its origin in the English-speaking nations. Examples of English measures are: weight, 1 pound; length, 1 foot; volume, 1 quart.

Equivalent denominate numbers are different measures of the same quantity or object using a different unit of measure (for example, 12 inches = 1 foot).

The *measure* of a denominate number is the numeral without its label. "5" is the measure of "5 miles."

English measures and their equivalents that you should know are listed in Table 4–1.

Using the equivalent measures in this table allows us to express equivalent denominate numbers. Consider converting 4 feet to inches:

$$4 \text{ feet} = 4 \cdot (1 \text{ foot})$$

$$= 4 \cdot (12 \text{ inches})$$

$$= 48 \text{ inches}$$

TABLE 4–1

Length	Time
12 inches (in) = 1 foot (ft)	60 seconds (sec) = 1 minute (min)
3 feet (ft) = 1 yard (yd)	60 minutes (min) = 1 hour (hr)
5280 feet (ft) = 1 mile (mi)	24 hours (hr) = 1 day
	7 days = 1 week

Liquid	Weight
3 teaspoons (tsp) = 1 tablespoon (tbs)	16 ounces (oz) = 1 pound (lb)
2 cups (c) = 1 pint (pt)	2000 pounds (lb) = 1 ton
2 pints (pt) = 1 quart (qt)	
4 quarts (qt) = 1 gallon (gal)	

> To express a denominate number in a new measure, replace the unit of measure by an equivalent measure and find the indicated product.

When 15 cents is added to 2 dollars, the sum is neither 17 cents nor 17 dollars. If we convert 2 dollars to cents:

$$2 \text{ dollars} = 2 (100 \text{ cents}) = 200 \text{ cents},$$

then we can add 15 cents and 200 cents and find the total measure to be 215 cents.

> The sum (or difference) of two denominate numbers with common units of measure is the sum (difference) of their *measures* followed by the unit of measure.

EXAMPLES

a) $13\frac{3}{4}$ minutes expressed in seconds is $13\frac{3}{4}$(1 minute) = $13\frac{3}{4}$ (60 seconds) = 825 seconds.

b)   6 feet + 7 feet + 11 feet = 24 feet.

c)   Add: 3 gal 2 qt
             5 gal 3 qt
         ─────────────
             8 gal 5 qt   = 8 gal + 4 qt + 1 qt
                          = 8 gal + 1 gal + 1 qt
                          = 9 gal 1 qt

d)   If a carpenter cuts off a board of length 2 ft 5 in from a board of length 8 ft 3 in, how much board is left?

8 ft   3 in   Since 5 cannot be subtracted from 3, borrow 1 ft from the 8 ft
2 ft   5 in

7 ft 15 in   (1 ft = 12 in)
2 ft   5 in
5 ft 10 in

Find the missing number.   **EXERCISES**

1) 2 ft = ? in   2) 2 yd = ? ft

3) 2 cups = ? pt   4) 2 pt = ? cups

5) 2 qt = ? gal   6) 2 qt = ? pt

7) $1\frac{1}{2}$ days = ? hours   8) $1\frac{1}{2}$ hours = ? minutes

9) $2\frac{1}{4}$ lb = ? oz   10) $3\frac{1}{2}$ tons = ? lb

11) 5 ft = ? inches   12) 2 miles = ? yd

13) 3.5 gal = ? qt   14) $4\frac{3}{16}$ lb = ? ounces

15) 6.5 yd = ? inches   16) 2.13 tons = ? pounds

17) $89\frac{2}{3}$ yd = ? ft

18) 4 ft 9 in = ? inches

19) 7 yd 2 ft = ? inches

20) 40 minutes = ? seconds

Add:

21) 2 ft 5 in
     6 ft 4 in

22)   6 ft 5 in
      2 ft 7 in
     10 ft 7 in

23) 9 gal 3 qt 1 pt
     4 gal 2 qt 1 pt
     3 gal 1 qt 1 pt

24)   5 lb 11 oz
     11 lb 9 oz

Subtract.

25) 6 yd 2 ft
     3 yd 1 ft

26) 21 min 39 sec
     14 min 47 sec

27) In the application at the beginning of this section, how much hamburger was sold?

28) If a car travels 55 miles in one hour, how far will it travel in 7 hours at the same speed?

29) A buyer purchased 52 quarts of a rare wine for four clients. If each client is to share equally, how many quarts of wine will they each get?

30) Paul pledged 7 dollars a month to the United Fund. What was his annual donation?

31) Carol bought one gallon of milk at the local store. When she got home, Dan drank 3 cups of milk, Larry drank $1\frac{1}{2}$ cups, Greg drank 2 cups, and Carol used $\frac{1}{2}$ cup in cooking. How much milk remained at the end of the day?

4.2 METRIC DENOMINATE NUMBERS

Paul used a meter stick to measure the distance from his receiver to **APPLICATION** the place where he wants to put his stereo speakers. He found that each speaker will be 1 meter and 15 centimeters from the receiver. How many meters of wire does he need to connect the two speakers?

44) Express a metric denominate number in an equivalent metric **OBJECTIVES** measure when the relationship between the two units of measure is known.

45) Express in specified units the sum or difference of metric denominate numbers.

Metric denominate numbers are expressed in basic units developed in **VOCABULARY** France almost 200 years ago. Most nations today use the metric system. In the United States, scientists and an increasing number of industries are using the metric system.

The standard units in the metric system are

length: 1 meter or 1 m

weight: 1 gram or 1 g

volume: 1 liter or 1 *l*

THE HOW AND WHY OF IT   Every measuring system has one or more standard units of length, weight, volume, and other quantities. The metric system was invented to take advantage of our base ten place value system in the same way that our monetary system does. (See Table 4—2.)

TABLE 4—2

1 mil	= .001 dollar	
10 mils = 1 cent	= .01 dollar	
10 cents = 1 dime	= .1 dollar	
10 dimes = 1 dollar	= 1 dollar	

For most purposes, the units of meter, gram, and liter along with their multiples are sufficient.

Some of the metric measures are shown in Table 4—3, which should be memorized.

TABLE 4—3

Length (basic unit is 1 m)

	1 millimeter	= .001 m
10 millimeters (mm)	= 1 centimeter	= .01 m
10 centimeters (cm)	= 1 decimeter	= .1 m
10 decimeters (dm)	= 1 METER	= 1 m
10 meters (m)	= 1 dekameter	= 10 m
10 dekameters (dam)	= 1 hectometer	= 100 m
10 hectometers (hm)	= 1 kilometer (km)	= 1000 m

Weight (basic unit is 1 g)

	1 milligram	= .001 g
10 milligrams (mg)	= 1 centigram	= .01 g
10 centigrams (cg)	= 1 decigram	= .1 g
10 decigrams (dg)	= 1 GRAM	= 1 g
10 grams (g)	= 1 dekagram	= 10 g
10 dekagrams (dag)	= 1 hectogram	= 100 g
10 hectograms (hg)	= 1 kilogram (kg)	= 1000 g
1000 kg	= 1 metric ton	

Liquid and Dry Measure (basic unit is 1 l)

	1 milliliter	= .001 l
10 milliliters (ml)	= 1 centiliter	= .01 l
10 centiliters (cl)	= 1 deciliter	= .1 l
10 deciliters (dl)	= 1 LITER	= 1 l
10 liters (l)	= 1 dekaliter	= 10 l
10 dekaliters (da$l$)	= 1 hectoliter	= 100 l
10 hectoliters (hl)	= 1 kiloliter (kl)	= 1000 l

Using the equivalent metric measures in Table 4–3, the necessary operations with metric denominate numbers can be performed in the same manner as they were performed for English denominate numbers.

EXAMPLES

a) **50 dm = ? cm**
$$50 \text{ dm} = 50 \text{ (1 dm)}$$
$$= 50 \text{ (10 cm), since 1 dm = 10 cm}$$
$$= 500 \text{ cm}$$

b) **How many meters are in .4 kilometers?**
$$.4 \text{ km} = .4 \text{ (1 km)}$$
$$= .4 \text{ (1000 m), since 1 km = 1000 m}$$
$$= 400 \text{ m}$$

c) **What is the measure of 8 meters + 45 cm in meters?**
$$8 \text{ meters} + 45 \text{ cm} = 8 \text{ m} + 45 \text{ (1 cm)}$$
$$= 8 \text{ m} + 45 \text{ (.01 m), since 1 cm = .01 m}$$
$$= 8 \text{ m} + .45 \text{ m}$$
$$= 8.45 \text{ m}$$

d) **What is the measure of 1.3 liters − 135 milliliters in milliliters?**
$$1.3 \text{ liters} - 135 \text{ milliliters} = 1.3 \text{ (1 } l\text{)} - 135 \text{ m}l$$
$$= 1.3 \text{ (1000 m}l\text{)} - 135 \text{ m}l$$
$$= 1300 \text{ m}l - 135 \text{ m}l$$
$$= 1165 \text{ m}l$$
$$\text{Since 1 } l = 1 \text{ (10 d}l\text{)} = 1 \cdot 10 \text{ (10 c}l\text{)} = 1 \cdot 10 \cdot 10 \text{ (10 m}l\text{)} = 1000 \text{ m}l$$

Find the missing number.   **EXERCISES**

1) 100 mils = ? cents   2) 100 mm = ? cm

3) 100 mg = ? g   4) 100 ml = ? cl

5) .5 km = ? m   6) .5 kg = ? g

7) 2 km = ? m   8) 3 kg = ? g

9) 2m 95 cm + 3m 10 cm = ?   10) 7 kg 80 g + 2 kg 60 g

11) 1.3 km = ? m   12) 1.3 kg = ? g

13) 1.3 kl = ? l 14) 244 mm = ? m

15) 245 ml = ? l 16) 246 mg = ? g

17) .4 km + 22 m = ? m   18) 2 kg + 45 hg = ? g

19) 2.6 km − 1900 m = ? m   20) 17l − 444 ml = ? ml

Add:

21) 3 km 250 m   22)   2 g 8 dg
     3 km 900 m   14 g 5 dg
     —————————   —————————
              = ? m   = ? dg

Subtract:

23) 3 kl 5 hl 24) 15 km 700 m
     1 kl 7 hl 3 km 850 m
     —————————   —————————
          = ? hl = ? m

25) In the application at the beginning of this section, how many meters
     of wire does Paul need?

26) If a can contains 298 grams of soup, how many grams of soup are contained in 7 cans?

27) Tim, who is a lab assistant, has 282 m*l* of acid that is to be divided among 24 students. How many m*l* will each student receive?

28) Gayle purchased a package of ground beef that cost $1.43 and weighed .5 kg. What was the price per kilogram?

4.3 CONVERSION OF UNITS

The French Concorde can fly at a speed of 3240 ft/sec, which is approximately three times the speed of sound. How many miles per hour does the Concorde fly?

APPLICATION

46) Convert a given denominate number to an equivalent denominate number with specified units.

OBJECTIVE

Recall that equivalent denominate numbers measure the same quantity but involve different basic units of measure. 6 feet and 2 yards are equivalent denominate numbers.

VOCABULARY

Since 12 inches = 1 foot, these two are equivalent denominate numbers. If we consider the indicated division, (12 inches) ÷ (1 foot), and think of these as numbers, the division asks "how many" units of measure 1 foot it will take to make 12 inches. Since they measure the same length, the answer is 1.

THE HOW AND WHY OF IT

$$\frac{12 \text{ inches}}{1 \text{ foot}} = 1$$

The indicated division of two equivalent denominate numbers can always be thought of as a name for 1. This concept, along with the multiplication property of one, will be used to convert from one measure to its equivalent in another measure. Consider converting 48 inches to feet.

$$48 \text{ inches} = (48 \text{ inches})(1)$$

$$= 48 \text{ inches} \cdot \frac{1 \text{ foot}}{12 \text{ inches}}$$

$$= \frac{48 \cdot 1 \text{ inch} \cdot 1 \text{ foot}}{12 \cdot 1 \text{ inch}}$$

$$= \frac{48 \cdot 1 \text{ foot}}{12} \cdot \frac{1 \text{ inch}}{1 \text{ inch}}$$

$$= 4 \cdot 1 \text{ foot} \cdot 1$$

$$= 4 \text{ feet}$$

Recall that, in working with fractions, $\dfrac{ab}{ac} = \dfrac{b}{c}$. Using this concept, the details of the problem can be simplified.

$$48 \text{ inches} = \frac{48 \text{ inches}}{1} \cdot \frac{1 \text{ foot}}{12 \text{ inches}}$$

$$= \frac{48}{12} \text{ feet}$$

$$= 4 \text{ feet}$$

In some cases it may be necessary to multiply by several different names of one. Convert 7200 seconds to hours.

$$7200 \text{ seconds} = \frac{7200 \text{ seconds}}{1} \cdot \frac{1 \text{ minute}}{60 \text{ seconds}} \cdot \frac{1 \text{ hour}}{60 \text{ minutes}}$$

$$= \frac{7200}{3600} \text{ hours}$$

$$= 2 \text{ hours}$$

The procedure for converting a denominate number is to multiply by one (in the form of an indicated division of equivalent denominate numbers) as many times as necessary until the correct unit of measure remains. Then perform the indicated operations with the numbers, retaining the desired unit of measure.

EXAMPLES

a)   Convert 4 gallons to pints.

$$4 \text{ gallons} = \frac{4 \text{ gallons}}{1} \cdot \frac{4 \text{ quarts}}{1 \text{ gallon}} \cdot \frac{2 \text{ pints}}{1 \text{ quart}}$$

$$= 4 \cdot 4 \cdot 2 \text{ pints}$$

$$= 32 \text{ pints}$$

b)   Convert 3 ft^2 to in^2. (See page 145 for an explanation of the units in this example). Since 1 ft^2 = (1 foot) (1 foot) and 1 in^2 = (1 inch) (1 inch) , we proceed as follows:

$$3 \text{ ft}^2 = 3 \cdot 1 \text{ foot} \cdot 1 \text{ foot} \cdot \frac{12 \text{ inches}}{1 \text{ foot}} \cdot \frac{12 \text{ inches}}{1 \text{ foot}}$$

$$= 3 \cdot 12 \cdot 12 \text{ in}^2$$

$$= 432 \text{ in}^2$$

c)   Convert 60 miles per hour to feet per second.

$$60 \text{ miles per hour} = \frac{60 \text{ miles}}{1 \text{ hour}} \cdot \frac{1 \text{ hour}}{60 \text{ min}} \cdot \frac{1 \text{ min}}{60 \text{ sec}} \cdot \frac{5280 \text{ ft}}{1 \text{ mile}}$$

$$= \frac{60 \cdot 5280}{60 \cdot 60} \frac{\text{ft}}{\text{sec}}$$

$$= 88 \text{ feet per second}$$

60 miles per hour is equivalent to 88 feet per second.

d)   Convert 480 grams to kilograms.

$$480 \text{ grams} = \frac{480 \text{ grams}}{1} \cdot \frac{1 \text{ dag}}{10 \text{ g}} \cdot \frac{1 \text{ hg}}{10 \text{ dag}} \cdot \frac{1 \text{ kg}}{10 \text{ hg}}$$

$$= \frac{480}{10 \cdot 10 \cdot 10} \text{kg}$$

$$= .48 \text{ kilogram}$$

EXERCISES   Convert each of the following to the indicated unit of measure.

1) 42 inches = ? feet

2) $1\frac{1}{3}$ feet = ? inches

3) 13 feet = ? yards

4) 1.6 kg = ? g

5) 13.1 kl = ? l

6) $2\frac{1}{2}$ gal = ? qt

7) 3m 9 cm = ? cm

8) 3 ft 9 in = ? in

9) 7 km 528 m = ? km

10) 7 miles 528 feet = ? miles

11) 1 mile to inches

12) 10,080 minutes to days

13) $\dfrac{44 \text{ feet}}{\text{second}}$ to $\dfrac{\text{miles}}{\text{hour}}$

14) 1000 mm to m

15) $\dfrac{9 \text{ tons}}{\text{ft}}$ to $\dfrac{\text{pounds}}{\text{in}}$

16) $\dfrac{10 \text{ ounces}}{\text{cup}}$ to $\dfrac{\text{pounds}}{\text{gallon}}$

17) $\dfrac{144 \text{ kilometers}}{\text{hour}}$ to $\dfrac{\text{meters}}{\text{second}}$

18) 3540 cm² to m²

19) 12 feet 9 inches to yards

20) In the application at the beginning of this section, what is the miles per hour rate of the Concorde?

21) Shirley is going on a diet that will cause her to lose 4 ounces every day. At this rate, how many pounds will she lose in 6 weeks?

22) The local candy manufacturer packs 30 pieces of candy in a box. To ship these to market he puts 40 boxes in a case. How many pieces of candy are in a shipment of 27 cases?

23) Larry's family normally eats 3 boxes of Yummy Snap and Pops, the latest breakfast craze, in a week. Each box contains 14 ounces of cereal. How many pounds of the cereal will Larry's family eat in one year?

24) If the cereal in Problem 23 costs 67¢ per box, what was the total cost of the cereal for the year?

25) Carol has a recipe that calls for 12 cubic centimeters of milk. She only has a liter measure. How much of a liter should she add? (1 cubic centimeter = 1 milliliter)

26) If Dan averages 50 miles per hour during an eight hour day of driving, how many days will it take him to drive 2000 miles?

27) Greg's Sport Shop decides to donate to charity 1 cent for every yard Joanne jogs during one week. Joanne jogs 2 miles every day. How much does Greg donate to charity?

28) If water weighs 64 pounds per ft^3, how many ounces (to the nearest tenth) does one in^3 weigh? Hint: 1 ft^3 = (1 foot) (1 foot) (1 foot) and 1 in^3 = (1 inch) (1 inch) (1 inch).

29) Mr. Smith's car averages 12.5 miles per gallon of gasoline. How many gallons of gas are needed to make a trip of 412 miles?

30) If a secretary can type 60 words per minute and the average page contains 500 words, how many pages can she type in 4 hours?

4.4 EVALUATING EXPRESSIONS AND FORMULAS

APPLICATION

An outdoor stage will be constructed of bricks with a concrete base. If it will be circular with a 30 ft diameter, how many square feet of stage area will be available? ($A = \pi r^2$, $\pi \approx 3.14$)

OBJECTIVE

47) Evaluate an algebraic expression or formula.

VOCABULARY

An *algebraic expression* consists of numerals, operation signs (for addition, subtraction, multiplication, and division) and letters of the alphabet used as placeholders for numbers. The placeholders (letters) are called *variables or unknowns*. When the variables have been replaced by numbers (substitution), the expression can be *evaluated* or *simplified*.

The order in which the operations should be carried out is the same as in Chapter 1.

$3 + 2t$ has the value 15 when t is replaced by 6 since

$$3 + 2 \cdot 6 = 3 + 12 = 15$$

The formula for the perimeter of a rectangle is

$$p = 2l + 2w$$

Therefore, a rectangle whose length is 13 cm and width is 7 cm has a perimeter of

$$p = 2l + 2w = 2 \cdot 13 + 2 \cdot 7$$
$$= 26 + 14$$
$$= 40 \text{ cm}$$

To evaluate an expression or formula, replace each variable by its given value and perform the indicated operation(s).

EXAMPLES

Evaluate the following expressions given $a = 2$, $b = 3$, $x = 5$ and $z = 0$.

a) $x + ab = 5 + 2(3) = 11$

b) $az + \dfrac{x}{b} = 2 \cdot 0 + \dfrac{5}{3} = 0 + \dfrac{5}{3} = \dfrac{5}{3}$

c) $ab - a + z = 6 - 2 + 0 = 4$

d) $ax + bx + xz = (10) + (15) + 0 = 25$

Evaluate the following formulas.

e) $A = \dfrac{1}{2}bc$ (area of a triangle); $b = 7$ and $c = 5$

$A = \dfrac{1}{2}(7)(5) = 17.5$

f) $S = \dfrac{1}{2}n(a + l)$ (sum of an arithmetic sequence). Given: $n = 100$, $a = 1$, and $l = 100$.

$S = \dfrac{1}{2} \cdot 100(1 + 100) = 50 \cdot 101 = 5050$

(The sum $1 + 2 + 3 + \ldots + 97 + 98 + 100 = 5050$)

g) $\quad R = \dfrac{1}{\dfrac{1}{R_1} + \dfrac{1}{R_2}}; R_1 = 2, R_2 = 1$

$\quad R = \dfrac{1}{\dfrac{1}{2} + \dfrac{1}{1}} = \dfrac{1}{\dfrac{3}{2}} = \dfrac{2}{3}$

EXERCISES

Evaluate the following expressions given $d = 10$, $f = 5$, $h = 2$, and $t = 1000$.

1) h^2

2) $6.3\,t$

3) $\dfrac{100\,d}{t} + h$

4) $\dfrac{2f}{d} \div h$

5) $2\,t - 10\,d$

6) $df + h$

7) $13.5\,t$

8) $2\,d + 3\,f + 4\,h$

9) $\dfrac{t}{2df} - h$

10) $t \div h \div f \div d$

Evaluate the following expressions given $a = 7$, $b = 10$, $c = 8$, $x = 1$, $y = 6$, and $z = 4$.

11) $a^2 x^2$

12) $a^2 + x^2$

13) $ab + yz$

14) $ay + bz$

15) $cx + cz$

16) $abz - abx$

17) $c(x + z)$

18) $ab(z - x)$

19) $(a + x)(c + y)$

20) $b + y(a + z)$

21) $2c + 4y$

22) $(3a + 5)z$

23) $\dfrac{6 - z}{b}$

24) $\dfrac{a + b + c}{x + y}$

25) $\dfrac{(x + y + z)^2}{c - x}$

26) $\dfrac{ax + by + cz}{3c - a}$

Evaluate the following formulas.

27) $A = \dfrac{1}{2}bc$ if $b = 32$ and $c = 29$

28) $S = \dfrac{1}{2}n(a + l)$ if $n = 100, a = 2,$ and $l = 200$

29) $R = \dfrac{1}{\dfrac{1}{R_1} + \dfrac{1}{R_2}}$ if $R_1 = .3$ and $R_2 = .6$

30) $V = \pi r^2 h$ if $\pi \approx 3.14, r = 2$, and $h = 3$

31) In the application at the beginning of this section, how many square feet of stage area will be available?

32) How much simple interest is earned on a savings account of $150 ($p = 150$) at the rate of $6\frac{1}{2}\%$ ($r = .065$) for 2 years ($t = 2$)? [Formula: $I = prt$]

33) How far does a plane travel at the rate of 425 mph ($r = 425$) in 9 hours ($t = 9$)? [Formula: $D = rt$]

34) The first ounce of first class mail costs 13¢ and each additional ounce costs 11¢. How much does a first class envelope weighing 9 ounces cost ($w = 8$)? [Formula: $C = 13 + 11w$]

4.5 PERIMETER AND CIRCUMFERENCE

A kitchen floor is in the shape of a rectangle with dimensions 9 feet by 11 feet. A mop board costing 38¢ per foot is to be installed around the perimeter of the room. Allowing 5 feet for door space, what will be the cost of the mop board?

48) Find the perimeter of geometric figures.

Circles, rectangles, triangles, and *squares* are four examples of *geometric figures.* An example of each is shown in Figure 4.1. Different parts of the figures are also shown.

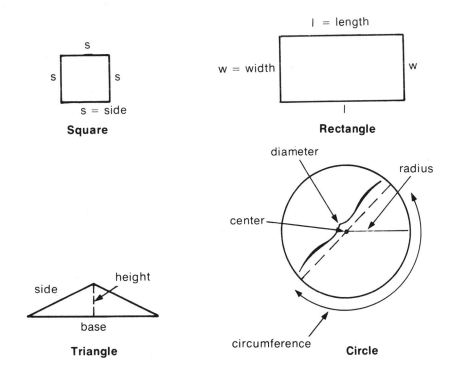

FIGURE 4.1.   **Square**   **Rectangle**   **Triangle**   **Circle**

The distance around any geometric figure is called the *perimeter* of that figure. The perimeter of a circle is called the *circumference.* The *radius* of the circle is the distance from the center to any point on the circle. The *diameter* is twice the radius.

The perimeter of any figure in which all of the sides are straight is the sum of the lengths of the sides. Recall that to add denominate numbers the unit of measure must be the same.

For example, find the perimeter of Figure 4.2.

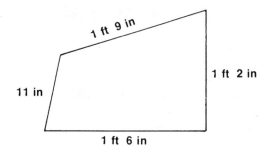

FIGURE 4.2.

P (the perimeter) is the sum of the lengths of the four sides, that is,

```
1 ft   9 in
1 ft   2 in
1 ft   6 in
       11 in
3 ft 28 in (28 in = 2 ft 4 in)
```

So P = 5 ft 4 in.

There are special formulas for the perimeters of squares, rectangles, and circles and for the length of a semicircle.

Square   If "P" is the perimeter and "s" is the length of one side, then the formula is

$$P = 4s$$

Rectangle   If "P" is the perimeter, "l" is the length, and "w" is the width, then the formula is

$$P = 2l + 2w$$

Circle   If "C" is the circumference and "d" is the diameter, then the formula is

$$C = \pi d = 2\pi r$$

(π is read "pi.")

Semicircle   If "L" is the length and "r" is the radius, then the formula is

$$L = \pi r = \frac{1}{2}\pi d$$

To find the perimeter of a geometric figure which is not one of the above, add the lengths of the sides.

There is no fraction or exact decimal that names the number π, so we will use 3.14 as an approximate value.

EXAMPLES

Find the perimeter of each of the following

a)

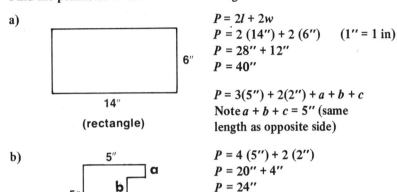

(rectangle)

$P = 2l + 2w$
$P = 2\ (14'') + 2\ (6'')$ $(1'' = 1\ \text{in})$
$P = 28'' + 12''$
$P = 40''$

$P = 3(5'') + 2(2'') + a + b + c$
Note $a + b + c = 5''$ (same length as opposite side)

b)

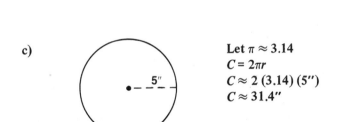

$P = 4\ (5'') + 2\ (2'')$
$P = 20'' + 4''$
$P = 24''$

c)

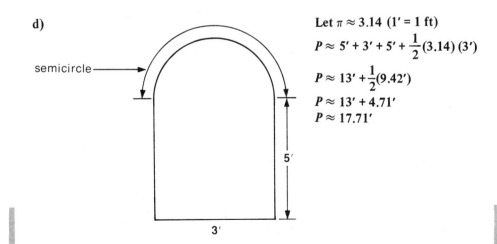

(circle)

Let $\pi \approx 3.14$
$C = 2\pi r$
$C \approx 2\ (3.14)\ (5'')$
$C \approx 31.4''$

d)

semicircle

Let $\pi \approx 3.14$ $(1' = 1\ \text{ft})$
$P \approx 5' + 3' + 5' + \dfrac{1}{2}(3.14)\ (3')$
$P \approx 13' + \dfrac{1}{2}(9.42')$
$P \approx 13' + 4.71'$
$P \approx 17.71'$

5'

3'

EXERCISES

Find the perimeter or circumference of each of the following. Let $\pi \approx 3.14$.

1) A square with side 3 cm.

2) A square with side 1.2 yd.

3) A rectangle that is 1 ft by 3 ft.

4) A rectangle that is 3 m by 2 m.

5) A triangle with sides 2 in, 4 in, and 5 in.

6) A triangle with sides 2.3 cm, 2.2 cm, and 3.1 cm.

7) A circle with radius 1 dm.

8) A circle with radius $\frac{1}{2}$ yd.

9) A four-sided figure (quadrilateral) with sides 12 in, 13 in, 15 in, and 17 in.

10) A five-sided figure (pentagon) with sides of 3 ft, 3 ft, 5 ft, 6 ft, and 8 ft.

Find the perimeter of each of the following figures. Let $\pi \approx 3.14$.

11)

12)

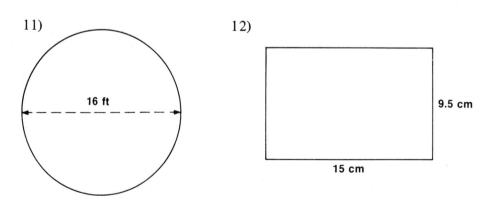

16 ft

9.5 cm

15 cm

13)

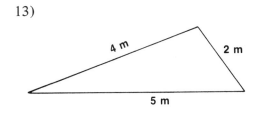

14)

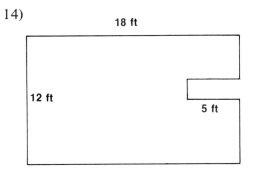

15)

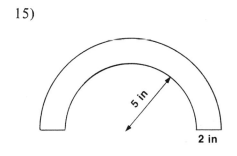

16)

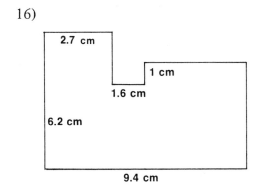

17)

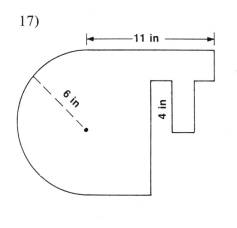

18)

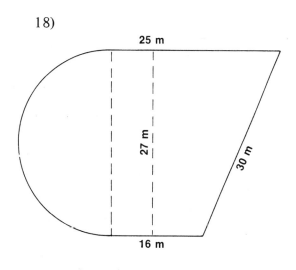

19)

20)

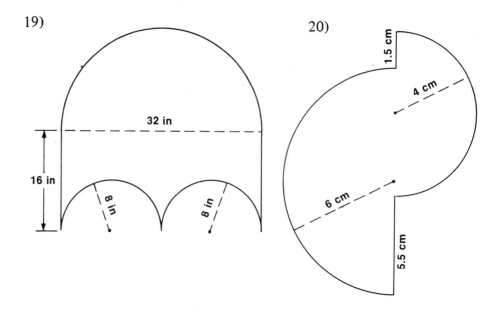

21) In the application at the beginning of this section, what is the cost of the mop board?

22) What is the distance around a baseball infield if it is 90 ft between the bases?

23) How many feet of picture framing does Bonnie need to frame 4 pictures, each measuring 8 inches by 10 inches?

24) How far will the tip of the minute hand of a clock travel in 15 minutes if the hand is one inch in length?

25) If fencing costs $5 per yard, what will be the cost of fencing a rectangular lot that is 407 ft long and 82 ft wide?

26) If Atha needs 2 minutes to put one foot of binding on a rug, how long will it take her to put the binding on a rug that is 15 ft by 12 ft?

27) A wheel on Mary's automobile is 16 inches in diameter. To the nearest whole number, how many revolutions will the wheel turn if the automobile travels one mile? ($\pi \approx 3.14$)

4.6 AREA OF COMMON GEOMETRIC FIGURES

A concrete retaining wall is 6 feet high and 596 feet long. Find the cost of paint for covering the wall with two coats of paint if one gallon of paint costs $13.25 and covers 75 square feet of concrete.   APPLICATION

49) Find the area of common geometric figures.   OBJECTIVE

Area is a measure of a surface, and surface is measured in square units. Two examples of surface measure are shown in Figure 4.3.   VOCABULARY

The unit of measure on the right is called a "*square inch*," since the square is one inch on each side. The "square inch" measures the surface that is contained within the square. "Square inch" is written "in^2."

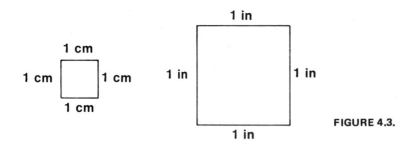

FIGURE 4.3.

The unit of measure on the left is called a *square centimeter,* since the square is one centimeter on each side. The "square centimeter" measures the surface that is contained within the square. "Square centimeter" is written "cm² ."

There are other units of surface measure, such as *"square foot," "square mile," "square meter,"* and *"square kilometer."* In each case the unit measures the amount of surface within a square having that length on each side.

In Figure 4.4, two different geometric figures are shown: the parallelogram and the trapezoid. Note that the trapezoid has two bases and one altitude, while the parallelogram has one base and one altitude.

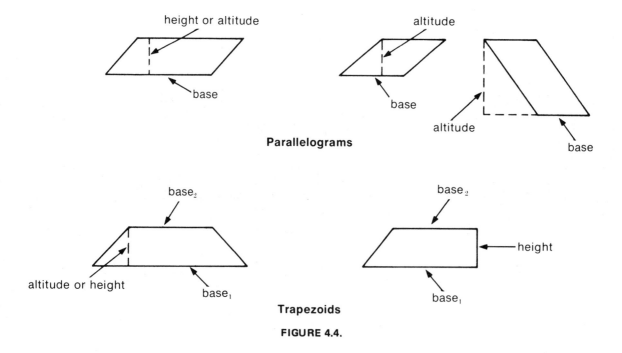

FIGURE 4.4.

THE HOW AND WHY OF IT   To find the area of a geometric figure means to determine how many square units of measure (surface units) are contained within that figure. Consider a square that is 2 inches on each side.

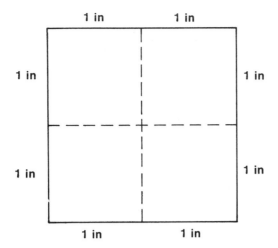

We can see from the drawing that this square can be divided into four squares, each of which is one inch on a side. This would indicate that the area of a square that is two inches on each side is four square inches. The area (4 in²) is the square of the length of a side. That is, $A = (2 \text{ in})^2 = (2 \text{ in})(2 \text{ in}) = 4 \text{ in}^2$. (Notice that "in" times "in" gives "in².")

> The area of a square can be found by squaring the length of one of its equal sides. That is,
>
> $$A = s^2$$

Consider a rectangle that has a length of 3 centimeters and a width of 2 centimeters. To find the area of this rectangle, we consider the following diagram.

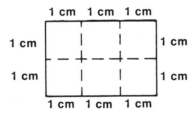

The diagram indicates that the rectangle can be divided into six squares, each of which is one centimeter on a side. This tells us that the area of the rectangle is 6 cm². Notice that if we multiply the length (3 cm) times the width (2 cm), we have (3 cm) × (2 cm), which is 6 cm².

> The area of a rectangle can be found by multiplying the length times the width. Stated as a formula,
>
> $$A = l \cdot w$$

There are formulas to find the area of a parallelogram, a triangle, a trapezoid, and a circle. We could consider each in a manner similar to the square and the rectangle, but instead the formulas will be stated.

The area of a parallelogram	$A = b \cdot h$
The area of a triangle	$A = \frac{1}{2}b \cdot h$
The area of a trapezoid	$A = \frac{1}{2}(b_1 + b_2) \cdot h$
The area of a circle	$A = \pi r^2$ or $A = \dfrac{\pi d^2}{4}$

Here, b indicates the base (or bases, in the case of the trapezoid), h indicates the height or altitude, r is the radius, and d is the diameter of the circle.

EXAMPLES

Find the area of each of the following.

a) A square that is 4 in on each side.

$A = s^2$
$A = (4 \text{ in})^2$
$A = 16 \text{ in}^2$

b)

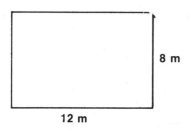

The figure is a rectangle with l = 12 m and w = 8 m.

$A = l \cdot w$
$A = (12\text{m})(8\text{ m})$
$A = 96 \text{ m}^2$

c)

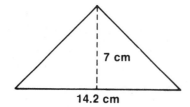

The figure is a triangle with $b = 14.2$ cm and $h = 7$ cm.

$$A = \frac{1}{2} \cdot b \cdot h$$

$$A = \frac{1}{2} \cdot (14.2 \text{ cm}) (7 \text{ cm})$$

$$A = 49.7 \text{ cm}^2$$

d)   A parallelogram whose base is 1 ft and whose height is 4 in.

$A = b \cdot h$

$A = (1 \text{ ft}) (4 \text{ in})$	**OR**	$A = (1 \text{ ft}) (4 \text{ in})$
$A = (12 \text{ in}) (4 \text{ in})$		$A = (1 \text{ ft}) \left(\frac{1}{3}\text{ft}\right)$
$A = 48 \text{ in}^2$		$A = \frac{1}{3}\text{ft}^2$

e)

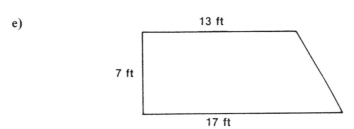

The figure is a trapezoid with $b_1 = 17$ ft, $b_2 = 13$ ft, and $h = 7$ ft.

$$A = \frac{(b_1 + b_2) \cdot h}{2}$$

$$A = \frac{(17 \text{ ft} + 13 \text{ ft}) \cdot 7 \text{ ft}}{2}$$

$$A = \frac{(30 \text{ ft}) (7 \text{ ft})}{2}$$

$$A = \frac{210 \text{ ft}^2}{2}$$

$$A = 105 \text{ ft}^2$$

f)   The figure is a circle with $r = 8$ cm.

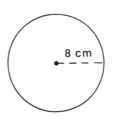

$A = \pi r^2$, where $\pi \approx 3.14$

$A \approx (3.14) (8 \text{ cm})^2$

$A \approx (3.14) (64 \text{ cm}^2)$

$A \approx 200.96 \text{ cm}^2$

EXERCISES   Find the area of each of the following. Let $\pi \approx 3.14$.

1) A square with side 8 cm.

2) A square with side $1\frac{1}{2}$ in.

3) A rectangle that is 4 ft by 11 ft.

4) A rectangle that is 1 ft by 6 in.

5) A parallelogram with base 16 cm and height 4 cm.

6) A triangle with base 16 cm and height 4 cm.

7) A circle with radius 1 dm.

8) A square with side 2.2 cm.

9) A parallelogram with base 4 in and height $1\frac{1}{2}$ in.

10) A triangle with base 4 in and height $1\frac{1}{4}$ in.

Find the area of each of the following figures. Let $\pi \approx 3.14$.

11)

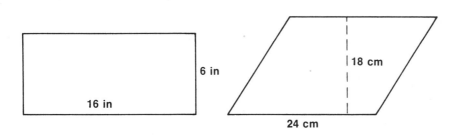

12)

13)

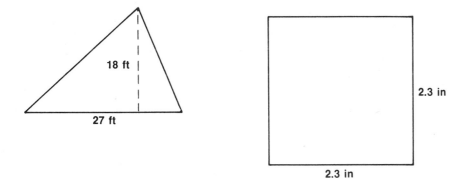

14)

15)

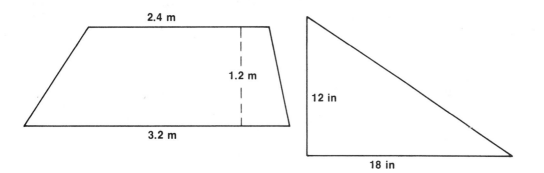

16)

17)

18)

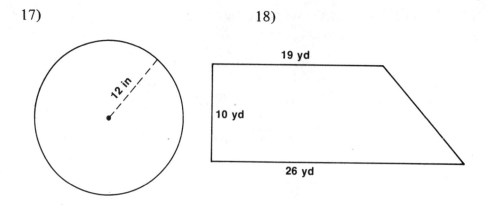

19) In the application at the beginning of this section, what is the cost of the paint?

20) Find the area of a rectangle whose length is 16 inches and whose width is 12 inches.

21) Find the area of a circle if it has a diameter of 16 centimeters.

22) Find the area of a parallelogram whose base is 18.9 meters and whose height is 17.3 meters.

23) Find the area of a trapezoid whose bases are 2 ft 4 in and 3 ft 2 in and whose height is 9 in.

24) How many square inches of glass are needed to make a circular mirror of diameter 36 inches?

25) How many square yards of carpet will Peggy need to carpet a rectangular floor that measures 21 ft by 30 ft?

26) If one ounce of weed killer treats one square meter of lawn, how many ounces of weed killer will Debbie need to treat a rectangular lawn that measures 30 m by 8 m?

27) How many acres are contained in a rectangular plot of ground if the length is 1320 ft and the width is 528 ft? ($43,560 \text{ ft}^2 = 1$ acre)

28) How many square tiles, each 12 in on a side, will Kirsten need to cover a floor that is 8 ft wide and 16 ft long?

4.7 A SECOND LOOK AT AREAS

The floor of a shop (floor plan shown below) is to be poured concrete. If concrete costs $3.35 a square yard, what will the floor cost? APPLICATION

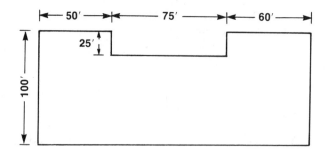

OBJECTIVE

50) Find the area of geometric figures that can be considered as a combination of two or more common geometric figures.

VOCABULARY

No new vocabulary.

THE HOW AND
WHY OF IT

It is possible to find the area of geometric figures that are not squares, rectangles, circles, or any of the other more common figures. Some of these figures can be divided into two or more of the common shapes. The sum of the areas of each of these common figures is the area of the entire region.

Consider the following shapes:

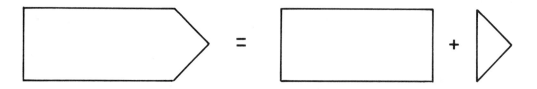

The figure on the left can be divided into a rectangle with a triangle attached; as shown on the right. If we can find the area of the rectangle and the area of the triangle, then the sum of those areas will be the area of the entire region.

Consider the following shapes:

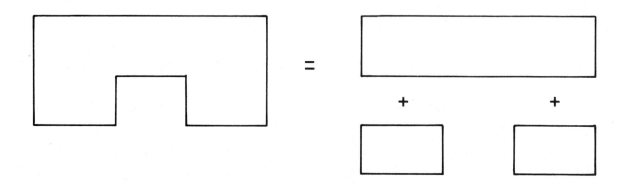

The figure on the left can be divided into three rectangles, as shown on the right. If we can find the area of each rectangle, the sum of those areas will be the area of the entire region.

In some figures it is helpful to append a region to the original figure so that it may be divided into the common figures. For example:

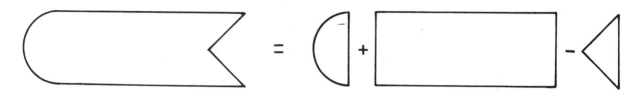

The figure on the left can be considered as a semicircle, a rectangle, and a triangle, as shown on the right. If we can find the area of the semicircle, the rectangle, and the triangle, then we find the sum of the area of the semicircle and the rectangle and subtract the area of the triangle to find the area of the entire region.

> To find the area of a geometric figure that is composed of the sum or the difference of two or more common geometric figures:
> 1) Divide the figure into common geometric figures or append a region or regions, and then divide into common geometric figures.
> 2) Find the area of each of the common figures.
> 3) Find the sum or difference of those areas.

EXAMPLES

Find the area of each of the following:

a)

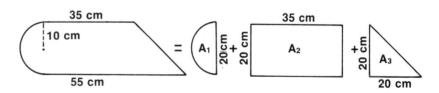

The area of the entire region is

$A =$ Area of semicircle + Area of rectangle + Area of triangle

$A = A_1 + A_2 + A_3$

$$A_1 = \frac{\pi r^2}{2} \approx \frac{(3.14)(10 \text{ cm})^2}{2} = 157 \text{ cm}^2$$

$$A_2 = l \cdot w = (35 \text{ cm})(20 \text{ cm}) = 700 \text{ cm}^2$$

$$A_3 = \frac{1}{2} \cdot b \cdot h = \frac{1}{2}(20 \text{ cm})(20 \text{ cm}) = 200 \text{ cm}^2$$

$$A \approx 157 \text{ cm}^2 + 700 \text{ cm}^2 + 200 \text{ cm}^2 = 1057 \text{ cm}^2$$

b)

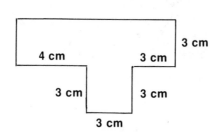

A = Area of semicircle + Area of rectangle − Area of triangle

A = $A_1 + A_2 - A_3$

$\approx$ 25.12 in² + 224 in² − 32 in²

$\approx$ 217.12 in²

EXERCISES

Find the area of each of the following figures. Let $\pi \approx 3.14$.

1)

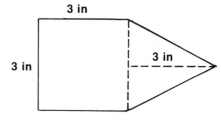

2)

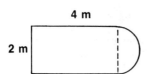

3)

4)

5)

6)

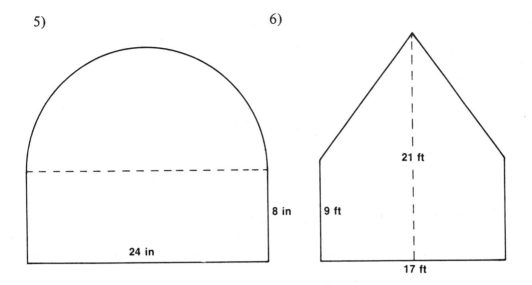

7)

8)

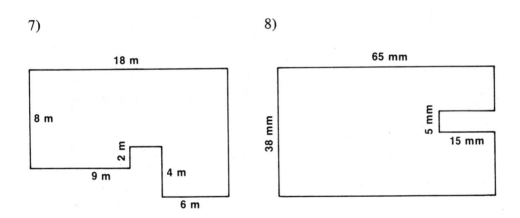

9)

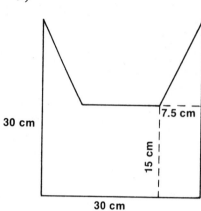

10) Round to the nearest hundredth.

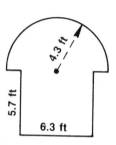

11)

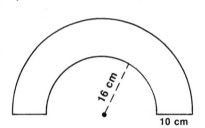

12) Round to the nearest tenth.

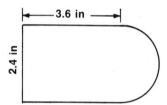

13)

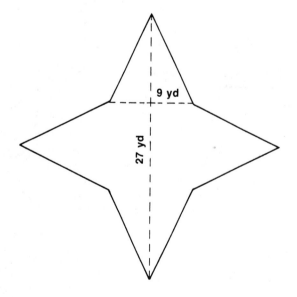

14) Round to the nearest hundredth.

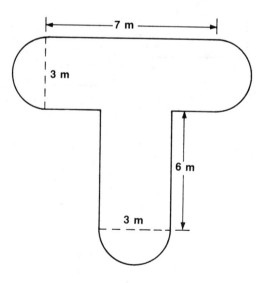

15) Find the area of the shaded portion.

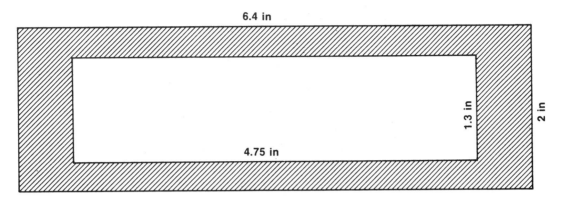

16) Find the area of the shaded portion.

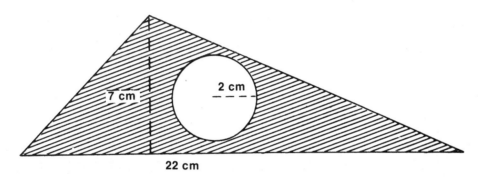

17) In the application at the beginning of this section, what will be the cost of the floor?

18) The following diagram shows the Jones' yard with respect to their house. How much grass seed is needed to sow the lawn if one pound of seed will sow 1000 ft²? Find the answer to the nearest pound.

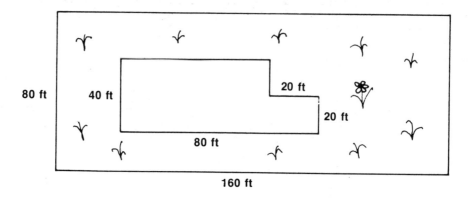

19) What is the area of the infield of a race track in which the straight-aways are 100 yards long and the ends are semicircles with 30 yard diameters?

4.8   VOLUME OF COMMON GEOMETRIC SOLIDS

APPLICATION   How many tons of coal will a boxcar hold if the inside dimensions of the car are 8 feet wide and 35 feet long? The car is to be filled to a depth of 5 feet and coal weighs 50 pounds per cubic foot.

OBJECTIVE   51)   Find the volume of common geometric solids.

VOCABULARY   *Volume* is the name given to the amount of space that is contained inside a three-dimensional figure. A box, for example, has volume. The question "How much does the box hold?" is referring to its volume. Volume is measured in terms of a cubic unit, that is, a *cube* (a box) that is one unit on each edge. Volume can be thought of as the number of cubic units needed to form the figure. An example of a unit of volume measure is a *cubic inch* (in³), which is shown in Figure 4.5.

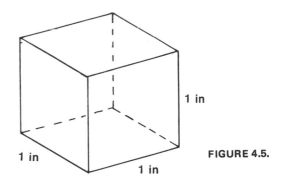

FIGURE 4.5.

A cube that is one inch on each edge is a cubic inch. A cube that is 1 centimeter on each edge is a cubic centimeter (cm^3). A cube that is one foot on each edge is a cubic foot (ft^3). Each of the cubes is a measure of volume. In fact, any cube may be thought of as a unit measure of volume.

THE HOW AND WHY OF IT

The volume of a rectangular solid (a box is a rectangular solid) is $V = l \cdot w \cdot h,$ where V is the volume, l is the length, h is the height, and w is the width. (See Figure 4.6.)

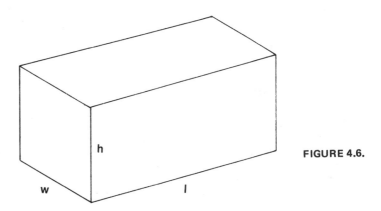

FIGURE 4.6.

The length, width, and height must all be measured with the same unit of measure or converted to the same unit.

A cube is a rectangular solid in which the length, width, and height are all equal.

To find the volume of a cube that is of length e on each edge, we use the formula $V = e^3$, that is: $V = e \cdot e \cdot e.$

Pictures of a cylinder, a sphere, a cone, and a pyramid are shown in Figure 4.7. The formula for the volume of each is as follows:

Cylinder	$V = \pi r^2 h$
Sphere	$V = \frac{4}{3}\pi r^3$
Cone	$V = \frac{1}{3}\pi r^2 h$
Pyramid	$V = \frac{1}{3}Bh$ (where B is the area of the base)

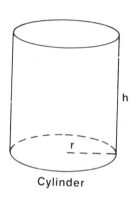

Cylinder

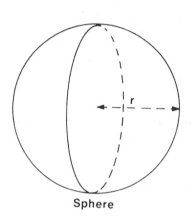

Sphere

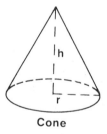

Cone

FIGURE 4.7.

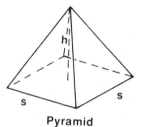

Pyramid

EXAMPLES

a)   The volume of a cube that is 6 inches on each edge is:

$$V = e^3 \text{ or } V = e \cdot e \cdot e$$
$$V = (6 \text{ in})^3$$
$$V = (6 \text{ in}) (6 \text{ in}) (6 \text{ in})$$
$$V = 216 \text{ in}^3$$

b)

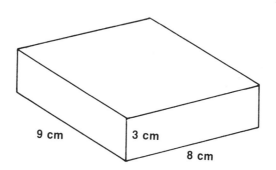

The volume of the rectangular solid (box) is:

$$V = l \cdot w \cdot h$$
$$V = (9 \text{ cm}) (8 \text{ cm}) (3 \text{ cm})$$
$$V = 216 \text{ cm}^3$$

c)

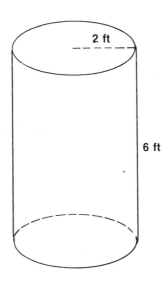

The volume of the cylinder is:

$$V = \pi r^2 h \qquad (\text{Let } \pi \approx 3.14.)$$
$$V \approx (3.14) (2 \text{ ft})^2 \ (6 \text{ ft})$$
$$V \approx 75.36 \text{ ft}^3$$

d) The volume of a sphere whose diameter is 18 centimeters is:

$$V = \frac{4}{3}\pi r^3 \qquad (\text{Let } \pi \approx 3.14.)$$
$$V \approx \frac{4}{3}(3.14) (9 \text{ cm})^3 \qquad (r = 9 \text{ cm since } d = 18 \text{ cm})$$
$$V \approx \frac{4}{3}(3.14) (729 \text{ cm}^3)$$
$$V \approx 3052.08 \text{ cm}^3$$

e)

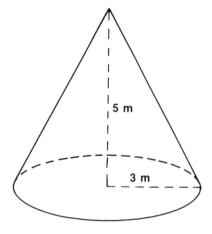

The volume of the cone is:

$$V = \frac{1}{3}\pi r^2 h \qquad \text{(Let } \pi \approx 3.14.\text{)}$$

$$V \approx \frac{1}{3}(3.14)\,(3\text{ m})^2\,(5\text{ m})$$

$$V \approx \frac{1}{3}(3.14)\,(45\text{ m}^3)$$

$$V \approx 47.1\text{ m}^3$$

f) Let us find the volume of a pyramid whose base is a rectangle with length 6 in and width 4 in and which has a height of 5 in. Since the base is a rectangle, the area of the base is $l\cdot w$, and we have

$$V = \frac{1}{3}(l\cdot w)h$$

$$V = \frac{1}{3}\,(6\text{ in})\,(4\text{ in})\,(5\text{ in})$$

$$V = 40\text{ in}^3$$

EXERCISES   Find the volume of each of the following. Let $\pi \approx 3.14$.

1) A cylinder with radius 1 ft and height 2 ft.

2) A cube with side 2 m.

3) A cone with radius 1 dm and height 3 dm.

4) A pyramid with base 10 in² and height 6 in.

5) A sphere with radius 1 m.

6)

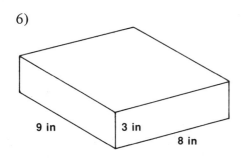

7)

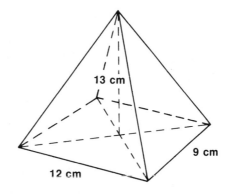

8) Round to the nearest tenth.

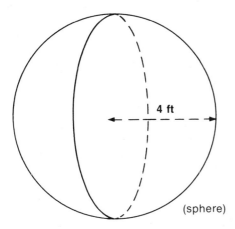

9)

10) Find the volume of a cube that is 7.3 cm on each edge.

11) Find the volume in cubic inches of a cylinder that has radius 9 inches and height 2 feet.

12) Find the volume of a pyramid whose base is a square with side 150 feet and whose height is 30 feet.

13) Find the volume of a sphere with diameter 30 meters.

14) In the application at the beginning of this section, how many tons of coal will the box car hold?

15) A water tank is a cylinder which is 22 inches in diameter and $3\frac{1}{2}$ feet high. If there are 231 in^3 in a gallon, how many gallons of water will the tank hold?

16) How many boxes that are 12 in wide, 8 in high, and 18 in long can be loaded into a truck bed that is 8 ft wide, 6 ft high, and 21 ft long?

17) An excavation is being made for a basement. The hole is 24 ft wide, 35 ft long, and 9 ft deep. If the bed of a truck holds 14 yd^3, how many truck loads of dirt will need to be hauled away?

18) A swimming pool that is 30 ft long and 10 ft wide is filled to a depth of 5 ft.
   a)   How many cubic feet of water are in the pool?
   b)   If one cubic foot of water is approximately 7.5 gallons, how many gallons of water are in the pool?

4.9 SURFACE AREA

How many square feet of sheet metal are needed to make a cylindrical heat duct that is 6″ in diameter and 15′ in length?   APPLICATION

52) Compute the surface area of given geometric solids.   OBJECTIVE

Lateral surface area is the area of the sides (lateral faces) of geometric solids such as cylinders, rectangular solids, cones, and pyramids. *Total surface area* is the lateral surface area plus the area of the top and/or bottom.   VOCABULARY

The sides of many geometric solids are actually simple geometric figures. The rectangular solid has six faces, each of which is a square or a rectangle. The cylinder has circles for top and bottom and a rectangle or square for the curved surface. The solids that we will consider, along with their formulas for surface area, follow.   THE HOW AND WHY OF IT

Rectangular Solid

$A = 2lw + 2hl + 2hw$ (Total area)

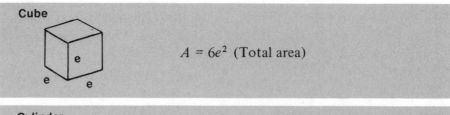

Cube

$$A = 6e^2 \text{ (Total area)}$$

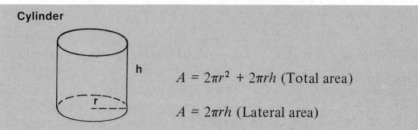

Cylinder

$$A = 2\pi r^2 + 2\pi rh \text{ (Total area)}$$

$$A = 2\pi rh \text{ (Lateral area)}$$

r is the radius of the base and *h* is the height of the cylinder.

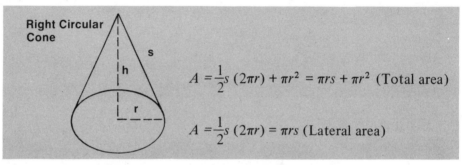

Right Circular Cone

$$A = \frac{1}{2}s\,(2\pi r) + \pi r^2 = \pi rs + \pi r^2 \text{ (Total area)}$$

$$A = \frac{1}{2}s\,(2\pi r) = \pi rs \text{ (Lateral area)}$$

s is the slant height and *r* is the radius of the base.

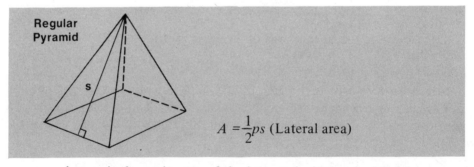

Regular Pyramid

$$A = \frac{1}{2}ps \text{ (Lateral area)}$$

where *p* is the perimeter of the base and *s* is the slant height.

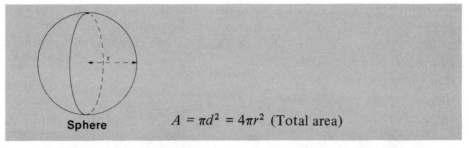

Sphere

$$A = \pi d^2 = 4\pi r^2 \text{ (Total area)}$$

d is the diameter of the sphere and *r* is the radius of the sphere.

EXAMPLES

a) Find the lateral surface area of a cylinder that is 14 cm high and has a radius of 4 cm.

$$A = 2\pi rh$$
$$A \approx 2(3.14)(4)(14)$$
$$A \approx 351.68 \text{ cm}^2$$

b) Find the total surface area of a cube that is 18″ on each edge.

$$A = 6e^2$$
$$A = 6 \cdot 18^2$$
$$A = 6 \cdot 324$$
$$A = 1944 \text{ in}^2$$

c) What is the lateral surface area of a pyramid whose base is a square that is 3 meters on each side and whose slant height is 2 meters?

$$A = \frac{1}{2}ps$$
$$A = \frac{1}{2}(4 \cdot 3) \cdot 2$$
$$A = \frac{1}{2}(12) \cdot 2$$
$$A = 12 \text{ m}^2$$

---

1) Find the total surface area of a cube whose edge is 50 cm.

EXERCISES

2) Find the lateral surface area of a cone whose radius is 2 feet and slant height is 3 feet.

3) Find the total surface area of a cylinder whose radius is 8 inches and height is 48 inches.

4) Find the lateral surface area of a pyramid whose base is a square 16 cm on each side and whose slant height is 20 cm.

5) What is the total surface area of a rectangular solid whose length, width, and height are 8 inches, 6 inches and 10 inches respectively?

6) What is the total surface area of a sphere whose radius is 2 m?

7) In the application at the beginning of this section, how many square feet of sheet metal are needed?

8) How many square feet of sheet metal will it take to make a cylindrical tank that is to be 20 inches in diameter and 4 feet in height? (Total surface area.)

9) A T.P. tent is in the shape of a cone. How many square feet of canvas are needed to make a T.P. tent (with floor) whose base has a radius of 8 feet and whose slant height is 15 feet? (Total surface area.)

10) The interior of a storage tank which is a rectangular solid is to be painted. If one gallon of paint will cover 300 square feet of surface, how many gallons are needed if the tank's length, width, and height are 20 feet, 18 feet, and 12 feet respectively? (Total surface area.)

1. (Obj. 48) Find the circumference of a circle whose radius is 6 inches. (Let $\pi \approx 3.14$.) _____

2. (Obj. 47) Evaluate the following formula if $a = 32$ and $t = 3$.

$$d = \frac{1}{2}at^2 - \frac{1}{2}a\,(t-1)^2$$

_____

3. (Obj. 50) Find the area of the following figure.

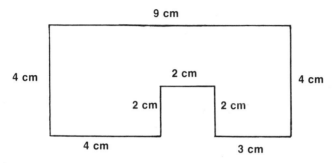

9 cm

4 cm   2 cm   4 cm

2 cm   2 cm

4 cm   3 cm

_____

4. (Obj. 46) Convert 480 minutes to hours. _____

5. (Obj. 42) 4 yd 1 ft 3 in = ? in. _____

6. (Obj. 44) 4682 ml = ? l. _____

7. (Obj. 43) Subtract.   5 yd 1 ft
                         2 yd 2 ft _____

8. (Obj. 51) The bed of a truck is 9 feet long, 6 feet wide, and 4 feet deep. How many scoops of gravel, each containing 1 cubic yard (27 cubic feet), will it take to fill the truck bed? _____

9. (Obj. 51) What is the weight of a metal ingot that is 2 ft long, 18 in wide, and 6 in high, if the metal weighs 58.75 lb per cubic foot? _____

10. (Obj. 43) Add.   6 ft 4 in
                     2 ft 5 in
                     3 ft 8 in _____

11. (Obj. 49) Find the area of a circle whose radius is 6 inches. (Let $\pi \approx 3.14$.) _____

12. (Obj. 45) Subtract.   28 km 450 m
                          6 km 580 m _____

13. (Obj. 42) 2.75 gal = ? quarts. _____

14. (Obj. 44) 2.3 km = ? m. _____

15. (Obj. 45) Add.   4 km 350 m

               <u>9 km 840 m</u> _____

16. (Obj. 46) Convert 50 miles per hour to feet per second. (To the nearest tenth.) _____

17. (Obj. 46) Norm's automobile averages 25 miles per gallon as he travels to and from work. If he travels 52 miles each day, round trip, how many gallons of gasoline does he use in one week? He works five days each week. _____

18. (Obj. 48) What is the perimeter of the following figure? _____

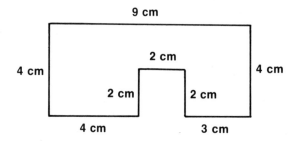

19. (Obj. 52) How many gallons of paint are needed to paint the walls and ceiling of a room that is 18 feet long, 15 feet wide, and 7.5 feet high if one gallon will cover 450 square feet? (Assume that an allowance has already been made for windows and doorways.) _____

20. (Obj. 50) What is the area of the shaded portion of the following figure? _____

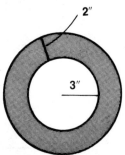

21. (Obj. 42) 6 ft. = ? inches.   _____

22. (Obj. 52) What is the lateral surface area of a cylinder that has a radius of 16 in and a height of 48 in. (Let $\pi \approx 3.14$.)   _____

23. (Obj. 47) Evaluate the following formula if $r = 2$ and $\pi \approx 3.14$. (Round to nearest tenth.)

$$V = \frac{4}{3}\pi r^3$$ _____

24. (Obj. 44) 462 cm = ? m.   _____

25. (Obj. 49) Find the area of a rectangle whose length is 18 inches and whose width is 12 inches.   _____

CHAPTER 5

SIGNED NUMBERS

5.1 SIGNED NUMBERS, OPPOSITES, AND ABSOLUTE VALUE

At the New York Stock Exchange positive and negative numbers are used to record changes in stock prices on the board. What is the opposite of a gain of three-eighths $(+\frac{3}{8})$?

53) Determine the opposite of a signed number.
54) Determine the absolute value of a signed number.

Positive numbers are numbers greater than zero. Negative numbers are numbers less than zero. Positive numbers and negative numbers together are called *signed numbers*.

The *opposite* of a signed number is that number which, on the number line, is the same number of units from zero but on the opposite side of zero. Zero is its own opposite. The symbol −5 can be read as the opposite of 5.

The absolute value of a signed number is the number of units (on the number line) between the number and zero. The symbol |7| is read as the absolute value of 7.

The application in this section involves numbers other than whole numbers, fractions, or decimals. These numbers are sometimes called signed numbers since there is another symbol (sign) that is used to identify them. Consider the following number line:

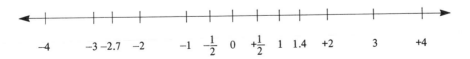

Notice that those symbols assigned to numbers that are to the right of zero are those that were used in arithmetic, while those to the left of zero have a dash, or minus sign, in front of them. The dash is read as negative and indicates that the number is to the left of zero on the number line. Those numbers to the right of zero are called positive (and can be written with a plus sign). Zero is neither positive nor negative.

The following symbols name the numbers indicated in the column on the right.

7	Seven or positive seven
$+\frac{1}{2}$	One-half or positive one-half
-3	Negative three
$-.12$	Negative twelve hundredths
0	Zero is neither positive nor negative

The opposite of a signed number is located on the opposite side of and the same number of units from zero on the number line. If a is a positive number, then

$-(a) = -a$ (The opposite of a positive number is negative.)
$-(-a) = a$ (The opposite of a negative number is positive.)

For instance: $-(-3) = 3$

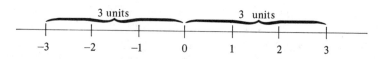

The absolute value of a signed number is determined by the number of units between the number and zero on the number line. If a is a positive number (or zero), then

$$|a| = a$$
$$|-a| = a$$

For instance, $|-5| = 5$

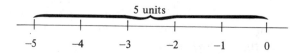

EXAMPLES

a)   The opposite of 8 is written -8, which is negative eight (on the opposite side and 8 units from 0).

b)   The opposite of negative 9 is written $-(-9)$, and this is 9 (on the opposite side and 9 units from 0).

c) -12 is read negative twelve or the opposite of twelve.

d) The absolute value of six is written $|6|$ and is 6.

e) The absolute value of negative sixteen is written $|-16|$ and is 16.

f) $|0| = 0$; the absolute value of zero is zero.

EXERCISES

Find the opposite of each of the following numbers.

1) -3 2) 5   3) $\dfrac{1}{2}$ 4) -2.1 5) $-\dfrac{1}{3}$

Find the absolute value of each of the following numbers.

6) $|-1|$ 7) $|-5|$ 8) $\left|\dfrac{1}{6}\right|$ 9) $|4|$ 10) $|-1.2|$

Find the opposite of each of the following numbers.

11) -31 12) 13   13) $-\dfrac{2}{3}$ 14) -2.35

15) $3\dfrac{1}{8}$ 16) $-22\dfrac{2}{3}$ 17) 0.23   18) -103.6

19) $-\left(4\dfrac{15}{16}\right)$ 20) -0.625

Find the absolute value of each of the following numbers.

21) $|-6|$ 22) $\left|-\dfrac{4}{5}\right|$ 23) $|.035|$ 24) $\left|-\dfrac{25}{8}\right|$

25) $|-21.75|$ 26) $|2.03|$ 27) $|6 - 1.5|$ 28) $|0|$

29) $|2 + 5|$ 30) $|-.003|$

31) In the application at the beginning of this section, what is the opposite of the gain?

32) On a thermometer temperatures above zero are listed as positive and those below zero as negative. What is the opposite of a reading of $-12°$ C?

33) The modern calendar counts the years after the birth of Christ as positive numbers (A.D. 1976 or +1976). Years before Christ are listed using negative numbers (2045 B.C. or −2045). What is the opposite of 1875 B.C. or −1875?

34) The empty weight center of gravity of an airplane is determined. A generator is installed at a moment of −300. At what moment could a weight be placed so that the center of gravity remains the same? (Moment is the product of a quantity, such as weight, and its distance to a fixed point. In this application the moments must be opposites to keep the same center of gravity.)

5.2 ADDITION OF SIGNED NUMBERS

APPLICATION   An airplane is being reloaded. 577 pounds of baggage and mail are removed (−577 pounds) and 482 pounds of baggage and mail are loaded on (+482 pounds). What net change in weight should the cargo master report?

OBJECTIVE   55) Find the sum of two signed numbers if one or both are negative.

VOCABULARY   Note that $-(x - y)$ is read "the opposite" of x minus y.

THE HOW AND WHY OF IT   Positive and negative numbers are used to show quantities having opposite characteristics.
+482 lb may show 482 pounds loaded.
−577 lb may show 577 pounds unloaded.
+27 dollars may show 27 dollars earned.
−19 dollars may show 19 dollars spent.
Using these characteristics we can find the sum of signed numbers. The various combinations follow.

$$(27) + (-19) = ?$$

If you think of this as 27 dollars earned (positive) and 19 dollars spent (negative), the end result is 8 dollars left in your pocket (positive). So $(27) + (-19) = 8$.

$$(-23) + (15) = ?$$

If you think of this as 23 dollars spent (negative) and 15 dollars earned (positive), the end result is that you still owe 8 dollars (negative). So $(-23) + (15) = -8$.

$$(-5) + (-2) = ?$$

If you think of this as 5 dollars spent (negative) and 2 dollars spent (negative), the end result is 7 dollars spent (negative). So $(-5) + (-2) = -7$. Shortcut: $-(5 + 2) = -7$.

The sum of two positive numbers is positive.

The sum of two negative numbers is negative.

The sum of a positive and a negative number is found by subtracting and

a) choosing a positive answer if the larger number (in absolute value) is positive;

b) choosing a negative answer if the larger number (in absolute value) is negative;

c) choosing zero if the numbers are opposites.

EXAMPLES

a) $(37) + (-15) = 22$

b) $(-48) + (29) = -19$

c) $(-.33) + (-.17) = -.50$

d) $(-8.3) + (25) = 16.7$

e) $\frac{4}{5} + -\frac{3}{2} = -\frac{7}{10}$

f) $(-13) + (72) + (-20) + (-23) = 16$

g) $(-.1) + (.5) + (-3.4) + (.8) = -2.2$

h) $-\frac{3}{4} + \frac{7}{8} + (-\frac{1}{2}) + (-\frac{7}{8}) = -\frac{5}{4}$

Add.

EXERCISES

1) $(-3) + (-2)$ 2) $(4) + (-5)$ 3) $(-6) + (2)$

4) $(-1) + (-1)$ 5) $(-10) + (-3)$ 6) $(7) + (-14)$

7) $(-8) + (3)$ 8) $(11) + (-3)$ 9) $(-15) + (-12)$

10) $(-17) + (9)$ 11) $(-39) + (17)$ 12) $(-87) + (94)$

13) $(-34) + (34)$ 14) $(-72) + (-72)$ 15) $(48) + (-39)$

16) $(-28) + (-47)$ 17) $(-432) + (439)$ 18) $(-239) + (225)$

19) $(-41) + (-144)$ 20) $(198) + (200)$ 21) $(-8.2) + (-3.2)$

22) $(6.3) + (-3.7)$ 23) $(-1.4) + (4.1)$ 24) $(1.66) + (-3.5)$

25) $(\frac{5}{12}) + (-\frac{7}{12})$ 26) $(-\frac{4}{15}) + (-\frac{6}{15})$

27) $(-31) + (18) + (-63) + (22)$ 28) $(-52) + (59) + (-26) + (-44)$

29) $(-19) + (54) + (-68) + (6)$ 30) $(32) + (11) + (-68) + (21)$

31) $(-17.3) + (14.6) + (6.9)$ 32) $(-11.23) + (15.36) + (-27.22)$

33) $(-65.52) + (19.51) + (-87.72)$ 34) $(16.49) + (-23.45) + (8.35)$

35) $(-\frac{1}{2}) + (-\frac{7}{8}) + (-\frac{5}{6})$

36) $(1\frac{2}{3}) + (-4\frac{5}{6}) + (-2\frac{1}{4})$

37) $(13\frac{2}{9}) + (-8\frac{2}{3}) + (-2\frac{2}{3})$

38) $(-15\frac{1}{2}) + (24\frac{3}{4}) + (-18\frac{5}{8})$

39) $(-.035) + (.751) + (-.111)$

40) In the application at the beginning of this section, what net change in weight should the cargo master report?

41) During the current fiscal year the LeBaroque Coffee House recorded the following quarterly earnings (positive numbers represent profit, negative numbers represent loss): $3456, −$507, −$498, $4007. What was the total profit (or loss) for the year?

42) The Pacific Northwest Book Depository handles most textbooks for local schools. On September 1, 1977 the inventory was 18,340 volumes. During the month the company had the following transactions (positive numbers represent volumes received, negative numbers represent shipments): 1800, −356, −843, −500, 250, −650. What is the inventory at the end of the month?

43) The change in altitude of a plane in flight was measured every ten minutes. The figures between 3 PM and 4 PM were:

3:00 PM	30,000 ft initially	(+30,000)
3:10 PM	increase of 220 ft	(+220)
3:20 PM	decrease of 200 ft	(−200)
3:30 PM	increase of 55 ft	(+55)
3:40 PM	decrease of 110 ft	(−110)
3:50 PM	decrease of 25 ft	(−25)
4:00 PM	increase of 40 ft	(+40)

What was the altitude of the plane at 4 PM? (Hint: Find the sum of the initial altitude and the six measured changes between 3 and 4 PM).

5.3 SUBTRACTION OF SIGNED NUMBERS

APPLICATION

Viking II recorded high and low temperatures of −22° and −107° for one day on the surface of Mars. What was the change in temperature for that day?

OBJECTIVE

56) Find the difference of two signed numbers.

VOCABULARY

Recall that subtraction is the inverse of addition, that is, $10 - 4 = 6$ because $4 + 6 = 10$.

THE HOW AND WHY OF IT

$(11) - (8) = ?$ asks what must be added to 8 to equal 11. $(11) - (8) = 3$, because $8 + 3 = 11$.

$(-3) - (+5) = ?$ asks what number must be added to +5 to equal −3. $(-3) - (+5) = -8$, because $5 + (-8) = -3$.

$(-4) - (-7) = ?$ asks what number must be added to −7 to equal −4. $(-4) - (-7) = 3$, because $(-7) + (3) = -4$.

Compare:

$$11 - 8 \text{ and } 11 + (-8) \qquad [\text{both equal } 3]$$
$$(-3) - (5) \text{ and } (-3) + (-5) \quad [\text{both equal } -8]$$
$$(-4) - (-7) \text{ and } (-4) + (7) \quad [\text{both equal } 3]$$

Every subtraction problem can be turned into an addition problem.

> To find the difference between two numbers, add the opposite of the number that is to be subtracted.
> If a and b are any two numbers,
>
> $$a - b = a + (-b)$$

EXAMPLES

a) $(45) - (33) = 12$ [because $45 + (-33) = 12$]

b) $(-47) - (29) = -76$ [because $(-47) + (-29) = -76$]

c) $(-38) - (-56) = 18$ [because $(-38) + (56) = 18$]

d) $(88) - (114) = -26$

e) $\dfrac{4}{5} - (-\dfrac{1}{2}) = 1\dfrac{3}{10}$

f) $(16) - (-4) - (7) = 13$

g) $(-.21) - (-3.4) - (-.15) - (.42) = 2.92$

h) $-\dfrac{3}{4} - \dfrac{7}{8} - (-\dfrac{1}{2}) - (-\dfrac{1}{8}) = -1$

Subtract.

EXERCISES

1) $(6) - (4)$

2) $(-6) - (4)$

3) $6 - (-4)$

4) $(-6) - (-4)$

5) $(-8) - (5)$

6) $(10) - (7)$

7) $(-15) - (-3)$

8) $(-15) - (3)$

9) $(11) + (-5)$

10) $(-9) - (4)$

11) $(-\frac{1}{4}) - (\frac{2}{4})$

12) $(-\frac{2}{3}) - (-\frac{1}{3})$

13) $(16) - (-25)$

14) $(56) - (77)$

15) $(-42) - (-33)$

16) $(-49) - (43)$

17) $(-45) - (45)$

18) $(-83) \div (-83)$

19) $(3.45) - (4.22)$

20) $(7.33) - (-2.48)$

21) $(-8.92) - (4.68)$

22) $(-43.8) - (-62.9)$

23) $(-25) - (-18.7)$

24) $(\frac{3}{4}) - (1\frac{1}{2})$

25) $(-\frac{2}{3}) - (\frac{1}{2})$

26) $(-\frac{5}{8}) - (-\frac{2}{5})$

27) $(\frac{6}{7}) - (\frac{9}{10})$

28) In the application at the beginning of this section, what was the change in temperature for the day?

29) At the beginning of the month, Joe's bank account had a balance of $315.65. At the end of the month, the account was overdrawn by $63.34 (−$63.34). If there were no deposits during the month, what was the total amount of the checks Joe wrote? (Hint: subtract the ending balance from the original balance.)

30) What is the difference in altitude between the highest point in the United States and the lowest point?

Highest point: Mt. McKinley is 20,300 ft above sea level (+20,300).

Lowest point: Death Valley is 282 ft below sea level (−282).

5.4 MULTIPLICATION OF SIGNED NUMBERS

The formula for converting a temperature measurement from **APPLICATION**
Fahrenheit to Celsius is $C = \frac{5}{9}(F - 32)$. What Celsius measure is equal to
18°F?

57) Find the product of two signed numbers. **OBJECTIVE**

When the sum of two numbers is zero, the numbers are opposites. So **VOCABULARY**
if $a + b = 0$, $a = -b$ (a is the opposite of b). The distributive law of
multiplication over addition is $a(b + c) = ab + ac$. So $2(3 + 4 = 2 \cdot 3 + 2 \cdot 4$.

Consider the following problem: **THE HOW AND WHY OF IT**

$3 + (-3) = 0$	Sum of opposites
$4[3 + (-3)] = 4 \cdot 0$	Multiply by 4
$4[3 + (-3)] = 0$	Multiplication property of zero
$4(3) + 4(-3) = 0$	Distributive law
$12 + 4(-3) = 0$	Multiply
So $4(-3) = -12$	$4(-3)$ must be the opposite of 12 because their sum is 0

This example can be repeated for any sum equal to zero and leads to the
conclusion that the product of a positive number and a negative number is
a negative number.

The product of two negatives can be found in a similar fashion.
Consider:

$(2) + (-2) = 0$	Sum of opposites
$(-3)[(2) + (-2)] = (-3)(0)$	Multiply by -3
$(-3)[(2) + (-2)] = 0$	Multiplication property of 0
$(-3)(2) + (-3)(-2) = 0$	Distributive law
$-6 + (-3)(-2) = 0$	Product of a positive and a negative
So $(-3)(-2) = -(-6) = 6$	$(-3)(-2)$ is added to -6 for a sum of 0

Therefore, we conclude that the product of two negative numbers is a
positive number.

In summary:

The product of two positive numbers is positive.
The product of two negative numbers is positive.
The product of a positive and a negative number is negative.

EXAMPLES

a) $(-6)(5) = -30$

b) $(2)(-3.3) = -6.6$

c) $(-\frac{1}{3})(\frac{1}{5}) = -\frac{1}{15}$

d) $(5)(-.7) = -3.5$

e) $(-6)(-7) = 42$

f) $(-1.5)(-.7) = 1.05$

g) $(-\frac{2}{7})(-\frac{3}{8}) = \frac{6}{56} = \frac{3}{28}$

h) $(-121)(-5) = 605$

i) $(3)(6) = 18$

j) $(-6)(-3)(-4) = (18)(-4) = -72$

EXERCISES   Find the products.

1) $(-3)(2)$

2) $(-5)(-6)$

3) $(9)(-6)$

4) $(-5)(-2)$

5) $(-1)(6)$

6) $(-17)(0)$

7) $(8)(-6)$

8) $(-6)(9)$

9) $(-1)(-1)(-1)$

10) $(-12)(-3)$

11) $(-7)(-2)(-3)$

12) $(2)(5)(-6)$

13) $(-3.02)(6)$

14) $(-\frac{3}{4})(-\frac{1}{2})$

15) $(-20.16)(-1.5)$

16) $(-.6)(-.4)(.2)$

17) $(5-3)(6-8)$

18) $|-3| \cdot |-5|$

19) (3) (−.7) (|−5|)   20) $(\frac{1}{2}+\frac{1}{3})\,(5-4\frac{1}{3})$ 21) (−2) (−3) (−4) (−5)

22) (−7) (8) (0)   23) (−3.18) (−1.6) (0.1) 24) (−1.7) (5.3) (−.2)

25) (−1) (−1) (−1) (−1) (−1) (3.07)   26) $(-1)^{11}(25)$

27) In the application at the beginning of this section, what Celsius measure is equal to 18°F?

28) For five consecutive weeks Ms. Obese recorded a loss of 2.5 lb. If each loss is represented by −2.5 lb, what was the total weight loss for the five weeks, expressed as·a signed number?

29) The Industrial Stock Average sustained twelve straight days of a 2.83 point decline (−2.83) each day. What was the total decline during the twelve-day period, expressed as a signed number?

30) Safeway Stores offered as a loss leader 10 lb of sugar at a loss of 12¢ per bag (−12¢). If 560 bags were sold during the sale, what was the total loss, expressed as a signed number?

5.5 DIVISION OF SIGNED NUMBERS

The coldest temperature in Eycee Northland for each of five days was −15°, −5°, −4°, 2°, and −3°. What was the average of these temperatures?   APPLICATION

OBJECTIVE   58) Find the quotient of two signed numbers.

VOCABULARY   Recall that division is the inverse of multiplication, that is, $12 \div 3 = 4$ because $3 \cdot 4 = 12$.

THE HOW AND WHY OF IT   To divide two signed numbers we must find the number that when multiplied by the divisor equals the dividend.

$(-8) \div (4) = ?$ asks what number when multiplied by 4 equals -8. So, $(-8) \div (4) = -2$, because $(4)(-2) = -8$.

$(-21) \div (-3) = ?$ asks what number when multiplied by -3 equals -21. So, $(-21) \div (-3) = 7$, because $(-3)(7) = -21$.

$15 \div (-5) = ?$ asks what number when multiplied by -5 equals 15. So, $15 \div (-5) = -3$, because $(-5)(-3) = 15$.

These examples lead to rules similar to those for multiplication.

The quotient of two positive numbers is positive.
The quotient of two negative numbers is positive.
The quotient of a positive and a negative number is negative.

EXAMPLES

a) $(-6) \div (3) = -2$ b) $(-8.6) \div (4.3) = -2$

c) $(-\frac{3}{4}) \div (-\frac{1}{2}) = \frac{3}{2}$ d) $(-7) \div (-\frac{4}{3}) = \frac{21}{4}$

e) $(1.44) \div (-.3) = -4.8$ f) $(68) \div (-4) = -17$

EXERCISES   Divide.

1) $(-10) \div (5)$ 2) $(10) \div (-2)$ 3) $(-10) \div (-5)$

4) $(8) \div (-2)$ 5) $(-16) \div (-4)$ 6) $(15) \div (-3)$

7) $(18) \div (-1)$ 8) $(-12) \div (-3)$ 9) $(15) \div (-3)$

10) $(-14) \div (-2)$ 11) $(-33) \div (11)$ 12) $(6.06) \div (-6)$

13) $(-4.95) \div (-.9)$ 14) $(-15.5) \div (.05)$ 15) $(.65) \div (-.13)$

16) $[(-5)(-2)] \div (-10)$ 17) $36 \div [(-3)(-6)]$ 18) $(+24) \div (-6)$

19) $[(28)(-1)] \div (-7)$ 20) $(-1.2) \div (.06)$ 21) $(100) \div [(-5)(-5)]$

22) $(-\frac{3}{4}) \div (-\frac{3}{5})$ 23) $(6-15) \div 3$ 24) $(3-8-5) \div (-2)$

25) $(-7+4-9) \div (4-2)$ 26) $[(-8.2)(3.6)] \div (-1.8)$

27) In the application at the beginning of this section, what was the average temperature in Eycee?

28) Mr. Gambit lost $836 (−$836) in 12 straight hands of poker at a casino in Reno. What was his average loss per hand, expressed as a signed number?

29) The membership of the Burlap Baggers Investment Club took a loss of $183.66 (−$183.66) on the sale of stock. If there are six coequal members in the club, what is each member's share of the loss, expressed as a signed number?

30) The temperature in Fairbanks, Alaska, dropped from 10° above zero (+10°) to 15° below zero (−15°) in an eight-hour period. This is a drop of 25° (−25°). What was the average drop per hour, expressed as a signed number?

5.6 REVIEW: ORDER OF OPERATIONS

APPLICATION   Does −40°F equal −40°C? To check, substitute −40 for F and also for C in the formula

$$F = \frac{9}{5}C + 32$$

and tell whether the statement is true.

OBJECTIVE   59) Perform any combination of operations (addition, subtraction, multiplication, or division) on signed numbers in the conventional order.

VOCABULARY   No new vocabulary.

THE HOW AND WHY OF IT   The conventional order of operations for signed numbers is the same as that for whole numbers.

EXAMPLES

a) $(−64) + (−22) \div (2) = (−64) + (−11)$ (division performed first)

$= −75$ (addition performed next)

b) $(−12)(5) − (54) \div (−3) = (−60) − (−18)$ (multiplication and division performed first)

$= −42$ (subtraction performed next)

c) $(3) \div (-.8) + 2(-1.7) = (-3.75) + (-3.4)$
$$= -7.15$$

d) $6 - (\frac{1}{2})(-5)$ $= 6 - (-2\frac{1}{2})$

$$= 6 + 2\frac{1}{2}$$

$$= 8\frac{1}{2}$$

Perform the indicated operations.

1) $(-8)(6) + (-18)$ 2) $25 + (-3)(-8)$

3) $(2)(-8) + 9$ 4) $19 - (4)(-2)$

5) $(-2)(3) + (5)(-2)$ 6) $(-7) + 2(-3)$

7) $(-5)(-3) - (2)(6)$ 8) $(-3)(-7) - 10$

9) $2(-6 + 5) - 3$ 10) $(-7)(6) \div (-3)$

11) $(-13)(-3) + (15)(-2)$ 12) $14(-2) - 8(-5)$

13) $6(-10 + 4) - 33 \div (-11)$ 14) $(-20) \div (-8)(-\frac{3}{4})$

15) $84 - (9)(9) + (-6)(2)$ 16) $[(-8)\,9 + 20] \cdot (-3) - 25$

17) $28 \div (-4) - 7 + (-1)(-8) - 6$ 18) $(-126) - [(-6)(6) - 6]$

19) In the application at the beginning of this section, does $-40°F$ equal $-40°C$?

20) The temperature at 5 PM was $18°$ C. If the temperature dropped $.6°$ every hour until midnight, we can find the midnight temperature by calculating the value of

$$T = 18 + (7)(-0.6)$$

What was the midnight temperature?

21) If the temperature at 6 AM was $-10°F$ and rose $3.4°F$ every hour until 11 AM, what was the temperature at 11 AM?

22) To check whether $x = -6$ is a solution to the equation

$$x^2 - 4x - 7 = 53$$

we substitute -6 for x and find the value of

$$(-6)(-6) - 4(-6) - 7$$

Check whether the above expression equals 53.

23) Substitute $x = -1\frac{1}{2}$ in the following equation and check whether it is a solution.

$$\frac{3}{4}(x + \frac{2}{3}) - \frac{3}{8} = -1$$

24) Does $-10°F$ equal $-10°C$? To check, substitute -10 for F and also for C in the formula in the application at the beginning of this section and tell whether the statement is true.

Perform the indicated operations.

1. (Obj. 55) $(-21) + (-17) + (42) + (-18)$ _____

2. (Obj. 58) $(-\frac{3}{8}) \div (\frac{3}{10})$ _____

3. (Obj. 56) $(-\frac{7}{15}) - (-\frac{3}{5})$ _____

4. (Obj. 59) $(36 - 42)(-18 + 6)$ _____

5. (Obj. 59) $(-3 + 18) - (21 - 7) + 3$ _____

6. (Obj. 55) $(-3.65) + (4.72)$ _____

7. (Obj. 53, 54) a. $-(-21) = ?$ a._____
                  b. $|-21| = ?$ b._____

8. (Obj. 56) $(-37) - (-41)$ _____

9. (Obj. 59) $(-16 + 4) \div 3 + 2(-6)(-3)$ _____

10. (Obj. 58) $(-88) \div (-22)$ _____

11. (Obj. 57) $(-7)(-9)(2)$ _____

12. (Obj. 57) $(|-6|)(-4)(-1)(-1)$ _____

13. (Obj. 56) $(-57.9) - (32.5)$ _____

14. (Obj. 59) $(-21)(3) - (-6)(-7)$ _____

15. (Obj. 58) $(30.66) \div (-0.6)$ _____

16. (Obj. 55) $(-\frac{1}{3}) + (\frac{5}{6}) + (-\frac{1}{2}) + (-\frac{1}{6})$ _____

17. (Obj. 58) $(-56) \div (-7)$ _____

18. (Obj. 56) $(18) - (-25)$ _____

19. (Obj. 59) $-7(3 - 11)(-2) - 4(-3 - 5) + 18 \div (-6)$ _____

20. (Obj. 57) $(-\frac{1}{3})(\frac{6}{7})$ _____

21. (Obj. 55) $(-17) + (-36)$ _____

22. (Obj. 59) 3 (−7) + 25   _____

23. (Obj. 56) The temperature in Chicago ranged from a high of 12°F to a low of −9°F within a twenty-four hour period. What was the drop in temperature expressed as a signed number?   _____

24. (Obj. 55) A stock on the West Coast Exchange opened at $6\frac{5}{8}$ on Monday. It recorded the following changes during the week: Monday, $+\frac{1}{8}$; Tuesday, $-\frac{3}{8}$; Wednesday, $+1\frac{1}{4}$; Thursday, $-\frac{7}{8}$; Friday, $+\frac{1}{4}$. What was its closing price on Friday?   _____

25. (Obj. 59) What Fahrenheit temperature is equal to a reading of −10°C? Use the formula
$$F = \frac{9}{5}C + 32$$
                                            _____

CHAPTER 6

EQUATIONS AND POLYNOMIALS

6.1 LINEAR EQUATIONS

Λ Fahrenheit thermometer scale shows a temperature of 95 degrees. Use the formula $F = \frac{9}{5}C + 32$ to find the corresponding temperature on the Celsius scale. (That is, solve for C in $95 = \frac{9}{5}C + 32$.)

APPLICATION

60) Solve a linear equation that can be written in the form $ax + b = c$ when a, b, and c are given.

OBJECTIVE

An *equation* is a statement that two expressions are equal.

An equation may be true ($3x = 15$ is true if $x = 5$). An equation may also be false ($2x = 7$ is false if $x = 3$).

In a *linear equation* the largest exponent of the variable is the number 1 ($3x^1 - 2 = 13$ or $3y - 7 = 2y$, but not $x^2 - 3x = 4$).

An *identity* is an equation that is true for all replacements of the variable ($2x = 2x$ or $3y + 4y = 7y$).

A *conditional equation* is true for some replacements, but not for all ($2x = 7$).

The *solution* or root of an equation is the set of all number replacements, that, when substituted for the variable, make the equation true. In the equation $2x = 6$, $x = 3$ is the solution.

VOCABULARY

To find the solution of a linear equation, we write a series of equations (usually simplifying as we go along) until we arrive at an equation of the form $x =$ (some number) or (some number) $= x$. For instance $x = 6$ or $-\frac{13}{4} = x$. (In some cases, explained later in the book, a form such as $x = a$ or $x = b - c$ is the last step.)

THE HOW AND WHY OF IT

This number or expression on the right is the solution of the linear equation. The series of equations leading up to this step comes from applying one or more of the following:

Fundamental Laws of Equations:

(1) Adding the same number to (or subtracting the same number from) both sides of an equation does not change the solution.

(2) Multiplying or dividing both sides of an equation by the same number (except 0) does not change the solution.

To solve the equation $3x - 2 = 13$, we can write this series of equations:

$$3x - 2 = 13 \qquad \text{original equation}$$
$$3x - 2 + 2 = 13 + 2 \qquad \text{add 2 to both sides}$$
$$3x = 15 \qquad \text{simplify}$$
$$\frac{3x}{3} = \frac{15}{3} \qquad \text{divide both sides by 3}$$
$$x = 5 \qquad \text{simplify (5 is the solution)}$$

$$3(5) - 2 = 13 \qquad \text{to check, replace } x \text{ with 5 in the}$$
$$15 - 2 = 13 \qquad\qquad \text{original equation}$$
$$13 = 13$$

The shortcuts for this process involve doing some of the steps mentally (see Example e).

> To find the solution of an equation that can be written in the form $ax + b = c$, apply the fundamental laws of equations until the variable is isolated on one side of the equation.

EXAMPLES

a) $6x = 372$

$$\frac{6x}{6} = \frac{372}{6} \qquad \text{divide both sides by 6}$$
$$x = 62 \qquad \text{substitution}$$

The solution is 62.

b) $$2x + \frac{5}{6} = \frac{2}{3}$$

$$2x + \frac{5}{6} - \frac{5}{6} = \frac{2}{3} - \frac{5}{6} \qquad \text{subtract } \frac{5}{6} \text{ from both sides}$$
$$2x = -\frac{1}{6} \qquad \text{substitution}$$
$$\frac{2x}{2} = -\frac{1}{6} \div 2 \qquad \text{divide both sides by 2}$$
$$x = -\frac{1}{12} \qquad \text{substitution}$$

The solution is $-\frac{1}{12}$.

c)
$$\frac{1}{2}x - \frac{1}{6} = \frac{1}{3}$$
$$\frac{6}{1}(\frac{1}{2}x - \frac{1}{6}) = \frac{6}{1} \cdot \frac{1}{3}$$ multiply both sides by 6
                    (6 is the LCM of 2, 3, and 6)
$$3x - 1 = 2$$ substitution
$$3x = 3$$ add 1 to both sides and substitute
$$x = 1$$ divide both sides by 3 and substitute

The solution is 1.

d)
$$7.4 = 1.8x - 16$$
$$7.4 + 16 = 1.8x - 16 + 16$$
$$23.4 = 1.8x$$
$$\frac{23.4}{1.8} = \frac{1.8x}{1.8}$$
$$13 = x$$

e)
$$13 - 6x = 5$$
$$-6x = -8$$
$$x = \frac{-8}{-6}$$
$$x = \frac{4}{3}$$

Solve.   **EXERCISES**

1) $5x = 30$ 2) $2x + 6 = 30$ 3) $x - 9 = 21$

4) $3x - 11 = -2$ 5) $x - \frac{1}{3} = 3$ 6) $3 = 7 + 2x$

7) $7x - 1 = 8$ 8) $4x + 5 = 21$ 9) $3x - 8 = 7$

10) $5a - 12 = -3$ 11) $6b + 16 = -2$ 12) $3y + 4 = 13$

13) $12z - 6 = -30$

14) $\frac{1}{2}x - 5 = 7$

15) $\frac{1}{4}y + \frac{1}{4} = \frac{1}{2}$

16) $\frac{5}{3}z - \frac{1}{2} = \frac{7}{6}$

17) $(3)(4) = 2x + 12$

18) $4a + 1 = 3$

19) $4 - 3b = 1$

20) $15 = 6 - 3z$

21) $.2x + .8 = 1.4$

22) $\frac{1}{4} - \frac{1}{2}a = \frac{2}{3}$

23) $9 = 16 - 21x$

24) In the application at the beginning of this section, what is the corresponding value on the Celsius scale?

25) The formula for the perimeter of a rectangle is $P = 2l + 2w$. If the perimeter of a certain rectangle is 40 cm and its length is 12 cm, what is the measure of its width?

26) The velocity of a falling object at any time t with initial velocity v_0 is given by $v_t = v_0 + 32.2\ t$. What is the initial velocity if the velocity after 5 seconds is 188 feet per second?

6.2 THE LAWS OF EXPONENTS

Sheetmetal trades usually refer to the area of a circle in terms of its diameter. If the formula for the area of a circle is $A = \pi r^2$ (r is the radius), what is the area in terms of the diameter? ($r = \dfrac{d}{2}$)

61) Multiply or divide powers with the same base.
62) Raise a power to a power.

See Chapter 1 for the meanings of *exponent, base,* and *power.*

The following laws of exponents are shortcuts for algebraic multiplication and division. Each law is followed by an example that is done first the long way and second by using the law.

$$\text{Law 1: } x^a \cdot x^b = x^{a+b}$$

Example: $x^3 \cdot x^5 = (xxx)(xxxxx)$
$= xxxxxxxx$
$= x^8$
or more quickly, $x^3 \cdot x^5 = x^{3+5} = x^8$

$$\text{Law 2: } \frac{x^a}{x^b} = x^{a-b}$$

Example: $\dfrac{x^9}{x^3} = \dfrac{xxxxxxxxx}{xxx}$
$= xxxxxx \text{ (by reducing)}$
$= x^6$
or more quickly, $\dfrac{x^9}{x^3} = x^{9-3} = x^6$

$$\text{Law 3: } (x^a)^b = x^{ab}$$

Example: $(x^3)^5 = x^3 \cdot x^3 \cdot x^3 \cdot x^3 \cdot x^3$
$= x^{3+3+3+3+3}$
$= x^{15}$
or more quickly, $(x^3)^5 = x^{3 \cdot 5} = x^{15}$

Law 4: $(xy)^a = x^a y^a$

Example:
$$(wz)^5 = wz \cdot wz \cdot wz \cdot wz \cdot wz$$
$$= wwwww \cdot zzzzz$$
$$= w^5 z^5$$
or more quickly, $(wz)^5 = w^5 z^5$

Law 5: $\left(\dfrac{x}{y}\right)^a = \dfrac{x^a}{y^a}$

Example:
$$\left(\frac{x}{4}\right)^3 = \frac{x}{4} \cdot \frac{x}{4} \cdot \frac{x}{4}$$
$$= \frac{x^3}{4^3} \text{ or } \frac{x^3}{64}$$
or more quickly, $\left(\dfrac{x}{4}\right)^3 = \dfrac{x^3}{4^3}$

EXAMPLES

a) $y^5 \cdot y^6 \cdot y = y^{5+6+1} = y^{12}$

b) $w^a \cdot w^2 = w^{a+2}$

c) $\dfrac{t^{10}}{t^2} = t^{10-2} = t^8$

d) $\dfrac{a^7}{a^5} = a^{7-5} = a^2$

e) $(3^2)^{10} = 3^{2 \cdot 10} = 3^{20}$

f) $(4w^3)^2 = 4^2 \cdot (w^3)^2 = 16 w^6$

g) $(x^2 y^5)^3 = (x^2)^3 (y^5)^3 = x^6 y^{15}$

h) $\left(\dfrac{2}{r^2}\right)^3 = \dfrac{2^3}{(r^2)^3} = \dfrac{8}{r^6}$

i) $\left(\dfrac{a^3}{b^4}\right)^5 = \dfrac{(a^3)^5}{(b^4)^5} = \dfrac{a^{15}}{b^{20}}$

EXERCISES   Evaluate.

1) $(2^2)^3$

2) $2^3 \cdot 2^2$

3) $\dfrac{2^5}{2^3}$

4) $\left(\dfrac{1}{2}\right)^2$

5) $\dfrac{3^4}{3^2}$

Express each of the following as a power or a product of powers of x and y.

6) $x^3 \cdot x^2$

7) $(y^2)^5$

8) $(x^2 y^3)^6$

9) $\dfrac{(x^2)^2}{y^3}$

10) $x^4 \cdot x^5 \cdot x^2$

Evaluate.

11) $(2^3 \cdot 3^2)^2$

12) $(6^3)^2$

13) $\left(\dfrac{2}{3}\right)^4$

Express each of the following as a power or a product of powers of a, b, x, or y.

14) $x^7 \cdot x^2$

15) $(y^3)^4$

16) $(a^3 b^2)^2$

17) $\dfrac{x^6}{x^3}$

18) $\left(\dfrac{a^2}{b}\right)^3$

19) $(2ab^2)^3$

20) $\dfrac{a^4 b^7}{a^2 b^6}$

21) $a^3 \cdot a^3 \cdot a^4$

22) $(3a^4 b^2 x^3)^3$

23) $\dfrac{(x^3 y^2)^2}{x^2 y^3}$

24) $(a^4 \cdot a^3)^2$

25) $x^{2a} \cdot x^{3a}$

26) $(3ax)^4 (b^2 y^2)^2$ 27) $\dfrac{(a^3 b^4)^2}{(ab^2)^3}$ 28) $(x^2)^3 (x)^4 (x^3)^5$

29) $(x^2)^{3m}$ 30) $\dfrac{x^{4a}}{x}$

31) In the application at the beginning of this section, what is the area in terms of the diameter?

32) The volume of a sphere is given by $V = \dfrac{1}{6}\pi d^3$ (d is the diameter). Express the volume in terms of the radius ($d = 2r$).

33) The volume of a box is given by $V = lwh$ (l is the length; w is the width; and h is the height). If the length is the square of the width ($l = w^2$) and the height is the cube of the width ($h = w^3$), express the volume in terms of w, using a single exponent, and then find the volume if $w = 2$ ft.

6.3 OPERATIONS ON MONOMIALS

APPLICATION   The primary applications are found in the solutions of equations and formulas.

OBJECTIVE   63) Add, subtract, multiply, and divide monomials.

VOCABULARY   The *terms* of an algebraic expression are connected by the symbols + and −. (The expression $3x - by + 2m^2 w$ has three terms: the first is $3x$, the second is by, and the third is $2m^2 w$.)

A single term is called a *monomial* and contains numerals (numerical coefficients), variable factors, and symbols for multiplication, but no indicated division by a variable $\left(-\frac{8}{3}x^2y\right)$.

Like terms are terms that have common variable factors. ($3xy$ and $4xy$ are like terms. $5xy$ and $5x^2y$ are not like terms because x^2 is a factor of only $5x^2y$.)

To add (or subtract) two or more monomials, we use a law of algebra:

THE HOW AND WHY OF IT

$$Distributive\ law:\quad ab + ac = a(b + c)$$
$$or\quad ab - ac = a(b - c)$$

This law states that each of these pairs of expressions

$$3 \cdot 5 + 3 \cdot 7 \quad and \quad 3(5 + 7)$$
$$2\frac{1}{2} \cdot 4 + 3\frac{1}{2} \cdot 4 \quad and \quad \left(2\frac{1}{2} + 3\frac{1}{2}\right) \cdot 4$$
$$6(8.7) - 6(3.8) \quad and \quad 6(8.7 - 3.8)$$

has the same value. Thus both sides of

$$3 \cdot 5 + 3 \cdot 7 = 3(5 + 7)$$
$$15 + 21 = 3 \cdot 12$$
$$36 = 36$$

have the value 36. Both sides of

$$2\frac{1}{2} \cdot 4 + 3\frac{1}{2} \cdot 4 = \left(2\frac{1}{2} + 3\frac{1}{2}\right) \cdot 4$$
$$10 + 14 = (6) \cdot 4$$
$$24 = 24$$

have the value 24, and both sides of the third example have the value 29.4.

To simplify (add) $6x + 3x$, we can use the distributive law to rewrite:

$$6x + 3x = (6 + 3)x \quad \text{distributive law}$$
$$= 9x \quad \text{substitution}$$

To add (or subtract) like terms, add (or subtract) the numerical coefficients and annex the common variable factors.

To multiply two monomials we use two laws of algebra. The associative law of multiplication permits the rearrangement of parentheses.

$$(3x^3)\,(7xy) = (3x^3 \cdot 7)\,(xy) = (3x^3)\,(7x)\,(y)$$

The commutative law of multiplication permits the interchanging of factors.

$$(3x^3 \cdot 7)\,(x7) = (3 \cdot 7 \cdot x^3)\,(x7)$$

By using these laws repeatedly the product of two monomials can be written so that the numerical coefficient precedes the variable factors.

$$(3x^3)\,(7xy) = (3 \cdot 7)\,(x^3 \cdot x)\,(y)$$
$$= 21x^4 y$$

> **To multiply monomials, multiply the numerical coefficient and multiply the variable factors.**

To divide two monomials we use the definition of division ($a \div b = c$, provided $bc = a$) and the law for dividing powers of the same base $\left(\dfrac{x^a}{x^b} = x^{a-b} \right)$. For instance, $(-51a^3 bc^2) \div (3abc)$ can be written

$$\frac{-51\,a^3 bc^2}{3abc} = \frac{-51}{3} \cdot \frac{a^3}{a} \cdot \frac{b}{b} \cdot \frac{c^2}{c}$$
$$= -17\,a^{3-1} \cdot 1 \cdot c^{2-1}$$
$$= -17\,a^2 c$$

> **To divide monomials, divide the numerical coefficients and divide the variable factors.**

When an equation contains more than one term involving an unknown, the terms can be combined, and then the equation can be solved as before.

$$6x - 14x + 3 = 51$$
$$-8x + 3 = 51 \qquad \text{combine the monomials}$$
$$\text{containing } x$$
$$-8x = 48$$
$$x = -6$$

EXAMPLES

a) $4d + 5d = 9d$ because $4d + 5d = (4 + 5)d$
                                    $= 9d$

b) $7bx - 19bx = -12bx$

c) $13x^2 + 4x^2 = 17x^2$

d) $8x^2 - 3x$ cannot be simplified (or subtracted) because they are not like terms

e) $8.3m^2 n + 5.2m^2 n - 12m^2 n = (8.3 + 5.2 - 12)m^2 n$
                                            $= 1.5m^2 n$

f) $(-3by)(by^2) = -3(bb)(yy^2) = -3b^2 y^3$

g) $3x(-4xy)(5xy^2) = (3 \cdot -4 \cdot 5)(xxx)(yy^2) = -60x^3 y^3$

$(-6cw^3)^3 = (-6 \cdot -6 \cdot -6)(ccc)(w^2 w^2 w^2) = -216c^3 w^6$
  or $= (-6)^3 (c)^3 (w^2)^3 = -216 c^3 w^6$

i) $\frac{7x^6}{3x^2} = 2\frac{1}{3}x^4$ or $\frac{7}{3}x^4$ or $\frac{7x^4}{3}$

j) $\frac{-5m^3 n^2}{-m} = \frac{-5m^3 n^2}{-1 \cdot m} = \frac{-5}{-1} \cdot \frac{m^3}{m} \cdot n^2 = 5m^2 n^2$

k) $\frac{6w^2 y}{-4w^2 y} = -1\frac{1}{2}$ or $-\frac{3}{2}$ or -1.5

l) $5x + 14 - 3x + 9x - 4 = 43$
        $11x + 10 = 43$
        $11x = 33$
        $x = 3$

m) $13x + 16 = 7x + 20$
    $13x + -7x + 16 = 7x + -7x + 20$ add $-7x$ to each side to isolate the terms with x on one side
        $6x + 16 = 20$
        $6x = 4$
        $x = \frac{4}{6} = \frac{2}{3}$

Perform the indicated operations.   **EXERCISES**

1) $6a + 3a$ 2) $3x - 5x$ 3) $4y^2 + 12y^2$

4) $5a - 7a$ 5) $(2x)(4x)$ 6) $(3ab)(-2a)(-b)$

7) $\dfrac{-18a^2 b}{9a}$ 8) $(-2xy^2)^2$ 9) $5a^2 b + 2ab - 3a^2 b$

10) $\dfrac{16a^{10}}{a^5}$ 11) $6y + 11y + (-y) + (-18y)$

12) $5a^2 + 13a^2 - 11a^2 + 4a^2$ 13) $st + 7st - 15st$

14) $-3xy^2 + 7xy^2 - 17xy^2 + 10xy^2$ 15) $3ab + 4a + 2ab$

16) $5xy^2 - 9xy^2 + 4x^2 y^2$ 17) $(3x)(2x)$

18) $(xy)(2y)$ 19) $(6ab)(-2a)(4b)$

20) $\dfrac{2ab^2}{2ab}$ 21) $\dfrac{-16x^2 y^2}{6xy}$

22) $\dfrac{-15ax^3}{5x^2}$ 23) $(3x)^3 - 4x(2x)^2 + 6x^3$

24) $(3abc)(2ab) - \dfrac{15a^2 b^2 c^2}{5c} + 12a^2 bc^2$

25) $25z^3 - \dfrac{12xyz^3}{-6xy} + \dfrac{18x^2 yz^4}{-9x^2 yz}$

26) $10x + 5 = 2x + 29$

27) $13a - 4 = 7a + 8$

28) $8y - 12y - 4 = 8$

29) $2a - 12 = 10a + 6$

30) $-2x + 5 = -8x + 6$

31) $\frac{1}{5}x - \frac{1}{2}x + \frac{3}{4}x = 9$

32) $\frac{3}{2}x - \frac{1}{4}x + \frac{5}{6}x = 5$

33) $x + 5 - 3x = 2x + 8 - 5x$

6.4 COMBINING POLYNOMIALS AND REMOVING PARENTHESES

The primary applications are found in the solution of equations and formulas.

APPLICATION

64) Add and subtract polynomials.
65) Remove parentheses and brackets in an algebraic expression.

OBJECTIVES

A *polynomial* is an algebraic expression containing monomials separated by + and/or − signs ($a - b + 2c - 4d$).
A *monomial* is a one-term polynomial ($4x$).
A *binomial* is a two-term polynomial ($3x + 4y$).
A *trinomial* is a three-term polynomial ($3x^2 - 4x + 7$).

VOCABULARY

To combine polynomials (add, subtract, or both) we use one or more of the following laws of algebra:

THE HOW AND WHY OF IT

Distributive law (see previous section)
Associative law of addition
    Example: $(6x + 3y) + (4x + 5y) = 6x + (3y + 4x) + 5y$
Commutative law of addition
    Example: $6x + (3y + 4x) + 5y = 6x + (4x + 3y) + 5y$

The following examples show how these laws are used to add polynomials. The examples do not show all the steps, but rather show how shortcuts may be used.

$$(6x + 3y) + (4x + 5y) = (6x + 4x) + (3y + 5y)$$

use both the commutative and associative laws of addition

$$=10x + 8y$$

add the monomials as before

The example can also be arranged vertically so that the monomials with a common variable are lined up.

$$
\begin{array}{r}
6x + 3y \\
4x + 5y \\
\hline
10x + 8y
\end{array}
$$

It is possible to combine more polynomials by similar procedures.

$$(-.2a + .4c) + (.6a + .5b) + (-.7a + .5b)$$
$$= [(-.2a) + .6a + (-.7a)] + (.5b + .5b) + .4c$$
commutative and associative laws of addition
$$= -.3a + b + .4c$$

A vertical arrangement may again be used.

$$
\begin{array}{r}
-.2a \quad\quad + .4c \\
.6a + .5b \\
-.7a + .5b \\
\hline
-.3a + \quad b + .4c
\end{array}
$$

To add two or more polynomials, add the like terms.

To subtract polynomials we use the same method as in the section on subtracting signed numbers. Every subtraction problem can be changed to an addition problem; then the laws of addition can be used.

$$(6x + 3y) - (4x + 5y) = (6x + 3y) + - (4x + 5y)$$
change to addition problem, that is, add the "opposite"
$$= (6x + 3y) + (-4x + -5y)$$
find the opposite
$$= (6x + -4x) + (3y + -5y)$$

use the commutative and associative laws
of addition

$= 2x + -2y$

$= 2x - 2y$

change the addition of the opposite to a
subtraction problem

When only one subtraction problem is involved, the work can be arranged vertically.

subtract	or	add
$6x + 3y$		$6x + 3y$
$4x + 5y$		$-4x - 5y$
$2x - 2y$		$2x - 2y$

To find the difference of two polynomials, add the opposite of the polynomial to be subtracted from the first.

Exercises with more than one addition or subtraction as indicated by braces{ }, brackets [ ], or parentheses ( ) can be done by using the method for each operation, working from inside out.

$10 + 3x - \{9 - [6x - (4x + 7)]\}$
$= 10 + 3x + - \{9 + -[6x + -(4x + 7)]\}$
$= 10 + 3x + - \{9 + -[6x + -4x + -7]\}$
$= 10 + 3x + - \{9 + -6x + 4x + 7\}$
$= 10 + 3x + -9 + 6x + -4x + -7$
$= (10 + -9 + -7) + (3x + 6x + -4x)$
$= -6 + 5x$ or $5x - 6$

To remove braces, brackets, and parentheses, perform the indicated operations, working from inside out.

When an equation contains polynomials within parentheses, the parentheses can be removed and the equation solved as before.

$(x + 3) + (2x - 5) - (x - 4) = 16$
$(x + 3) + (2x + -5) + -(x + -4) = 16$ change all subtractions to additions

$(x + 3) + (2x + -5) + (-x + 4) = 16$ find the opposite
$2x + 2 = 16$ remove parentheses and combine monomials

$2x = 14$
$x = 7$

EXAMPLES

a)   Add. $6x, 11x, -x, -12x$

$$6x + 11x + -1x + -12x = (6 + 11 + -1 + 12)\,x$$
$$= 4x$$

b)   Simplify $(4a - 5b) + (a - 2b) + (3 - 3a)$

$$(4a - 5b) + (a - 2b) + (3 - 3a) = (4a + -5b) + (a + -2b) + (3 + -3a)$$
$$= (4a + a + -3a) + (-5b + -2b) + 3$$
$$= 2a + -7b + 3 \text{ or } 2a - 7b + 3$$

c)   Add. $2a - 3x + 7y, 4x - 2a + 8y$, and $3y - 5x + 3a$

$$
\begin{array}{r}
2a - 3x + 7y \\
-2a + 4x + 8y \\
\underline{3a - 5x + 3y} \\
3a - 4x + 18y
\end{array}
\quad \text{or} \quad
\begin{array}{r}
2a + -3x + 7y \\
-2a + 4x + 8y \\
\underline{3a + -5x + 3y} \\
3a + -4x + 18y
\end{array}
$$

or

$$3a - 4x + 18y$$

d)   Subtract. $-6x + y^2$ from $4y^2 - 3x$

$$(4y^2 - 3x) - (-6x + y^2) = (4y^2 + -3x) + -(-6x + y^2)$$
$$= (4y^2 + -3x) + (6x + -y^2)$$
$$= 3y^2 + 3x$$

e)   Simplify. $m + [5t - (2t + 4m)]$

$$m + [5t - (2t + 4m)] = m + [5t + -(2t + 4m)]$$
$$= m + [5t + -2t + -4m]$$
$$= m + 5t + -2t + -4m$$
$$= -3m + 3t$$

f)

$$5x - (3x + 5) = x + 13$$
$$5x + -(3x + 5) = x + 13$$
$$5x + (-3x + -5) = x + 13$$
$$2x + -5 = x + 13$$
$$2x + -x + -5 + 5 = x + -x + 13 + 5 \quad \text{add } -x \text{ and } 5 \text{ to each side to isolate}$$
$$x = 18 \qquad \qquad \text{the terms with } x \text{ on one side}$$

g)

$$(3x - 7) - (4x + 9) = 3x - (2 - 4x)$$
$$(3x + -7) + -(4x + 9) = 3x + -(2 + -4x)$$
$$(3x + -7) + (-4x + -9) = 3x + (-2 + 4x)$$
$$-x + -16 = 7x + -2$$
$$-x + -7x + -16 + 16 = 7x + -7x + -2 + 16$$
$$-8x = 14$$
$$x = -\frac{14}{8} = -\frac{7}{4}$$

Simplify.

1) $3x - (2x + 6)$

2) $(a - 4) + (a - 5)$

3) $(2x + 6) + (3x + 9) + (x - 1)$

4) $(2a - 4) - (a + 2)$

5) $(5a - 2) - 3a$

6) $4 + [2 - (3 - x)]$

7) $2 - (x + 3) - (x - 2)$

Add.

8) $\begin{array}{r} 7a + 3w \\ -5a + \ \ w \end{array}$

9) $\begin{array}{r} x + \ \ y - \ \ z \\ 3x - 4y + 5z \end{array}$

10) $\begin{array}{r} 6x^2 - 7x - 8 \\ -7x^2 + 9x - 2 \end{array}$

11) $\begin{array}{r} 10r + 9s - 8t \\ -r + 5s + 2t \\ -12r - 7s - 3t \end{array}$

12) $\begin{array}{r} -5a^2 - 3b^2 + 2ac \\ -4a^2 + \ \ b^2 - 5ac \\ -\ \ a^2 + \ \ b^2 - 5ac \end{array}$

13) $\begin{array}{r} 3p \qquad + 5r \\ -6p + 7q \\ -12q - 11r \end{array}$

14) $\begin{array}{r} \dfrac{2}{3}mn - \dfrac{1}{2}n \\ \dfrac{3}{4}mn + \dfrac{5}{6}n \end{array}$

15) $\begin{array}{r} .8bc - 3.5ad \\ -1.1bc - \ \ .8ad \end{array}$

16) $\begin{array}{r} 6.22w - 8.91 \\ -9.13w + 5.83 \end{array}$

Simplify.

17) $(6x - 2y) + (7x - 8y) + 3y$

18) $(9a - b + c) - (3a + 2b - 2c) + (8b + 2c)$

19) $(\frac{1}{4}x + 2y - 1\frac{1}{2}z) + (1\frac{1}{4}x - \frac{3}{4}y + z) - (-\frac{1}{2}x - \frac{1}{2}y - \frac{1}{2}z)$

20) $(.34r - .55s - .17t) + (.28r + .32s - .83t) - (-.12r + s + .19t)$

21) $(5x^2 - 6x + 7) - (3x^2 + 4x - 8)$

22) $16 - [3x - (4x - 1)]$

23) $[8at + 3 - (-2a - 4b)] - [(a + 2b) - 4a]$

24) $2 - \{[4x - (2x + 2y)] + 6y\} + (3x - 4y)$

25) $y - \{2 - [2y - (y + 4)]\}$

Solve.

26) $(2b + 8) - 3 = (4b - 8) + 1$

27) $(y - 16) - (6 - 3y) = 6$

28) $5 - (6x + 18) = 2 - (2x - 12)$

29) $22 - (4x + 1) + 5x = -13$

30) $(x - 1) - (2x - 4) - (3x + 5) = 9$

31) $6x - (2x - 6) + (x - 5) = 30 - (2x + 4)$

6.5 SOLVING WORD PROBLEMS USING EQUATIONS

Three lamps are connected in series. The first has a resistance of 150 ohms more than the second, and the third has a resistance of 100 ohms less than the second. Their total resistance (when connected in series, it is the sum of the individual resistances) is 800 ohms. What is the resistance of each of the lamps?

APPLICATION

66) Solve word problems using equations.

OBJECTIVE

A *word phrase*, such as "one number is six larger than another number," can be translated into a *mathematical expression* (x is the smaller number and $x + 6$ is the larger number) by letting a letter (x) represent one of the unknowns.

VOCABULARY

Applications of mathematics are often written in word form. The word form usually contains an implied equation that can be solved. In order to set up the equation it is necessary to translate the word phrases into mathematical expressions.

THE HOW AND WHY OF IT

Consider the following phrases and the mathematical translations:

one number is five less than another | x is one number
$x - 5$ is the other

one number is three more than twice another number | x is one number
$2x + 3$ is the other

the cost of a TV set is twice the cost of a hi-fi set | x is the cost of the hi-fi set
$2x$ is the cost of the TV set

Once the unknowns have been established and represented by mathematical expressions, a simpler word form of the equation can be found. If two different ways of expressing the same relationship can be found, the desired equation can be written.

Consider the following problems and the simpler word forms with the corresponding equations:

The sum of two numbers is 35. One number is five less than the other. Find the two numbers.

(first number) + (second number) is 35   simpler word form

$$x \quad + \quad x - 5 \quad = 35 \qquad \text{equation}$$

One number is three more than twice a second number. If the smaller number is subtracted from the larger, the result is 23. Find the two numbers.

(larger number) − (smaller number) is 23   simpler word form

$$2x + 3 \quad - \quad x \quad = 23 \qquad \text{equation}$$

The cost of a new TV set is $678.50. The cost of the TV is twice the cost of a hi-fi set. What is the cost of the hi-fi set?

2(cost of hi-fi) is (cost of TV)   simpler word form

$$2(x) \quad = 678.50 \qquad \text{equation}$$

After solving the equation a check should be made into the original wording of the problem rather than into the equation. The solution is then written in terms of the problem. For example, solving the second problem above we have:

$$
\begin{aligned}
2x + 3 - x &= 23 \\
x + 3 &= 23 \\
x &= 20 \quad \text{(one number)} \\
2x + 3 &= 43 \quad \text{(the other number)}
\end{aligned}
$$

Check: The difference is 23; $43 - 20 = 23$.
The two numbers are 20 and 43.

The five basic steps for solving word problems are:
1. Represent one of the unknowns with a letter (x) and all the others in terms of the same letter. (In some problems a sketch can be helpful.)
2. Write a simpler word form of the problem. (Find two ways of saying the same thing.)
3. Write the equation for the problem by replacing the words in the simpler word form by the mathematical expressions.
4. Solve the equation.
5. Check the solution(s) in the original wording of the problem. The correct solution is then written in terms of the problem.

EXAMPLES

a) The sum of two numbers is 35. One number is five less than the other. Find the two numbers.

Let x represent one number and
$x - 5$ represent the other

(first number) + (second number) is 35   simpler word form

$$
\begin{aligned}
x + x - 5 &= 35 \qquad \text{equation} \\
2x - 5 &= 35 \\
2x &= 40 \\
x &= 20 \quad \text{(one number)} \\
x - 5 &= 15 \quad \text{(the other number)}
\end{aligned}
$$

Check: The sum is 35; $20 + 15 = 35$. The two numbers are 20 and 15.

b) The cost of a new TV set is $678.50. The cost of the TV is twice the cost of a hi-fi set. What is the cost of the hi-fi set?

Let x represent the cost of the hi-fi and
$2x$ represent the cost of the TV

2(cost of hi-fi) is (cost of the TV)

$$2x = 678.50$$
$$x = 339.25 \text{ (cost of hi-fi)}$$

Check: Cost of TV ($678.50) is twice the cost of the hi-fi;
($339.25) (2) = $678.50. The hi-fi set cost $339.25.

c) If the sum of three consecutive odd numbers is 129, what are the numbers?
(Hint: consecutive numbers are in order of value and can be represented by x,
$x + 1, x + 2, \ldots$; consecutive odd or even numbers differ by 2, and can be represented by $x, x + 2, x + 4, \ldots$)

Let x represent the first number, then
$x + 2$ represents the second number and
$x + 4$ represents the third number

(first number) + (second number) + (third number) is 129
$$x \quad + \quad x + 2 \quad + \quad x + 4 \quad = 129$$

$$3x + 6 = 129$$
$$3x = 123$$
$$x = 41 \quad \text{(first number)}$$
$$x + 2 = 43 \quad \text{(second number)}$$
$$x + 4 = 45 \quad \text{(third number)}$$

Check: The sum is 129; 41 + 43 + 45 = 129.
The three consecutive odd numbers are 41, 43, and 45.

EXERCISES

1) In the application at the beginning of this section, what is the resistance of each lamp?

2) If a father is five years older than twice his son's age and the sum of
their ages is 52, what is the age of each?

3) Four resistors in series have a total resistance of 1940 ohms. If each one after the first is 150 ohms more than the previous one, find the ohms for each resistor.

4) A student bought two books; one cost $3.85 more than the other. If the total cost was $14.75, what was the cost of each book?

5) A roofer laid shingles for three days. The second day he was able to lay two bundles more than on the first day. The third day was cut short by rain and he laid five bundles less than on the first day. If he laid a total of 27 bundles, how many bundles did he lay each day?

6) The difference between two numbers is 123. If the larger number is five less than three times the smaller, find the two numbers.

7) Find five consecutive numbers whose sum is 125.

8) The sum of four consecutive odd numbers is 224. Find the numbers.

9) Four lamps connected in series have a total resistance of 1050 ohms. One lamp has a resistance of 250 ohms, and a second of 200 ohms. The third has a resistance that is twice the fourth. Find the resistance of the third and fourth lamps.

10) The perimeter of a triangle is 48 ft. If the first side is three feet longer than the second, and the second is the same length as the third, what is the length of each of the sides? (See the diagram below.)

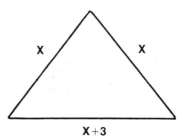

11) A farmer has 890 ft of fencing. He wants to use it to fence in a rectangular lot where the length of the lot is 175 ft more than the width. What are the dimensions of the lot he can fence in?

12) What are the length and width of a rectangle with perimeter of 48 in and width four inches less than its length?

13) At an agricultural test station three tests were conducted on germination of corn seed. The second test had twenty-five fewer germinations than the third test and five less than the first. If a total of 3000 seeds germinated, how many germinations were there in each test?

14) The current through one branch of an electrical circuit is 2.45 amperes more than through another circuit. When the circuits are joined, the total amperage is 8.75. What is the current in each of the circuits?

15) On a vacation trip John Snyder traveled twice as far the first day as he did the second day. On the third day he traveled fifty miles less than on the first day. If he traveled 1075 miles during the three days, how far did he go each day?

16) The total cost of a suit and a pair of shoes was $165. If the cost of the suit was $15 less than twice the cost of the shoes, find the cost of each article.

17) A strip of plate metal is to be cut into six pieces so that each successive piece is $\frac{1}{2}$ in longer than the preceding one. What is the length of each piece, if the original length of the plate was 36 in?

18) A 350 acre farm has three times as many acres in pasture as it has in timber. The remaining acreage is in wheat and is thirty acres less than the number in pasture. How many acres of wheat, timber, and pasture are there?

19) A metal bar $9\frac{5}{16}$ inches long is to be cut into two lengths so that one is $\frac{3}{16}$ inch shorter than the other. What is the length of each piece if $\frac{3}{16}$ inch allowance is made for the cut?

20) An alloy called "antifriction metal" contains twice as much antimony as copper, and six times as much tin as antimony. If a sample weighs 12 lb, how much of each metal does it contain?

6.6 MULTIPLICATION OF POLYNOMIALS

APPLICATION

A motorist drives from town A to town B at a rate of 45 mph. On the return trip he drives at a rate of 55 mph. If the round trip took five hours to complete, what is the distance between the two towns?

OBJECTIVE

67) Find the product of polynomials.

VOCABULARY

No new vocabulary.

THE HOW AND WHY OF IT

Products of polynomials can be found by repeated use of the distributive law and the other laws stated in previous sections. Study these three examples:

monomial times binomial
$$-6x\,(2x + 4y) = (-6x)\,(2x) + (-6x)\,(4y)$$
$$= -12x - 24xy$$

binomial times binomial

$$(3x + 4)(2x - 5) = (3x + 4)(2x + -5)$$
$$= (3x + 4)2x + (3x + 4)(-5)$$
$$= (3x)(2x) + (4)(2x) + (3x)(-5) + (4)(-5)$$
$$= 6x^2 + 8x - 15x - 20$$
$$= 6x^2 - 7x - 20$$

binomial times trinomial

$$(2x - 5y)(3x^2 - 4xy + 1) = (2x + -5y)(3x^2 + -4xy + 1)$$
$$= (2x + -5y)(3x^2) + (2x + -5y)(-4xy)$$
$$+ (2x + -5y)(1)$$
$$= (2x)(3x^2) + (-5y)(3x^2) + (2x)(-4xy)$$
$$+ (-5y)(-4xy) + (2x)(1) + (-5y)(1)$$
$$= 6x^3 - 23x^2y + 2x + 20xy^2 - 5y$$

In each example we first changed subtraction problems to addition of opposites (see section on combining polynomials) and then repeatedly applied the distributive law. In each case the last step shown is the sum of monomial products.

The steps can be cut short by realizing that every term of one polynomial is multiplied by each term of the second polynomial. A particular example of this is the use of the FOIL shortcut for two binomials.

$$(2x + 7)(5 + 2y)$$

F(irst)	+	O(uter)	+	I(nner)	+	L(ast)
$(2x)(5)$	+	$(2x)(2y)$	+	$(7)(5)$	+	$(7)(2y)$

or
$$10x + 4xy + 35 + 14y$$

The examples can also be written vertically.

$$\begin{array}{r} 2x + 4y \\ -6x \\ \hline \end{array}$$
$$(-6x)(2x) + (-6x)(4y)$$
$$= -12x - 24xy$$

$$\begin{array}{r} 2x - 5 \\ 3x + 4 \\ \hline \end{array}$$
$$(3x)(2x) + (3x)(-5)$$
$$(4)(2x) + (4)(-5)$$

or

$$\begin{array}{r} 6x^2 - 15x \\ + 8x + -20 \\ \hline 6x^2 - 7x - 20 \end{array}$$

$$\begin{array}{r} 3x^2 + -4xy + 1 \\ 2x + -5y \\ \hline 6x^3 + -8x^2y + 2x \\ -15x^2y + 20xy^2 + -5y \\ \hline 6x^3 - 23x^2y + 2x + 20xy^2 - 5y \end{array}$$

In general, to find the product of two polynomials,

1. change subtraction to addition $[a - b = a + {-}b]$;
2. write the product of each term of one polynomial and the terms of the other;
3. simplify (combine the terms of step two).

When an equation contains the product of polynomials, the product can be found and the equation solved as before if it is a linear equation. For example:

$$3(x + 2) - 2(x + 6) = 18$$
$$(3x + 6) - (2x + 12) = 18$$
$$(3x + 6) + (-2x + {-}12) = 18$$
$$x + {-}6 = 18$$
$$x = 24$$

To determine if the equation is linear it might be necessary to isolate the variable and combine monomials. (See Examples d and e.)

EXAMPLES

a) $3x\,(3a^2 + 4ab - 2b^2) = 3x\,(3a^2 + 4ab + {-}2b^2)$
$\qquad\qquad = (3x)\,(3a^2) + (3x)\,(4ab) + (3x)\,(-2b^2)$
$\qquad\qquad = 9a^2x + 12abx - 6b^2x$

b) $(3x + 2)\,(5x - 4) = (3x)\,(5x) + (3x)\,(-4) + (2)\,(5x) + (2)\,(-4)$
$\qquad\qquad\qquad = 15x^2 + {-}12x + 10x + {-}8$
$\qquad\qquad\qquad = 15x^2 + {-}2x + {-}8$
$\qquad\qquad\qquad = 15x^2 - 2x - 8$

c) $(.5y^2 + .3y - .5)\,(.2y - .75)$

$$
\begin{array}{r}
.5y^2 + \quad .3y\ + {-}.5 \\
.2\,y + {-}.75 \\
\hline
.1y^3 + \quad .06y^2 + {-}.1y \\
+ {-}.375y^2 + {-}.225y + .375 \\
\hline
.1y^3 + {-}.315y^2 + {-}.325y + .375
\end{array}
$$

$$\text{or}$$
$$.1y^3 - .315y^2 - .325y + .375$$

d) $$
\begin{aligned}
x(x - 4) + 6 - x(x + 7) &= 12 \\
(x^2 - 4x) + 6 - (x^2 + 7x) &= 12 \\
(x^2 - 4x) + 6 + (-x^2 + {-}7x) &= 12 \\
-11x + 6 &= 12 \\
-11x &= 6 \\
x &= -\frac{6}{11}
\end{aligned}
$$

e)
$$(x + 2)(x - 3) = (x + 7)(x - 10)$$
$$x^2 + 2x - 3x - 6 = x^2 + 7x - 10x - 70$$
$$x^2 - x - 6 = x^2 - 3x - 70$$
$$2x = -64$$
$$x = -32$$

Multiply.

1) $3(2a - 1)$

2) $4(x + 6)$

3) $-2(3y - 1)$

4) $(x + 1)(x + 2)$

5) $(x + 2)(x + 4)$

6) $(x - 2)(x + 3)$

7) $7x + 3$
$\underline{\quad -5}$

8) $-5a + 2b$
$\underline{\qquad 3a}$

9) $4x + 2xy$
$\underline{\quad -7x}$

10) $-5w(6w - 5z)$

11) $10xy(4x - 7y)$

12) $-mn(mn - m^2n - mn^2)$

13) $3x - 7$
$\underline{2x + 8}$

14) $\quad 8y + 2w$
$\underline{\;-3y + 4}$

15) $-r + 4s$
$\underline{2r - 3s}$

16) $(x - 3)(x + 7)$

17) $(x + 5)(x + 5)$

18) $(x + 2)(x - 1)$

19) $(x - 6)(x - 9)$ 20) $(2x + 1)(3x - 1)$ 21) $(2x + 3)(x - 6)$

22) $(3x + 5)(4x - 3)$ 23) $(5x - 6)(5x + 6)$ 24) $(x - 6)(x + 6)$

25) $(3y - 4)(3y + 4)$ 26) $(3x - 9y)(5x - 6y)$ 27) $(p - q)(p + q)$

28) $(y + b)(y + c)$ 29) $\begin{array}{r} 4x + 2y \\ \underline{3x - 5y} \end{array}$ 30) $\begin{array}{r} 3w - 4x + 2y \\ \underline{3x - 5y} \end{array}$

31) $(x + 4)(2x^2 - 7x - 7)$ 32) $(2a - 3b)(4a - 8ab + 5b)$

Solve.

33) $4(y - 7) = y + 2$ 34) $2(b + 4) - 3 = 4(b - 2) + 1$

35) $(x + 2)(x + 3) = x(x + 3)$ 36) $x(2x - 6) - x(x + 3) = x(x + 7) + 20$

37) $(x + 4)(x + 3) = (x - 6)(x + 2)$

Simplify.

38) $16 - 2[3x - 5(4x - 1)]$

39) $2 - 3 \{-2[4x - (2x + 2y)] + 6y\} + 2 (3y - 4y)$

40) $y - \{2-3 [2y - 4(y + 4)]\}$

Distance problems: Distance (d) in terms of rate (r) and time (t) is given by the formula $d = rt$. Equations for solving these problems usually indicate that the sum (or difference) of two distances is equal to the total distance. Solve the following.

41) In the application at the beginning of this section, what is the distance between the two towns?

42) Two cyclists start in opposite directions from the same point, one at 8 mph and the other at 10 mph. How long will it take until they are 54 miles apart? Hint: the simpler word form is

(distance of rider one) + (distance of rider two) is 54.

43) One automobile starts out from town at 6 A.M. and travels at an average speed of 45 mph. Two hours later a second automobile starts out to overtake the first, traveling at an average speed of 60 mph. How long will it take the second to overtake the first?

44) Two cars, 580 miles apart, start out toward each other at 1 P.M. One car is averaging 40 mph and the other 60 mph. The one traveling at 60 mph takes a one-hour lunch break. At what time will the two cars meet?

45) One car starts out at 3 P.M. traveling at 40 mph. If a second car starts out to overtake the first at 5 P.M., how fast will it have to travel in order to overtake the first by 10 P.M.?

6.7 DIVIDING POLYNOMIALS

APPLICATION   A cylindrical tank is to be painted two colors, the top half red and the bottom half blue. In order to purchase the correct amount of paint, the surface area for half the tank must be found. If $S = 2\pi r^2 + 2\pi rh$ is the formula for the surface area of the tank, what is the formula for half the area?

OBJECTIVE   68) Find the quotient of a polynomial divided by a monomial.

VOCABULARY   No new vocabulary.

THE HOW AND WHY OF IT
    To divide any polynomial by a monomial, change the subtraction signs in the polynomial to addition of opposites and then divide each term by the monomial divisor.

The exercise may be written as a fraction or as a long division problem.

$$(3x^4 - 6x^3 + 12x^2 - 9x) \div -3x$$

$$\frac{3x^4 + -6x^3 + 12x^2 + -9x}{-3x} = \frac{3x^4}{-3x} + \frac{-6x^3}{-3x} + \frac{12x^2}{-3x} + \frac{-9x}{-3x}$$

$$= -x^3 + 2x^2 - 4x + 3$$

or

$$
\begin{array}{r}
x^3 + 2x^2 - 4x + 3 \\
-3x \overline{\smash{)}\, 3x^4 + -6x^3 + 12x^2 + -9x} \\
\underline{3x^4} \\
0 + -6x^3 \\
\underline{-6x^3} \\
0 + 12x^2 \\
\underline{+ 12x^2} \\
0 + -9x \\
\underline{-9x} \\
0
\end{array}
$$

EXAMPLES

a) $(8y^2 - 4y) \div 2y$

$$\frac{8y^2 + -4y}{2y} = \frac{8y^2}{2y} + \frac{-4y}{2y} = 4y - 2$$

b) $(36a^3 b - 27a^2 b^2 + 18ab^3) \div -9ab$

$$\frac{36a^3 b + -27a^2 b^2 + 18ab^3}{-9ab} = \frac{36a^3 b}{-9ab} + \frac{-27a^2 b^2}{-9ab} + \frac{18ab^3}{-9ab}$$

$$= -4a^2 + 3ab - 2b^2$$

c) $(25x^3 - 20x^2 - 55x) \div -5x$

$$
\begin{array}{r}
-5x^2 + 4x + 11 \\
-5x \overline{\smash{)}\, 25x^3 + -20x^2 + -55x} \\
\underline{25x^3} \\
0 + -20x^2 \\
\underline{-20x^2} \\
0 + -55x \\
\underline{-55x} \\
0
\end{array}
$$

EXERCISES   Divide.

1) $(2x + 6) \div 2$

2) $(7y - 21) \div 7$

3) $(3x^2 - 6x - 9) \div 3$

4) $(x^2 - x) \div x$

5) $(x^3 - x^2 + x) \div x$

6) $(3a^2 b - 3ab^2) \div 3ab$

7) $5x \overline{)\, 10x^2 - 5x}$

8) $(4x^3 + 8x^2) \div 4x$

9) $(3a^3 - 6a) \div 3a$

10) $(15y^2 + 5y) \div 5y$

11) $(12ab^2 c - 24a^2 b^2 c^2) \div 6ab$

12) $(14x^3 - 21x) \div -7x$

13) $(36p^2 q - 18p^2 q^2 - 24p^2 q^3) \div 6p^2$

14) $(14x^3 - 28x^2 - 42x) \div 14x$

15) $(5x^2y^2 + 10xy^2 - 15xy) \div 5xy$

16) $(-8abc + 16a^2bc - 4ab^2c) \div -4abc$

17) $\dfrac{12r^2s^2t^2 - 16r^2st^2 - 24rs^2t}{-4rst}$

18) $-15ab \overline{)-30abc - 45a^2b^2c^2 - 15ab^2c^2}$

19) $(12x^5 - 15x^4 + 18x^3 - 6x^2) \div 3x$

20) $(14y^2z + 16yz^2 + 24yz - 30y^2z^2) \div 24z$

21) $(a^3b^2c^4 + a^2bc^3 + ab^2c^2 - a^2bc - abc) \div -abc$

22) In the application at the beginning of this section, what is the formula for half of the area?

6.8 FACTORING POLYNOMIALS: COMMON MONOMIAL FACTOR

APPLICATION

Equations that have a polynomial on one side and zero on the other can often be solved by factoring. An example of such an equation is $s = 144t - 16t^2$. When s (height of falling object) is known, t (the time the object falls) can be found. If $s = 0$, t is the time it takes the object to hit the ground.

OBJECTIVE

69) Factor a polynomial by writing it as the product of a monomial and another polynomial.

VOCABULARY

To *factor* a number or polynomial means to write it in the form of a multiplication problem.

THE HOW AND WHY OF IT

In an earlier section the distributive law was used to multiply monomials and polynomials.

$$3x \, (x^2 - 4y) = 3x^3 - 12xy$$

To factor a polynomial is to reverse this process.

$$4y^3 - 12xy = 4y \, (y^2 - 3x)$$

To *factor a polynomial* we look for all of the common factors of the terms. The common monomial factor contains every factor that each of the terms of the polynomial has in common. This monomial factor is called the HCF (Highest Common Factor) of the terms of the polynomial. The HCF may often be seen by inspecting the polynomial. If it is not easily discovered, it can also be found by writing each term in factored form.

$$34x^2y - 51xy^2 - 68x^2y^2$$

$$= (2 \cdot 17 \cdot x \cdot x \cdot y) - (3 \cdot 17 \cdot x \cdot y \cdot y) - (2 \cdot 2 \cdot 17 \cdot x \cdot x \cdot y \cdot y)$$

$$= 17xy \, (2x) - 17xy \, (3y) - 17xy \, (4xy)$$

By rearranging the terms so that the common factors are written first, you can see that the HCF is $17xy$. The polynomial can then be written as a product (using the distributive law).

$$17xy \, (2x - 3y - 4xy)$$

If you can find the HCF by inspection, you can short-cut the factoring steps by doing them mentally.

To factor a polynomial as the product of a monomial and another polynomial,
1) determine the HCF of the terms of the polynomial;
2) if the HCF is not one (1), use the distributive law to write it as a product;
3) if the HCF is one (1), we say that the polynomial cannot be factored.

EXAMPLES

a) $4x + 16y = 4(x + 4y)$

b) $4x + 17xy = x(4 + 17y)$

c) $5x + 6y = 5x + 6y$ since HCF = 1, the polynomial cannot be factored.

d) $2xyz - 2wxy + 4wyz = 2y\,(xz - wx + 2wz)$

e) $15a^2 b - 35b^2 c - 25abc = 5b(3a^2 - 7bc - 5ac)$

f) $-4r^2 st - 20rs^2 t - 36rst^2 = -4rst(r + 5s + 9t)$

Factor each of the following:

EXERCISES

1) $2x + 6$

2) $8x - 12$

3) $6xy - 5y$

4) $2ab - 4a$

5) $2x^2 - 3x$

6) $3ab - 15b$

7) $2a + 8b$

8) $4x^2 + 5x$

9) $15xyz - 7wxz$

10) $12bc - 16bd$

11) $6a + 5b$

12) $\pi r^2 h - \pi R^2 h$

13) $4ab + 8bc - 6cd$

14) $25rst - 35st + 5stx$

15) $16x^2 y^2 z + 24x^2 yz^2 + 32x^2 yz$

16) $\pi r^2 h + \pi r^2 + 2\pi r$

17) $15x^3 y - 25x^2 y + 40xy$

18) $5abc + 15bcd - 10abd - 5a^2 b^2$

19) $3xy + 4yz - 5xz$

20) $3m^2 n^2 - 6mn - 12m^2 n - 15mn^2$

21) $x^2 y^2 + x^3 y^3 - x^4 y^4$

22) $36ac - 8ab - 12ac$

23) $8x^3y^2 + 9x^2y^3 - 7x^2y^2$ 24) $14ax + 10ay + 6az$

25) $24x^4 - 12x^3 - 36x^2 - 48x$ 26) $2\pi rh + 2\pi r^2$

27) In the application at the beginning of this section, factor the right side of the equation.

6.9 FACTORING POLYNOMIALS: TRINOMIALS

The time it takes a rock to fall to the bottom of a gully 176 feet deep if it is thrown *upward* from the rim at the rate of 56 ft/sec is given by the equation $16t^2 - 56t - 176 = 0$. Factor the left member of this equation.

APPLICATION

70) Factor a trinomial.

OBJECTIVE

No new vocabulary.

VOCABULARY

A trinomial (with no common factor) that can be factored using whole numbers and/or their opposites is the product of two binomials. The factors can be found by looking at all possible products of the numerical coefficients of the first and last terms (F and L from FOIL). For instance, study the polynomial

THE HOW AND WHY OF IT

$$x^2 + 6x + 8$$

The first term (F) has coefficient 1.

$$1 = 1 \cdot 1$$

The last term (L) is 8.

$$8 = 1 \cdot 8 \text{ or } 2 \cdot 4 \text{ or } -1 \cdot -8 \text{ or } -2 \cdot -4$$

We ignore $-1 \cdot -8$ and $-2 \cdot -4$, since the coefficient of the middle term (6) is positive.

The possible factors are

$$(x + 2)(x + 4) \text{ and } (x + 1)(x + 8)$$

We determine the correct factoring by multiplication (the FOIL shortcut can be used) and find

$$(x + 2)(x + 4) = x^2 + 6x + 8$$

$$(x + 1)(x + 8) = x^2 + 9x + 8$$

and conclude that the correct factoring is

$$x^2 + 6x + 8 = (x + 2)(x + 4)$$

If the coefficient of the first term is not 1, there are more possibilities.

$$6x^2 - 19x + 10$$

First (F) is 6.

$$1 \cdot 6 \text{ or } 2 \cdot 3$$

Last (L) is 10.

$$-1 \cdot -10 \text{ or } -2 \cdot -5$$

(since the middle term is -19, we ignore the products $1 \cdot 10$ and $2 \cdot 5$)

The possibilities are

$$(x - 10)(6x - 1)$$
$$(x - 1)(6x - 10)$$
$$(x - 2)(6x - 5)$$
$$(x - 5)(6x - 2)$$
$$(2x - 10)(3x - 1)$$
$$(2x - 1)(3x - 10)$$
$$(2x - 2)(3x - 5)$$
$$(2x - 5)(3x - 2)$$

Multiply each pair of binomials until the correct factoring is found.

$$6x^2 - 19x + 10 = (2x - 5)(3x - 2)$$

In some cases there is a common monomial factor in the trinomial, and it should be factored first. This will reduce the number of possibilities in factoring the remaining trinomial. (See Example f.)

EXAMPLES

a) $m^2 + 20m + 64$ F: $1 = 1 \cdot 1$
         L: $64 = 1 \cdot 64$ or $2 \cdot 32$ or
            $4 \cdot 16$ or $8 \cdot 8$

$(m + 1)(m + 64) = m^2 + 65m + 64$ no
$(m + 2)(m + 32) = m^2 + 34m + 64$ no
$(m + 4)(m + 16) = m^2 + 20m + 64$ yes

b) $x^2 - 11xy + 28y^2$ F: $1 = 1 \cdot 1$
         L: $28 = -1 \cdot -28$ or
            $-2 \cdot -14$ or
            $-4 \cdot -7$

$(x - 1y)(x - 28y) = x^2 - 29xy + 28y^2$ no
$(x - 2y)(x - 14y) = x^2 - 16xy + 28y^2$ no
$(x - 4y)(x - 7y) = x^2 - 11xy + 28y^2$ yes

c) $c^2 - 15c + 48$ F: $1 = 1 \cdot 1$
         L: $48 = -48 \cdot -1$ or $-24 \cdot -2$ or
            $-16 \cdot -3$ or $-12 \cdot -4$ or
            $-8 \cdot -6$

$(c - 48)(c - 1) = c^2 - 49c + 48$ no
$(c - 24)(c - 2) = c^2 - 26c + 48$ no
$(c - 16)(c - 3) = c^2 - 19c + 48$ no
$(c - 12)(c - 4) = c^2 - 16c + 48$ no
$(c - 8)(c - 6) = c^2 - 14c + 48$ no
Cannot be factored

d) $4y^2 - 3y - 10$ F: $4 = 1 \cdot 4$ or $2 \cdot 2$
         L: $-10 = -1 \cdot 10$ or $1 \cdot -10$ or $-2 \cdot 5$
            or $2 \cdot -5$

$(y - 1)(4y + 10) = 4y^2 + 6y - 10$ no
$(y + 1)(4y - 10) = 4y^2 - 6y - 10$ no
$(y - 10)(4y + 1) = 4y^2 - 39y - 10$ no
$(y + 10)(4y - 1) = 4y^2 + 49y - 10$ no
$(y - 2)(4y + 5) = 4y^2 - 3y - 10$ yes

e) $18w^2 + 66w + 36 = 6\,(3w^2 + 11w + 6)$

In $3w^2 + 11w + 6$ F: $3 = 1 \cdot 3$
                        L: $6 = 1 \cdot 6$ or $2 \cdot 3$

$(w + 1)\,(3w + 6)\; = 3w^2 + 9w + 6$ no
$(w + 6)\,(3w + 1)\; = 3w^2 + 19w + 6$ no
$(w + 2)\,(3w + 3)\; = 3w^2 + 9w + 6$ no
$(w + 3)\,(3w + 2)\; = 3w^2 + 11w + 6$ yes

So $18w^2 + 66w + 36 = 6\,(w + 3)\,(3w + 2)$.

EXERCISES   Factor the following, if possible.

1) $a^2 + a - 6$ 2) $b^2 + 10b + 24$

3) $x^2 - 5x + 6$ 4) $y^2 - 12y + 35$

5) $x^2 + x - 2$ 6) $x^2 - x - 42$

7) $y^2 - 6y + 8$ 8) $x^2 - 2x - 80$

9) $x^2 + 7x + 12$ 10) $a^2 - 5a + 4$

11) $c^2 + 4c - 21$ 12) $bx^2 - 8bx - 20b$

13) $2y^2 - 3y - 5$

14) $3x^2 + 5x + 2$

15) $3a^2 - 10a + 8$

16) $2y^2 + 7y - 20$

17) $3x^2 - 10xy + 3y^2$

18) $x^2 + 6x + 9$

19) $6x^2 + 13x + 7$

20) $4y^2 - 12ay + 9a^2$

21) $x^2 + x - 90$

22) $6x^2 + 3x - 7$

23) $4y^2 - 6y - 10$

24) $25x^2 - 40xy + 16y^2$

25) $3a^2 - 13ab + 4b^2$

26) $6x^2 + 2x + 1$

27) $7a^2 - 17a - 12$

28) In the application at the beginning of this section, factor the left member.

29) The area of a rectangle whose length is 4 m longer than its width is 117 m^2. This can be expressed by the equation $x(x + 4) = 117$. This in turn can be rewritten as $x^2 + 4x - 117 = 0$. Factor the left member.

6.10 FACTORING POLYNOMIALS: DIFFERENCE OF TWO SQUARES

APPLICATION

In arithmetic $(50 - 2)(50 + 2) = 2500 - 4 = 2496$ is a short cut for multiplying $48 \cdot 52$. $(50 - 2 = 48$ and $50 + 2 = 52$.) $1600 - 49 = 1551$ is a short cut for multiplying what two numbers?

OBJECTIVE

71) Factor a binomial that is the difference of two squares.

VOCABULARY

No new vocabulary.

THE HOW AND WHY OF IT

Binomials that are the difference of two squares are factored by observation (they can also be factored by the method discussed in the last section). By multiplication of the sum and difference of the terms x and 5

$$(x + 5)(x - 5)$$

we obtain

$$x^2 - 25$$

which is the difference of two squares. Also, since

$$(y + 5)(y - 5) = y^2 - 25$$

$$(m + 5)(m - 5) = m^2 - 25$$

we judge that, given $r^2 - 25$, we may factor as follows:

$$r^2 - 25 = (r + 5)(r - 5)$$

The factoring is correct as we can check by multiplying. Similarly,

$$9x^2 - 100$$

is the difference of two squares, since $3x \cdot 3x = (3x)^2 = 9x^2$ and $10 \cdot 10 = 10^2 = 100$. We try the factors that are the sum and difference of $3x$ and 10. Does

$$9x^2 - 100 = (3x + 10)(3x - 10)?$$

After multiplying (to check) we can answer yes.

In general the difference of two squares is the product of two binomials. One is the sum and the second is the difference of the two terms whose squares are the same as those in the original. $a^2 - b^2 = (a + b)(a - b)$

EXAMPLES

a) $a^2 - 144 = (a + 12)(a - 12)$

b) $16t^2 - 1 = (4t + 1)(4t - 1)$

c) $x^2 - 10$ is not factored, since 10 is not the square of a whole number.

d) $196y^2 - 81 = (14y + 9)(14y - 9)$

e) $4x^2 - 36 = 4(x^2 - 9) = 4(x + 3)(x - 3)$

Factor each of the following if possible.   **EXERCISES**

1) $x^2 - 1$ 2) $x^2 - 4$ 3) $x^2 - 9$

238 / CHAPTER 6—EQUATIONS AND POLYNOMIALS

4) $a^2 - 16$ 5) $y^2 - 25$ 6) $4x^2 - 9$

7) $9a^2 - 1$ 8) $a^2 - 64$ 9) $x^2 - 36$

10) $a^2 - 100$ 11) $4x^2 - 9$ 12) $9a^2 - 64b^2$

13) $25x^2 - 36y^2$ 14) $\frac{9}{16}b^2 - \frac{16}{25}$ 15) $48a^2 - 75$

16) $100y^2 - 64z^2$ 17) $9x^2 - 36$ 18) $4ax^2 - 4a$

19) $36a^3 - 64a$ 20) $-36 + 25a^2$ 21) $25b^2 - 16$

22) $2\pi r^2 - 2\pi R^2$ 23) $14a^3 - 56ab^2$ 24) $9a^2 - 25b^2$

25) In the application at the beginning of this section, what are the two numbers?

6.11 QUADRATIC EQUATIONS—SOLVED BY FACTORING

The time it takes a rock to fall to the bottom of a gully 176 ft deep **APPLICATION**
if it is thrown upward from the rim at the rate of 56 ft/sec is given by the
equation $16t^2 - 56t - 176 = 0$. How long will it take the rock to reach the
bottom? (The negative value of t has no physical meaning.)

72) Solve a quadratic equation by factoring.   **OBJECTIVE**

In a *quadratic equation* the variable has the number 2 as the largest **VOCABULARY**
exponent ($2x^2 + x - 3 = 0$ or $4x^2 - 1 = 0$, but not $x^3 - x^2 + 1 = 7x$). The
solutions (there are usually two) are the number replacements for the
variable which make the equation true.

By the use of another fundamental law of equations we can write a **THE HOW AND**
series of equations to find the solution of a quadratic equation (recall the **WHY OF IT**
two laws used for linear equations).

Third law: If two expressions have a product of 0, then one or both of
them must equal 0.
Examples: If $ab = 0$, then $a = 0$ or $b = 0$.
 If $(x - c)(x - d) = 0$, then $x - c = 0$ or $x - d = 0$.

To solve the equation $2x^2 - 9x + 4 = 0$, we can write this series of
equations:

$2x^2 - 9x + 4 = 0$ original equation

$(2x - 1)(x - 4) = 0$ factor the left side so the product of the two factors
 equals 0

$2x - 1 = 0$ or $x - 4 = 0$ one or both factors must equal 0

$2x = 1$ or $x = 4$ solve the two linear equations as
 usual

$x = \dfrac{1}{2}$ or $x = 4$

Check if $x = \frac{1}{2}$.

$$2 \left(\frac{1}{2}\right)^2 - 9 \left(\frac{1}{2}\right) + 4 = 0$$

$$2 \left(\frac{1}{4}\right) - \frac{9}{2} + 4 = 0$$

$$\frac{1}{2} - \frac{9}{2} + 4 = 0$$

$$-4 + 4 = 0$$

$$0 = 0$$

If $x = 4$,

$$2 (4)^2 - 9 (4) + 4 = 0$$

$$2 (16) - 36 + 4 = 0$$

$$32 - 36 + 4 = 0$$

$$-4 + 4 = 0$$

$$0 = 0$$

If the original quadratic equation does not have zero as its right member, the fundamental laws of equations are used to change it. (See Example c.)

> To solve a quadratic equation by factoring,
> 1) Rewrite the equation so the right member equals zero.
> 2) Factor the left member.
> 3) Set both factors equal to zero.
> 4) Solve both linear equations.

EXAMPLES

a) $x^2 - 4x - 77 = 0$
    $(x + 7)(x - 11) = 0$
    $x + 7 = 0$ or $x - 11 = 0$
    $x = -7$ or $x = 11$

b) $15x^2 - 16x + 4 = 0$
    $(3x - 2)(5x - 2) = 0$
    $3x - 2 = 0$ or $5x - 2 = 0$
    $3x = 2$ or $5x = 2$
    $x = \frac{2}{3}$ or $x = \frac{2}{5}$

c) $4x + 12 = x^2$
    $-x^2 + 4x + 12 = -x^2 + x^2$
    $-x^2 + 4x + 12 = 0$
    $-1(-x^2 + 4x + 12) = -1(0)$
    $x^2 - 4x - 12 = 0$
    $(x + 2)(x - 6) = 0$
    $x + 2 = 0$ or $x - 6 = 0$
    $x = -2$ or $x = 6$

d) The product of two consecutive positive numbers is 56. What are the numbers?

Let x represent the first number; then
$x + 1$ represents the second number.

(first number) (second number) is 56

$$(x)(x + 1) = 56$$
$$x^2 + x = 56$$
$$x^2 + x - 56 = 0$$
$$(x + 8)(x - 7) = 0$$

$x + 8 = 0$ or $x - 7 = 0$
$x = -8$ or $x = 7$
$x + 1 = -7$ $x + 1 = 8$

Check: The product is 56. (7)(8) = 56. (–8)(–7) = 56. So the numbers are –8 and –7 or 7 and 8.

Solve the following.   **EXERCISES**

1) $(x + 7)(x - 8) = 0$ 2) $(x + 5)(x + 6) = 0$

3) $(x - 9)(x - 4) = 0$ 4) $(x + 8)(x - 11) = 0$

5) $x(x - 1)(x - 3) = 0$ 6) $x^2 - x - 72 = 0$

7) $x^2 - 2x - 48 = 0$

8) $x^2 + 5x + 6 = 0$

9) $x^2 - x - 12 = 0$

10) $a^2 + 4a - 21 = 0$

11) $y^2 - 8y = 48$

12) $x^2 - 35 = -2x$

13) $x^2 - 9 = 8x$

14) $10a = 24 - a^2$

15) $0 = b^2 + 3b - 10$

16) $y^2 = 49$

17) $3x^2 + 11x + 6 = 0$

18) $4x^2 = 36$

19) $7x^2 - 17x - 12 = 0$

20) $3x^2 - 10x + 3 = 0$

21) $6x^2 + 1 = -5x$

22) $4x^2 + 4x + 1 = 0$

23) $16s^2 = 9$

24) $-90 = -e^2 - e$

25) $x^2 + 3x = 0$

26) In the application at the beginning of this section, how long did it take for the rock to reach the bottom?

27) The area of a triangle whose altitude is x and whose base is $x + 7$ is 60. What is the measure of each if the formula for the area of a triangle is $\frac{1}{2}ba = A$.

28) The product of two consecutive positive odd numbers is 63. Find the numbers.

29) One positive number is six more than another number. The product of the two numbers is 72. Find the numbers.

30) A rectangular lot is 3 ft longer than it is wide. It contains 88 sq ft in area. Find the dimensions.

31) The perimeter of a rectangle is 40 in. Its area is 96 sq in. Find its dimensions.

32) If the height (s) of a falling object is given by the formula $s = 144t - 16t^2$, where t is the time the object is falling, how long will it take the object to hit the ground ($s = 0$)?

1. (Obj. 60) Solve for x. $4x + 5 = 33$ _____

2. (Obj. 68) Divide. $(-9ab^2c + 15a^2bc - 3abc^2) \div 3abc$ _____

3. (Obj. 63) Add. $4x + 2x + 5x$ _____

4. (Obj. 64) Simplify. $(6a + 2b - 7) - (3a - 2b + 4)$ _____

5. (Obj. 65) Solve for x. $(4x + 5) - (2x + 6) = 18$ _____

6. (Obj. 72) Solve for x. $x^2 - 13x + 42 = 0$ _____

7. (Obj. 67) Multiply. $(x - 7)(x + 5)$ _____

8. (Obj. 61) Express as a power of x. $\dfrac{(x^7)^2}{x^3}$ _____

9. (Obj. 62) Evaluate. $(4^3)^2$ _____

10. (Obj. 63) Perform the indicated operations.

$$3x + \frac{6x^3}{2x^2} - 2x$$ _____

11. (Obj. 67) Multiply. $3x(2x - 7)$ _____

12. (Obj. 64) Add. $3x + 3y + 6$, $7x - 5y + 9$, and $2x + 4y - 16$ _____

13. (Obj. 64) Solve for y. $4y + 8 = 2y - 10$ _____

14. (Obj. 70) Factor. $2y^2 - 15y + 7$ _____

15. (Obj. 71) Factor. $16x^2 - 49$ _____

16. (Obj. 67) Multiply. $(2x + 5)(3x + 2)$ _____

17. (Obj. 61) Express as a power of a. $a^3 \cdot a^5 \cdot a^6$ _____

18. (Obj. 69) Factor. $8xy + 12xyz - 4xyz^2$ _____

19. (Obj. 67) Solve for x. $3(x - 7) + 2(x + 5) = 27$ _____

20. (Obj. 70) Factor. $x^2 - 7x - 30$ _____

21. (Obj. 72) Solve for x. $x^2 - 43x + 42 = 0$ _____

22. (Obj. 67) Solve for x. $3(x + 2) - 5(2x + 5) = 7(3 - 2x)$ _____

23. (Obj. 72) One positive number is three less than twice another. The product of the two numbers is 20. Find the numbers.

    _____

24. (Obj. 66) The sum of four consecutive even numbers is 100. Find the numbers.

    _____

25. (Obj. 66) A landscaper planted six more mugho pines than he did azaleas. He also planted twice as many roses as he did mugho pines. If he planted 62 plants in all, how many of each did he plant?

    _____

CHAPTER 7

RATIONAL EXPRESSIONS

7.1 MULTIPLICATION OF RATIONAL EXPRESSIONS

The main application of this procedure is in the solution of fractional equations.

APPLICATION

73) Multiply two or more rational expressions.

OBJECTIVE

A *rational expression* is a fraction that has a polynomial for its numerator and/or denominator. A rational expression is also an indicated quotient of two polynomials. Some examples are

VOCABULARY

$$\frac{1}{x}, \frac{3x}{4y} \text{ and } \frac{2a + b}{x + 3}$$

For this chapter it will be assumed that no variable will have a value that will make any denominator zero.

Since rational expressions look like fractions in arithmetic, one might expect that the operations performed on fractions would be used in a similar manner when working with rational expressions. This is exactly the case.

THE HOW AND WHY OF IT

To multiply two rational expressions, multiply the numerators and multiply the denominators.

$$\frac{A}{B} \cdot \frac{C}{D} = \frac{AC}{BD}, \; BD \neq 0$$

So, $\dfrac{3}{x} \cdot \dfrac{1}{2} = \dfrac{3 \cdot 1}{x \cdot 2} = \dfrac{3}{2x}$.

EXAMPLES

a) $\dfrac{2}{a} \cdot \dfrac{3}{b} = \dfrac{2 \cdot 3}{a \cdot b} = \dfrac{6}{ab}$

b) $\dfrac{4x}{5} \cdot \dfrac{3a}{5b} = \dfrac{4x \cdot 3a}{5 \cdot 5b} = \dfrac{12ax}{25b}$

c) $\dfrac{2x + y}{4a} \cdot \dfrac{b}{x} = \dfrac{(2x + y)\,b}{4ax} = \dfrac{2bx + by}{4ax}$

EXERCISES   Multiply.

1) $\dfrac{3}{2} \cdot \dfrac{w}{z}$

2) $\dfrac{6}{m} \cdot \dfrac{p}{7}$

3) $\dfrac{12}{x} \cdot \dfrac{2}{y}$

4) $\dfrac{8a}{5b} \cdot \dfrac{9c}{7d}$

5) $\dfrac{-5x}{7u} \cdot \dfrac{12xy}{11uv}$

6) $\dfrac{-2}{x} \cdot \dfrac{-y}{7x}$

7) $\dfrac{3x}{5} \cdot \dfrac{2a}{5}$

8) $\dfrac{6ab}{7x} \cdot \dfrac{3ad}{5y}$

9) $\dfrac{x}{4} \cdot \dfrac{3x - 2}{5}$

10) $\dfrac{26 + z}{m} \cdot \dfrac{2y}{7m}$

11) $\dfrac{19x}{13y} \cdot \dfrac{-8a}{15b}$

12) $\dfrac{15rs}{7t} \cdot \dfrac{18r}{7v}$

13) $\dfrac{5a + 3b}{x} \cdot \dfrac{9ab}{3a + c}$

14) $\dfrac{4x - 2y}{3a - b} \cdot \dfrac{7}{4a}$

15) $\dfrac{z + y}{z} \cdot \dfrac{y}{z - y}$

16) $\dfrac{x-4}{x+3} \cdot \dfrac{x-4}{x+3}$ 17) $\dfrac{y+3}{2t+1} \cdot \dfrac{y-7}{3t-2}$ 18) $\dfrac{7p-2q}{18r} \cdot \dfrac{9q}{6r+7s}$

19) $\dfrac{x}{-5y} \cdot \dfrac{-11x}{15y} \cdot \dfrac{7}{-2y}$ 20) $\dfrac{3}{x} \cdot \dfrac{x+4}{2x} \cdot \dfrac{x-1}{x+1}$

7.2 RENAMING RATIONAL EXPRESSIONS

The main application of this procedure is in adding and subtracting rational expressions.   **APPLICATION**

74) Reduce rational expressions.
75) Write an equivalent rational expression for a given rational expression.   **OBJECTIVES**

Reducing rational expressions has the same meaning as reducing fractions in arithmetic. Eliminate like factors from the numerator and denominator. The reduced and the given rational expressions are said to be *equivalent.* $\dfrac{4xy}{6xz}$ and $\dfrac{2y}{3z}$ are equivalent.   **VOCABULARY**

The process of building or reducing algebraic expressions is similar to that used with fractions in arithmetic.   **THE HOW AND WHY OF IT**

To build an algebraic expression, introduce a common factor in both the numerator and the denominator.

$$\frac{A}{B} \cdot \frac{C}{C} = \frac{AC}{BC}, \ BC \neq 0$$

To reduce an algebraic expression, eliminate a common factor in both the numerator and the denominator.

$$\frac{AC}{BC} = \frac{A}{B} \cdot \frac{C}{C} = \frac{A}{B} \cdot 1 = \frac{A}{B}, \ BC \neq 0$$

$$\frac{2}{4} = \frac{2 \cdot 1}{2 \cdot 2} = \frac{2}{2} \cdot \frac{1}{2} = 1 \cdot \frac{1}{2} = \frac{1}{2}$$

We can use the same method for rational expressions.

$$\frac{4x}{12y} = \frac{4 \cdot x}{4 \cdot 3y} = \frac{4}{4} \cdot \frac{x}{3y} = 1 \cdot \frac{x}{3y} = \frac{x}{3y}$$

EXAMPLES

a) $\dfrac{7a}{21b} = \dfrac{7 \cdot a}{7 \cdot 3b} = \dfrac{7}{7} \cdot \dfrac{a}{3b} = 1 \cdot \dfrac{a}{3b} = \dfrac{a}{3b}$

b) $\dfrac{4x}{5x^2} = \dfrac{4 \cdot x}{5x \cdot x} = \dfrac{4}{5x} \cdot \dfrac{x}{x} = \dfrac{4}{5x} \cdot 1 = \dfrac{4}{5x}$

c) $\dfrac{2x + 4}{x^2 + 5x + 6} = \dfrac{2(x + 2)}{(x + 3)(x + 2)} = \dfrac{2}{x + 3} \cdot \dfrac{x + 2}{x + 2} = \dfrac{2}{x + 3}$

d) $\dfrac{3}{4} = \dfrac{?}{8} , \quad \dfrac{3}{4} \cdot \dfrac{2}{2} = \dfrac{6}{8}$

e) $\dfrac{4x}{5y} = \dfrac{?}{20y^2} , \quad \dfrac{4x}{5y} \cdot \dfrac{4y}{4y} = \dfrac{16xy}{20y^2}$

f) $\dfrac{x + y}{5} = \dfrac{?}{10x + 15y}$

$$\dfrac{x + y}{5} \cdot \dfrac{2x + 3y}{2x + 3y} = \dfrac{2x^2 + 5xy + 3y^2}{10x + 15y}$$

EXERCISES   Reduce.

1) $\dfrac{6a}{18b}$ 2) $\dfrac{5xyz}{10xz}$ 3) $\dfrac{8ab^2}{12ab}$

4) $\dfrac{9(2x + 3)}{12(2x + 3)}$ 5) $\dfrac{2(x - y)^2}{4(x - y)}$ 6) $\dfrac{8(x + 4)}{4(x + 8)}$

Find the missing numerator.

7) $\dfrac{8y}{9x} = \dfrac{?}{9x^2}$

8) $\dfrac{7p}{13q} = \dfrac{?}{26pq}$

9) $\dfrac{a}{x} = \dfrac{?}{x^2 y}$

10) $\dfrac{3}{7x} = \dfrac{?}{14x}$

11) $\dfrac{3x}{a + b} = \dfrac{?}{2a + 2b}$

12) $\dfrac{5}{x + 3} = \dfrac{?}{x^2 + 5x + 6}$

Reduce.

13) $\dfrac{16x^2 y^2}{24x}$

14) $\dfrac{21b^2 d}{35bd^2}$

15) $\dfrac{x + 2}{2x + 4}$

16) $\dfrac{2a + 6b}{a + 3b}$

17) $\dfrac{x^2 + 4x}{x^2 + 8x + 16}$

18) $\dfrac{6x + 12}{x^2 + 5x + 6}$

Find the missing numerator.

19) $\dfrac{5x}{12y} = \dfrac{?}{24xy}$

20) $\dfrac{a + b}{6} = \dfrac{?}{6a + 6b}$

21) $\dfrac{4x + 2y}{x + y} = \dfrac{?}{(x + y)^2}$

22) $\dfrac{x + 2}{x + 3} = \dfrac{?}{x^2 + 5x + 6}$

23) $\dfrac{5ac}{9xz} - \dfrac{?}{45bxz}$

24) $\dfrac{x^2 + 2x + 3}{x - 4} = \dfrac{?}{x^2 - 16}$

7.3 A SECOND LOOK AT MULTIPLYING RATIONAL EXPRESSIONS

APPLICATION

Floyd can mow his lawn in 3 hours and his grandson Norm can mow the same lawn in 2 hours. If they work together, the formula for the time (t) it takes them together is given by $\frac{1}{3} + \frac{1}{2} = \frac{1}{t}$. Simplify this equation by multiplying each term of the equation by $\frac{6t}{1}$.

OBJECTIVE

76) Find the product of two rational expressions and reduce the product to lowest terms.

VOCABULARY

No new vocabulary.

THE HOW AND WHY OF IT

Recall from multiplication of fractions in arithmetic that there were times that it was possible to do the work more quickly by reducing before performing the multiplication. For example,

$$\frac{3}{4} \cdot \frac{4}{5} = \frac{3}{\overset{}{\underset{1}{\cancel{4}}}} \cdot \frac{\overset{1}{\cancel{4}}}{5} = \frac{3}{5}$$

The same holds true when multiplying rational expressions. If a factor in a numerator is the same as a factor in a denominator, they may be eliminated (divided out) and the multiplication performed as before. Following the example in arithmetic just used, we can proceed as follows:

$$\frac{5}{x} \cdot \frac{x}{7} = \frac{5}{\underset{1}{\cancel{x}}} \cdot \frac{\overset{1}{\cancel{x}}}{7} = \frac{5}{7}$$

EXAMPLES

a) $\dfrac{x}{2a + 2b} \cdot \dfrac{a + b}{y} = \dfrac{x}{2\underset{1}{\cancel{(a+b)}}} \cdot \dfrac{\overset{1}{\cancel{(a+b)}}}{y} = \dfrac{x}{2y}$

b) $\dfrac{ax}{x+y} \cdot \dfrac{x^2 - y^2}{bx} = \dfrac{a\cancel{x}}{\underset{1}{\cancel{x+y}}} \cdot \dfrac{\overset{1}{\cancel{(x+y)}}(x-y)}{\underset{1}{b\cancel{x}}} = \dfrac{a(x-y)}{b} = \dfrac{ax - ay}{b}$

Multiply and reduce to lowest terms.   **EXERCISES**

1) $\dfrac{2x}{3y} \cdot \dfrac{12}{5x}$

2) $\dfrac{10u^3}{12z^2} \cdot \dfrac{6z}{5u}$

3) $\dfrac{3(x+4)}{6(x-2)} \cdot \dfrac{14(x-2)}{4(x+4)}$

4) $\dfrac{2x+2}{5x} \cdot \dfrac{15x}{x+1}$

5) $\dfrac{33}{x^2-9} \cdot \dfrac{x+3}{3}$

6) $\dfrac{x^2+5x+6}{10y} \cdot \dfrac{2y}{x+2}$

7) $\dfrac{6(x+4)}{3(x-4)} \cdot \dfrac{4y}{8y}$

8) $\dfrac{y}{x+z} \cdot \dfrac{x+z}{z}$

9) $\dfrac{a+b}{5} \cdot \dfrac{6a}{a+b}$

10) $\dfrac{x+y}{x+3} \cdot \dfrac{x+3}{x+y}$

11) $\dfrac{ax}{y^2} \cdot \dfrac{y^2+2y}{a^2+a}$

12) $\dfrac{2a+3b}{x+y} \cdot \dfrac{x^2+xy}{2ax+3bx}$

13) $\dfrac{(x+2)(x+3)}{x+5} \cdot \dfrac{x+7}{(x+3)(x+7)}$ 14) $\dfrac{(x+9)(x-9)}{(x-6)(x+9)} \cdot \dfrac{(x-6)(x+4)}{(x-4)(x-9)}$

15) $\dfrac{x^2+2x+3}{x+1} \cdot \dfrac{1}{x+2}$ 16) $\dfrac{a^2x^2y^2}{a+b} \cdot \dfrac{a^2+ab}{ab}$

17) $\dfrac{x^2-y^2}{15} \cdot \dfrac{40}{x-y}$ 18) $\dfrac{b^2+4b}{ab} \cdot \dfrac{a}{b^2-16}$

19) $\dfrac{4bc^2}{x^2+4x+4} \cdot \dfrac{x^2+6x+8}{4x+16}$ 20) $\dfrac{12a^2-48}{a^2-6a-8} \cdot \dfrac{a^2-a-12}{a+3}$

21) $\dfrac{x}{y} \cdot \dfrac{x+y}{x-y} \cdot \dfrac{3}{4}$

22) In the application at the beginning of the section, what is the equation after simplifying?

7.4 LEAST COMMON MULTIPLE

Find the least common multiple of the denominators of the follow- **APPLICATION**
ing fractions: $\dfrac{3}{x}, \dfrac{4}{x^2}, \dfrac{5}{2x}$. The least common multiple will be used to add
and subtract such fractions.

77) Find the least common multiple of two or more polynomials. **OBJECTIVE**

Recall that the *least common multiple* (LCM) of two or more whole **VOCABULARY**
numbers is the smallest number other than zero that is a multiple of each
of the given numbers. The *LCM of two or more polynomials* is similar in
that it is the polynomial with the least number of factors that is a
multiple of each of the given polynomials. The LCM of $6x$ and $9x^2$ is
$18x^2$.

Recall that when finding the LCM of two or more whole numbers, **THE HOW AND**
each of the whole numbers was prime factored. The LCM was then con- **WHY OF IT**
structed from these prime factors. To find the LCM of 15, 50, and 75, we
did the following:

$$15 = 3 \cdot 5$$
$$50 = 2 \cdot 5 \cdot 5 = 2 \cdot 5^2$$
$$75 = 3 \cdot 5 \cdot 5 = 3 \cdot 5^2$$

The LCM is the product of 2, 3, and 5^2 ($2 \cdot 3 \cdot 25 = 150$). 5^2 was chosen
since the largest exponent of 5 is 2. (See Chapter 2.)

> To find the LCM of two or more polynomials, each polynomial
> is factored completely. The LCM is the product of the highest power
> of each different factor.

To find the LCM of x^2, y^2, and x^3y^3 we have:

$$x^2 = x^2$$
$$y^2 = y^2$$
$$x^3y^3 = x^3 \cdot y^3$$

The different factors are x and y, and the highest power of each is 3;
therefore, the LCM is x^3y^3.

EXAMPLES

a) Find the LCM of $4x^2$, $3y^2$, and $8xy$.

$$4x^2 = 2^2 x^2$$
$$3y^2 = 3y^2$$
$$8xy = 2^3 xy$$
$$\text{LCM} = 2^3 \cdot 3x^2 y^2 = 24x^2 y^2$$

b) Find the LCM of x, $x+y$, and $x^2 + xy$.

$$x = x$$
$$x+y = x+y$$
$$x^2 + xy = x(x+y)$$
$$\text{LCM} = x(x+y) = x^2 + xy$$

c) Find the LCM of $x^2 - y^2$, $x+y$, and $x-y$.

$$x^2 - y^2 = (x+y)(x-y)$$
$$x+y = x+y$$
$$x-y = x-y$$
$$\text{LCM} = (x+y)(x-y) = x^2 - y^2$$

EXERCISES

Find the LCM of the following. LCM may be written in factored form.

1) $2x, 3x$

2) $3x^2, 2x$

3) $5x^2, 3x^2$

4) $6x^2, 10x, 15$

5) $2a, 3b, 5c$

6) $2(x+3), 3(x+3)$

7) $6(3a+7y), 8(3a+7y)$

8) $4x(a+1), 5y(a+1)$

9) $3x + 3y, 2x + 2y$

10) $x + 5, x^2 - 25$

11) $2ab, 4a^2, 8b^2$

12) $12z, 16xy, 4x$

13) $6a^2 b, 8ab, 12ab^2$

14) $x + 4, x + 6$

15) $x - y, x^2 - y^2$

16) $a + b, a - b$

17) $4a^2 - 9, 14a + 21$

18) $y^2 - 9, (y - 3)^2$

19) $x + y, x^2 + 2xy + y^2$

20) $x, x + y$

21) $b + c, b - c, b^2 - c^2$

22) $3x, 12x^2, x + y$

23) $y + 5, y + 2, y + 1$

24) $2y + 6, 2y - 8$

25) In the application at the beginning of this section, what is the LCM?

7.5 ADDITION OF RATIONAL EXPRESSIONS

APPLICATION

The formula for finding the total resistance (R) for two resistors wired in parallel is given by $\frac{1}{R} = \frac{1}{r_1} + \frac{1}{r_2}$. If r_1 has resistance x and r_2 has a resistance of 15 ohms greater than r_1 $(x + 15)$, the formula becomes $\frac{1}{R} = \frac{1}{x} + \frac{1}{x + 15}$. Simplify the right side of the formula.

OBJECTIVE

78) Add rational expressions.

VOCABULARY

No new vocabulary.

THE HOW AND WHY OF IT

The sum of two or more rational expressions is found by using the same process as was used when adding fractions in arithmetic.

To add two rational expressions that have a common denominator, add the numerators and put that sum over the common denominator.

$$\frac{2}{x} + \frac{5}{x} = \frac{7}{x}$$

> To add two rational expressions that do not have a common denominator, write each as an equivalent expression in which the denominators are the same (use the LCM for the common denominator) and then add.

To add $\frac{3}{2x} + \frac{5}{x^2}$, the LCM of x^2 and $2x$, which is $2x^2$, is picked for the common denominator. We then have

$$\frac{3}{2x} \cdot \frac{x}{x} + \frac{5}{x^2} \cdot \frac{2}{2} = \frac{3x}{2x^2} + \frac{10}{2x^2} = \frac{3x + 10}{2x^2}$$

EXAMPLES

a) $\frac{4}{x} + \frac{5}{x} = \frac{4+5}{x} = \frac{9}{x}$

b) $\frac{3x}{a+b} + \frac{4y}{a+b} = \frac{3x + 4y}{a+b}$

c) $\frac{3}{a} + \frac{4}{b} = \frac{3}{a} \cdot \frac{b}{b} + \frac{4}{b} \cdot \frac{a}{a} = \frac{3b}{ab} + \frac{4a}{ab} = \frac{4a + 3b}{ab}$, ab is the LCM of a and b.

d) $\dfrac{x}{a} + \dfrac{y}{a+b} = \dfrac{x}{a} \cdot \dfrac{a+b}{a+b} + \dfrac{y}{a+b} \cdot \dfrac{a}{a}$

$a(a+b)$ is the LCM of a and $a+b$.

$$= \dfrac{ax+bx}{a(a+b)} + \dfrac{ay}{a(a+b)}$$

$$= \dfrac{ax+bx+ay}{a(a+b)} = \dfrac{ax+bx+ay}{a^2+ab}$$

e) $\dfrac{3}{x} + \dfrac{4}{x^2+x} = \dfrac{3}{x} + \dfrac{4}{x(x+1)}$

$x(x+1)$ or x^2+x is the LCM of x and x^2+x.

$$\dfrac{3}{x} + \dfrac{4}{x(x+1)} = \dfrac{3}{x} \cdot \dfrac{(x+1)}{(x+1)} + \dfrac{4}{x(x+1)}$$

$$= \dfrac{3x+3}{x(x+1)} + \dfrac{4}{x(x+1)} = \dfrac{3x+7}{x(x+1)} = \dfrac{3x+7}{x^2+x}$$

Add and reduce where possible.

EXERCISES

1) $\dfrac{y}{5} + \dfrac{y}{5}$

2) $\dfrac{z}{6} + \dfrac{z}{6}$

3) $\dfrac{3}{14w} + \dfrac{4}{14w}$

4) $\dfrac{3}{15a} + \dfrac{5}{15a}$

5) $\dfrac{x}{2} + \dfrac{x}{6}$

6) $\dfrac{3x}{x+2} + \dfrac{7}{x+2}$

7) $\dfrac{m}{2y} + \dfrac{3m}{8y}$

8) $\dfrac{3}{p^2} + \dfrac{4}{3p}$

9) $\dfrac{4x-3}{7} + \dfrac{2x-1}{7}$

10) $\dfrac{x+2}{5} + \dfrac{x+3}{10}$ 11) $\dfrac{7}{y} + \dfrac{3}{z}$ 12) $\dfrac{x}{y} + \dfrac{x}{3y} + \dfrac{x}{3}$

13) $\dfrac{1}{4ab} + \dfrac{b}{6a} + \dfrac{a}{2b}$ 14) $\dfrac{3}{xy} + \dfrac{5}{x^2} + \dfrac{8}{y^2}$ 15) $\dfrac{8a}{9b} + \dfrac{a}{6b}$

16) $\dfrac{3}{5a} + 7$ 17) $\dfrac{2}{3a} + \dfrac{5}{a} + \dfrac{3}{6a}$ 18) $x + 3 + \dfrac{1}{x}$

19) $\dfrac{4}{a+b} + \dfrac{7}{a-b}$ 20) $\dfrac{2}{x+5} + \dfrac{7}{x-1}$ 21) $\dfrac{4}{y} + \dfrac{3}{y+1}$

22) $\dfrac{x}{x+y} + \dfrac{y}{x-y} + \dfrac{y}{x^2-y^2}$ 23) $\dfrac{a}{a+b} + \dfrac{a-b}{a}$

24) $\dfrac{1}{x} + (x+3)$ 25) $\dfrac{a}{b} + \dfrac{a+b}{a} + \dfrac{b}{a+b}$

26) In the application at the beginning of this section, what is the simplified formula?

7.6 SUBTRACTION OF RATIONAL EXPRESSIONS

The difference in the time it takes two automobiles, one traveling 10 mph faster $(R + 10)$ than the other (R), to cover 100 miles is given by $T = \dfrac{100}{R} - \dfrac{100}{R + 10}$. Simplify the right member of the formula by subtracting.

APPLICATION

79) Subtract rational expressions.

OBJECTIVE

No new vocabulary.

VOCABULARY

The difference of two rational expressions is found using the same process as was used when subtracting fractions in arithmetic.

THE HOW AND WHY OF IT

To subtract two rational expressions that have a common denominator, subtract the numerators in the order indicated and put that difference over the common denominator.

$$\frac{x}{a + b} - \frac{3}{a + b} = \frac{x - 3}{a + b}.$$

To subtract two rational expressions that do not have a common denominator, write each one as an equivalent expression in which the denominators are the same and then subtract. Use the LCM of the denominators as the common denominator.

$$\frac{4}{3} - \frac{5}{x} = \frac{4x}{3x} - \frac{15}{3x} = \frac{4x - 15}{3x}.$$

EXAMPLES

a) $\dfrac{5}{x} - \dfrac{2}{x} = \dfrac{5 - 2}{x} = \dfrac{3}{x}$

b) $\dfrac{9a}{a + b} - \dfrac{2a}{a + b} = \dfrac{9a - 2a}{a + b} = \dfrac{7a}{a + b}$

c) $\dfrac{8y}{x} - \dfrac{6b}{a} = \dfrac{8y}{x} \cdot \dfrac{a}{a} - \dfrac{6b}{a} \cdot \dfrac{x}{x} = \dfrac{8ay}{ax} - \dfrac{6bx}{ax} = \dfrac{8ay - 6bx}{ax}$

d) $\dfrac{5x}{x+y} - \dfrac{4y}{x-y} = \dfrac{5x}{x+y} \cdot \dfrac{x-y}{x-y} - \dfrac{4y}{x-y} \cdot \dfrac{x+y}{x+y}$

$$= \dfrac{5x(x-y)}{(x+y)(x-y)} - \dfrac{4y(x+y)}{(x-y)(x+y)}$$

$$= \dfrac{5x^2 - 5xy - (4xy + 4y^2)}{(x+y)(x-y)}$$

$$= \dfrac{5x^2 - 5xy - 4xy - 4y^2}{(x+y)(x-y)} = \dfrac{5x^2 - 9xy - 4y^2}{x^2 - y^2}$$

e) $\dfrac{2a}{a^2 + ab} - \dfrac{3}{a+b} = \dfrac{2a}{a(a+b)} - \dfrac{3}{a+b}$

$a(a+b)$ is the LCM of $a^2 + ab$ and $a+b$.

$$\dfrac{2a}{a(a+b)} - \dfrac{3}{a+b} = \dfrac{2a}{a(a+b)} - \dfrac{3}{a+b} \cdot \dfrac{a}{a}$$

$$= \dfrac{2a}{a(a+b)} - \dfrac{3a}{a(a+b)} = \dfrac{-a}{a(a+b)} = \dfrac{-1}{a+b}$$

EXERCISES   Subtract and reduce where possible.

1) $\dfrac{5a}{12} - \dfrac{a}{12}$

2) $\dfrac{4}{a^2} - \dfrac{7}{a^2}$

3) $\dfrac{7a}{x-y} - \dfrac{3}{x-y}$

4) $\dfrac{7}{y} - \dfrac{3}{z}$

5) $\dfrac{4}{b} - \dfrac{3}{b^2}$

6) $\dfrac{3}{x} - \dfrac{5}{x^3}$

7) $\dfrac{x}{y} - \dfrac{x}{3y}$

8) $4 - \dfrac{2}{x}$

9) $\dfrac{3}{xy} - \dfrac{5}{x^2}$

10) $\dfrac{1}{4a} - \dfrac{2}{3b}$

11) $\dfrac{6}{5y} - \dfrac{3}{10y}$

12) $\dfrac{x}{2y} - \dfrac{z}{5y}$

13) $\dfrac{8a}{9b} - \dfrac{a}{6b}$

14) $\dfrac{3}{5a} - 7$

15) $\dfrac{2}{3a} - \dfrac{5}{a} - \dfrac{3}{6a}$

16) $x - 3 - \dfrac{1}{x}$

17) $\dfrac{4}{a+b} - \dfrac{7}{a-b}$

18) $\dfrac{2}{x+5} - \dfrac{7}{x-1}$

19) $\dfrac{4}{y} - \dfrac{3}{y+1}$

20) $\dfrac{x}{x+y} - \dfrac{y}{x-y}$

21) $\dfrac{a}{a+b} - \dfrac{a+b}{a}$

22) $\dfrac{a}{a+b} - \dfrac{a-b}{a}$

23) $\dfrac{1}{x} - (x-3)$

24) $\dfrac{a}{b} - \dfrac{a+b}{a} - \dfrac{b}{a+b}$

25) In the application at the beginning of this section, what is the simplified formula?

7.7 DIVISION OF RATIONAL EXPRESSIONS

In time-motion problems, average velocity (v) is given by the formula APPLICATION
$v = \dfrac{D}{t}$, where D represents distance and t represents time. Acceleration (a)
is given by $a = \dfrac{v}{t} = v \div t$. By substituting, acceleration can be expressed in
terms of D and t by $a = \dfrac{D}{t} \div t$. Simplify the right side of the formula.

OBJECTIVE   80) Divide rational expressions.

VOCABULARY   Two rational expressions whose product is 1 are said to be *reciprocals* of each other. $\dfrac{3}{x} \cdot \dfrac{x}{3} = 1$.

THE HOW AND   Recall from division of fractions that an equivalent multiplication
WHY OF IT   problem was solved.

> To divide rational expressions, multiply by the reciprocal of the divisor.
>
> $$\frac{A}{B} \div \frac{C}{D} = \frac{A}{B} \cdot \frac{D}{C}, \; BCD \neq 0$$

EXAMPLES

a) $\dfrac{2x}{3y} \div \dfrac{3}{4} = \dfrac{2x}{3y} \cdot \dfrac{4}{3} = \dfrac{8x}{9y}$

b) $\dfrac{4a}{5b} \div \dfrac{10a^2}{9b} = \dfrac{\overset{2}{\cancel{4a}}}{\underset{1}{\cancel{5b}}} \cdot \dfrac{\overset{1}{\cancel{9b}}}{\underset{5a}{\cancel{10a^2}}} = \dfrac{18}{25a}$

c) $\dfrac{ab}{2a + 4b} \div \dfrac{a^2 - ab}{a^2 + 2ab} = \dfrac{ab}{2a + 4b} \cdot \dfrac{a^2 + 2ab}{a^2 - ab} = \dfrac{ab}{2\cancel{(a + 2b)}} \cdot \dfrac{\overset{1}{\cancel{a}}\overset{1}{\cancel{(a + 2b)}}}{\underset{1}{\cancel{a}}(a - b)} = \dfrac{ab}{2(a - b)}$

EXERCISES   Divide and reduce where possible.

1) $\dfrac{x}{y} \div \dfrac{a}{b}$ 　　　　　 2) $\dfrac{3}{4} \div \dfrac{p}{q}$ 　　　　　 3) $\dfrac{r}{s} \div \dfrac{4}{7}$

4) $\dfrac{2x}{3} \div \dfrac{1}{3}$ 　　　　　 5) $\dfrac{6}{5w} \div \dfrac{3}{5}$ 　　　　　 6) $\dfrac{x + 4}{6} \div \dfrac{x + 4}{5}$

7) $\dfrac{x^2 - 9}{2} \div \dfrac{x + 3}{1}$ 8) $\dfrac{1}{x^2 + 5x + 6} \div \dfrac{1}{x + 2}$ 9) $\dfrac{2x + 6}{x - 5} \div \dfrac{3x + 9}{x - 5}$

10) $\dfrac{ax - x}{36} \div \dfrac{bx - x}{12}$ 11) $\dfrac{4a}{5b} \div \dfrac{1}{2}$ 12) $\dfrac{6x}{5y} \div \dfrac{3}{4}$

13) $\dfrac{2a}{3b} \div \dfrac{6a^2}{5b^3}$ 14) $\dfrac{25xy}{3a} \div \dfrac{5x}{9a^2}$ 15) $\dfrac{a + 2}{15} \div \dfrac{a - 2}{10a}$

16) $\dfrac{4x - 6y}{3ay} \div \dfrac{2x - 3y}{15y^2}$ 17) $\dfrac{a + b}{x + y} \div \dfrac{a - b}{x - y}$ 18) $\dfrac{x + 2}{x^2 + 6x + 5} \div \dfrac{x^2 - 4}{x + 5}$

19) $\dfrac{3x + 9}{6a^2 - 3a} \div \dfrac{x + 3}{8ax - 4x}$ 20) $\dfrac{25ab^2}{b^2 - 16} \div \dfrac{15abx}{b^2 - b - 12}$

21) $\dfrac{x + 4y}{2a + 1} \div \dfrac{x^2 - 16y^2}{4a^2 - 1}$ 22) $\dfrac{3a + 4b}{5x + 2} \div \dfrac{9a^2 - 16b^2}{25x^2 - 4}$

23) $\dfrac{b^2 - b - 12}{b^2 + 5b + 6} \div \dfrac{b^2 - 4}{b^2 - 5b - 6}$ 24) $\dfrac{c^2 - c - 12}{c^2 + c - 6} \div \dfrac{c^2 - 2c - 8}{c^2 - 4}$

25) In the application at the beginning of this section, what is the simplified formula for acceleration?

7.8 COMPLEX RATIONAL EXPRESSIONS

APPLICATION

The formula for current produced by cells in parallel is $I = \dfrac{E}{r + \dfrac{R}{n}}$.

Simplify the right side of the formula.

OBJECTIVE

81) Simplify complex rational expressions.

VOCABULARY

Complex rational expressions are those rational expressions that contain rational expressions within their numerator and/or denominator. $\dfrac{\dfrac{1}{x} + 2}{\dfrac{2}{x} + 3}$ is a complex fraction.

THE HOW AND WHY OF IT

To simplify a complex rational expression,
(1) Find the LCM of the denominators that are within the numerator and/or denominator of the complex fraction.
(2) Multiply the numerator *and* denominator of the complex fraction by the LCM and simplify.

To simplify $\dfrac{\dfrac{1}{x}}{\dfrac{1}{y}}$ find the LCM of x and y, which is xy. Now multiply

both the numerator and denominator by xy and simplify.

$$\frac{\dfrac{1}{x} \cdot \dfrac{xy}{1}}{\dfrac{1}{y} \cdot \dfrac{xy}{1}} = \frac{\dfrac{xy}{x}}{\dfrac{xy}{y}} = \frac{y}{x}$$

EXAMPLES

a)   Simplify $\dfrac{\dfrac{1}{2}+\dfrac{1}{y}}{5}$.

The LCM of 2 and y is $2y$; therefore, we have

$$\dfrac{\left(\dfrac{1}{2}+\dfrac{1}{y}\right)\ \cdot\dfrac{2y}{1}}{\dfrac{5}{1}\ \cdot\dfrac{2y}{1}}=\dfrac{\dfrac{2y}{2}+\dfrac{2y}{y}}{10y}$$

$$=\dfrac{y+2}{10y}$$

b)   Simplify $\dfrac{x+\dfrac{1}{y}}{\dfrac{x}{y}}$.

The LCM of y and y is y; therefore, we get

$$\dfrac{\left(x+\dfrac{1}{y}\right)\ \cdot\dfrac{y}{1}}{\dfrac{x}{y}\ \cdot\dfrac{y}{1}}=\dfrac{\dfrac{xy}{1}+\dfrac{y}{y}}{\dfrac{xy}{y}}$$

$$=\dfrac{xy+1}{x}$$

Simplify.

EXERCISES

1) $\dfrac{\dfrac{1}{4}}{\dfrac{3}{5}}$

2) $\dfrac{\dfrac{1}{a}}{\dfrac{1}{b}}$

3) $\dfrac{\dfrac{2}{x}}{\dfrac{3}{x^2}}$

4) $\dfrac{\dfrac{1}{x}+\dfrac{1}{y}}{6}$

5) $\dfrac{\dfrac{2}{3}+\dfrac{3}{4}}{\dfrac{1}{2}}$

6) $\dfrac{x+\dfrac{1}{2}}{\dfrac{2}{3}}$

7) $\dfrac{a + \dfrac{b}{c}}{\dfrac{1}{a}}$

8) $\dfrac{\dfrac{1}{x} + \dfrac{1}{y}}{\dfrac{1}{x} - \dfrac{1}{y}}$

9) $\dfrac{\dfrac{1}{x} + \dfrac{1}{y}}{\dfrac{1}{w}}$

10) $\dfrac{a + b}{\dfrac{1}{x} + \dfrac{1}{y}}$

11) $\dfrac{a + \dfrac{1}{b}}{a - \dfrac{1}{b}}$

12) $\dfrac{\dfrac{a}{b} - \dfrac{b}{a}}{a - b}$

13) $\dfrac{\dfrac{1}{y} - \dfrac{1}{z}}{\dfrac{y - z}{z}}$

14) $\dfrac{1 + \dfrac{1}{b}}{b + \dfrac{1}{b}}$

15) $\dfrac{\dfrac{x}{y} + \dfrac{x}{z}}{\dfrac{y}{x} + z}$

16) $\dfrac{\dfrac{1}{a} - \dfrac{2}{b}}{\dfrac{3}{a} + 4}$

17) $\dfrac{\dfrac{1}{5} + \dfrac{x^2}{2}}{\dfrac{x}{4}}$

18) $\dfrac{\dfrac{x}{5}}{\dfrac{1}{4} + \dfrac{x^2}{2}}$

19) $\dfrac{\dfrac{5}{x} + \dfrac{2}{x^2}}{\dfrac{x}{4}}$

20) $\dfrac{\dfrac{5}{x}}{\dfrac{2}{x^2} - \dfrac{x}{4}}$

21) $\dfrac{\dfrac{1 + \dfrac{1}{x}}{x}}{\dfrac{1}{x} - 1}$

22) In the application at the beginning of this section, what is the simplified formula?

7.9 FRACTIONAL EQUATIONS

The formula for finding the total resistance (R) for two resistors wired in parallel is given by $\frac{1}{R} = \frac{1}{r_1} + \frac{1}{r_2}$. If $r_1 = 5$ ohms and $r_2 = 8$ ohms, what is the total resistance?

APPLICATION

82) Solve equations involving rational expressions (fractions).

OBJECTIVE

No new vocabulary.

VOCABULARY

To solve an equation such as $\frac{x}{9} + \frac{x}{3} = 5$, we can perform an operation so that it will have the same form as equations previously encountered. Using the multiplication law of equality and multiplying each member of the equation by 9 (the LCM of 3 and 9), the equation becomes:

THE HOW AND WHY OF IT

$$9\left(\frac{x}{9} + \frac{x}{3}\right) = 9 \cdot 5$$

$$9 \cdot \frac{x}{9} + 9 \cdot \frac{x}{3} = 45$$

$$x + 3x = 45$$

$$4x = 45$$

$$x = \frac{45}{4}$$

To solve an equation that contains rational expressions, multiply both members of the equation by the LCM of the denominators. This will eliminate all of the denominators, and the resulting equation can then be solved by methods previously discussed.

EXAMPLES

a) $\quad \frac{x}{3} + \frac{x}{4} = \frac{1}{2}$

Since 12 is the LCM of 2, 3, and 4, multiply both members by 12.

$$12\left(\frac{x}{3} + \frac{x}{4}\right) = 12 \cdot \frac{1}{2}$$

$$12 \cdot \frac{x}{3} + 12 \cdot \frac{x}{4} = 6$$

$$4x + 3x = 6$$

$$7x = 6$$

$$x = \frac{6}{7}$$

b) $\frac{18}{x} + \frac{6}{x} = 12$

. x is the LCM; therefore, multiply both members by x.

$$x\left(\frac{18}{x} + \frac{6}{x}\right) = x \cdot 12$$

$$x \cdot \frac{18}{x} + x \cdot \frac{6}{x} = 12x$$

$$18 + 6 = 12x$$

$$24 = 12x$$

$$2 = x$$

c) $\frac{5}{x} = \frac{3}{x + 2}$

$x(x + 2)$ is the LCM; therefore, multiply both members by $x(x + 2)$.

$$x(x + 2) \cdot \frac{5}{x} = x(x + 2) \cdot \frac{3}{x + 2}$$

$$5(x + 2) = x \cdot 3$$

$$5x + 10 = 3x$$

$$5x - 3x = -10$$

$$2x = -10$$

$$x = -5$$

EXERCISES Solve.

1) $\frac{8}{x} + \frac{6}{x} = 7$

2) $\frac{a}{3} + \frac{a}{4} = \frac{7}{2}$

3) $\frac{b}{14} + \frac{b}{7} + \frac{b}{21} = 10$

4) $\dfrac{1}{8a} - \dfrac{1}{6a} = \dfrac{1}{4}$

5) $\dfrac{2}{x} + \dfrac{1}{2x} = 6$

6) $\dfrac{3}{x-2} = \dfrac{2}{x+2}$

7) $\dfrac{2}{a+3} = \dfrac{5}{a-4}$

8) $\dfrac{2}{a} + 1 = \dfrac{4}{a}$

9) $\dfrac{2}{x-5} = \dfrac{5}{x-2}$

10) $\dfrac{3}{x+2} = \dfrac{6}{x+4}$

11) $\dfrac{5}{x} + \dfrac{6}{x} = 33$

12) $\dfrac{x}{4} = x - 3$

13) $\dfrac{4a-3}{6} + 2 = \dfrac{a-7}{4} + a$

14) $\dfrac{4}{b+2} + \dfrac{3}{b} = \dfrac{5}{b^2 + 2b}$

15) $\dfrac{10}{x+4} - \dfrac{3}{x-2} = 0$

16) In the application at the beginning of this section, what is the total resistance?

17) One seventh of a certain number is three more than one eighth of the same number. Find the number.

18) The denominator of a fraction is 5 more than twice the numerator. If 4 is subtracted from the numerator and 6 is added to the denominator, the resulting fraction is $\frac{3}{25}$. What is the fraction?

19) Jim drove 500 miles on Wednesday at a certain speed. On Thursday he drove 15 mph faster and went 180 miles farther. If Jim drove the same number of hours each day, find the number of hours he drove each day.

20) Flight 182A flew 800 miles in the same time it took Flight 291B to fly 625 miles. If Flight 291B averaged 75 mph less than Flight 182A, what was the rate of both planes?

21) Dan can row his rubber life raft 4 mph in still water. He rows up a stream 4 miles and then rows down stream 6 miles. The trip upstream takes the same time as the trip down stream. Find the rate of the stream.

22) The formula for finding the total resistance (R) for two resistors wired in parallel is given by $\frac{1}{R} = \frac{1}{r_1} + \frac{1}{r_2}$. If r_1 has resistance that is 15 ohms greater than r_2, and the total resistance is 4 ohms, find the resistance of each resistor.

23) The difference in time it takes two automobiles, one traveling 10 mph faster than the other, to cover 100 miles is one-half hour. Find the rate of each automobile.

Work Problems: Work problems can be set up by determining the part of the job accomplished per day, per hour, or during some other unit of time (rate of work). If it takes 6 hours to do a certain job, the rate of work would be $\frac{1}{6}$ of the job per hour. The sum of all the rates is the rate of those working together.

24) Floyd can mow his lawn in 3 hours and his grandson Norm can mow the same lawn in 2 hours. If they work together, how long will it take to mow the lawn?

Let x = the time it takes to mow the lawn together.

(Floyd's rate) + (Norm's rate) is (rate working together)

$$\frac{1}{3} \quad + \quad \frac{1}{2} \quad = \quad \frac{1}{x}$$

25) Larry and Greg can hoe an acre of strawberries in 3 hours working together. Working alone Larry can hoe the acre in 5 hours. How long would it take Greg to hoe the acre by himself?

26) A water tank for a small city can be filled in four days. During the summer months, the average rate of water use drains the tank in six days. If the tank is empty at the beginning of the summer, how many days will it take the tank to fill?

27) A tank can be filled through two pipes in 2 hours. One pipe alone can fill the tank in 3 hours less than the other can fill the tank alone. How long does it take each pipe to fill the tank?

7.10 SOLVING FORMULAS

APPLICATION

Fred wants to make a horsepower table or chart for use in his diesel shop. The formula for engine torque is $T = \dfrac{5252H}{S}$, where H is horsepower and S is engine speed in revolutions per minute. What formula can Fred use for horsepower in terms of torque and engine speed?

OBJECTIVES

83) Evaluate a formula for any of its variables when given the value of the others.

84) Solve or rewrite a formula for any one of its variables.

VOCABULARY

Recall that variables are often called unknowns when working with formulas.

THE HOW AND WHY OF IT

Formulas can be evaluated by substituting the known values for the variables in the formula and using the techniques of previous sections.

Find r in $A = p + prt$ given $p = 20,000$, $A = 45,500$, and $t = 15$

$45,500 = 20,000 + 20,000 \cdot r \cdot 15$ substitute the known values

$45,500 = 20,000 + 300,000r$ simplify

$45,500 + -20,000 = 20,000 + -20,000 + 300,000r$ add −20,000 to both sides

$25,500 = 300,000r$ simplify

$\dfrac{17}{200} = .085 = r$ divide both sides by 300,000; the solution is .085

When a formula is to be rewritten to solve for a different variable or unknown, the general steps are similar. Now, however, there are no known values to substitute, so we treat the variables as if they were numbers.

Solve for r in $A = p + prt$.

$$A + -p = p + -p + prt \qquad \text{add } -p \text{ to both sides}$$

$$A + -p = prt \qquad \text{simplify}$$

$$\frac{A + -p}{pt} = \frac{A - p}{pt} = r \qquad \text{divide both sides by } p \text{ and } t$$

The *new* formula, $r = \dfrac{A - p}{pt}$, is handier when using a calculator to find r when the values of A, p, and t are known.

EXAMPLES

a) $\quad C = \dfrac{5}{9}(F - 32) \quad$ Find F if $C = 60$.

$$60 = \frac{5}{9}(F - 32)$$

$$60 = \frac{5}{9}F - \frac{160}{9}$$

$$540 = 5F - 160$$

$$700 = 5F$$

$$140 = F$$

b) $\quad \dfrac{1}{T} = \dfrac{1}{a} + \dfrac{1}{b} + \dfrac{1}{c} \quad$ Find b if $T = .2$, $a = 7$, and $c = 3$.

$$\frac{1}{.2} = \frac{1}{7} + \frac{1}{b} + \frac{1}{3}$$

$$4.2b\left(\frac{1}{.2}\right) = 4.2b\left(\frac{1}{7} + \frac{1}{b} + \frac{1}{3}\right)$$

$$21b = .6b + 4.2 + 1.4b$$

$$21b = 2b + 4.2$$

$$19b = 4.2$$

$$b = \frac{4.2}{19} \approx .22$$

c) $\quad I = \dfrac{E}{R + r}$ Solve for r.

$$(R + r)I = (R + r)\dfrac{E}{R + r}$$

$$RI + rI = E$$

$$RI + -RI + rI = E + -RI$$

$$rI = E - RI$$

$$r = \dfrac{E - RI}{I} = \dfrac{E}{I} - R$$

d) $\quad \dfrac{1}{T} = \dfrac{1}{a} + \dfrac{1}{b} + \dfrac{1}{c}$ Solve for a.

$$abcT \cdot \dfrac{1}{T} = abcT \cdot \left(\dfrac{1}{a} + \dfrac{1}{b} + \dfrac{1}{c} \right)$$

$$abc = bcT + acT + abT$$

$$abc - acT - abT = bcT + acT - acT + abT - abT$$

$$abc - acT - abT = bcT$$

$$a(bc - cT - bT) = bcT$$

$$a = \dfrac{bcT}{bc - cT - bT}$$

EXERCISES

1) $P = 3a$ Find P if $a = 7$.

2) $V = \dfrac{1}{3}\pi r^2 h$ Find h if $\pi \approx 3.14$, $r = 3$ cm, and $V = 37.68$ cm^3.

3) $S = \dfrac{1}{2}gt^2$ Find g if $t = 4$ and $S = 128$.

4) $A = p + prt$ Find p if $r = .09$, $t = 3$, and $A = 2430$.

5) $S = \dfrac{n}{2}(a + \ell)$ Find ℓ if $n = 8$, $a = 6$, and $S = 106$.

6) $A = p(1 + r)^n$ Find p if $r = .08$, $n = 2$, and $A = 5832$.

7) $A = \dfrac{bh}{2}$ Find h if $A = 60$ and $b = 15$.

8) $S - M = C$ Find S if $M = 12$ and $C = -6$.

9) $\dfrac{c}{d} = \pi$ Find d if $c = 66$ and $\pi \approx \dfrac{22}{7}$.

10) $i = prt$ Find r if $i = 770$, $p = 7000$, and $t = 1$.

11) $V = \ell w h$ Find w if $V = 64$, $\ell = 10$, and $h = 4$.

12) $S = gt - \dfrac{1}{2}gt^2$ Find t if $g = 32$ and $S = 0$.

13) $\dfrac{c}{d} = \pi$ Find d if $\pi \approx 3.14$ and $c = 62.8$.

14) $I = \dfrac{E - e}{R}$ Solve for R.

15) $S = gt - \frac{1}{2}gt^2$ Solve for g.

16) $V = \ell wh$ Solve for w.

17) $S = ab + bh + ah$ Solve for a.

18) $F = \frac{kmM}{r^2}$ Solve for k.

19) $\frac{1}{f} = \frac{1}{s} + \frac{1}{t}$ Solve for f.

20) $A = \frac{h}{2}(B + b)$ Solve for B.

21) $S = T - \frac{k}{N}$ Solve for N.

22) $V = \frac{D_2 - D_1}{t_2 - t_1}$ Solve for D_2.

23) $y = mx + b$ Solve for m.

24) In the application at the beginning of this section, what is the formula?

1. (Obj. 83) If $A = p(1 + r)^n$, find p if $r = .05$, $n = 2$, and $A = 882$. _____

2. (Obj. 77) Find the LCM of the following:

    $3x^2$, $5xy$, and $10xy^2$ _____

3. (Obj. 73) Multiply. $\dfrac{23w}{4p} \cdot \dfrac{5w}{37p}$ _____

4. (Obj. 76) Multiply and reduce to lowest terms.

    $$\dfrac{x^2 - x}{10x^2} \cdot \dfrac{6x}{x^2 - 1}$$ _____

5. (Obj. 74) Reduce. $\dfrac{32a^2 b^3}{40a^3 b^2}$ _____

6. (Obj. 80) Divide and reduce.

    $$\dfrac{4a^2 b}{6a} \div \dfrac{8ab}{3ab^2}$$ _____

7. (Obj. 84) If $I = \dfrac{E - e}{R}$, solve for e. _____

8. (Obj. 75) Find the missing numerator.

    $$\dfrac{x - 1}{x + 3} = \dfrac{?}{x^2 - 9}$$ _____

9. (Obj. 79) Subtract and reduce.

    $$\dfrac{3x}{x^2 - 4} - \dfrac{6}{x^2 - 4}$$ _____

10. (Obj. 73) Multiply. $\dfrac{3x}{5} \cdot \dfrac{12x + 5}{17}$ _____

11. (Obj. 79) Subtract. $\dfrac{a}{a + 4} - \dfrac{2}{a + 7}$ _____

12. (Obj. 82) Solve. $\dfrac{x}{5} - \dfrac{3}{4} = \dfrac{x}{2}$ _____

13. (Obj. 81) Simplify. $\dfrac{\dfrac{3}{x + 3} + 1}{\dfrac{4}{x - 3} - 2}$ _____

14. (Obj. 78) Add. $\dfrac{x}{x+2} + \dfrac{3}{x-4}$

_____

15. (Obj. 81) Simplify. $\dfrac{\dfrac{x}{3} + \dfrac{y}{2}}{\dfrac{x}{4} - \dfrac{y}{6}}$

_____

16. (Obj. 75) Find the missing numerator.

$$\frac{7xy}{6w} = \frac{?}{72w^2 z}$$

_____

17. (Obj. 80) Divide and reduce.

$$\frac{y+5}{y^2-4} \div \frac{y^2+6y+5}{y+2}$$

_____

18. (Obj. 74) Reduce. $\dfrac{m^2-25}{2m^2-6m-20}$

_____

19. (Obj. 78) Add. $\dfrac{3y-5}{xy} + \dfrac{5y-3}{xy}$

_____

20. (Obj. 84) If $S = 2ab + 2ah + 2bh$, solve for h.

_____

21. (Obj. 77) Find the LCM of the following:

$$3a + 6b, \, 2a + 4b, \, 4$$

_____

22. (Obj. 82) Solve. $\dfrac{2}{x+3} + \dfrac{1}{2} = \dfrac{2}{3}$

_____

23. (Obj. 83) If $A = \dfrac{bh}{2}$, find h if $b = 2.1$ and $A = 7.56$.

_____

24. (Obj. 82) A carpenter and his apprentice can frame a house in 10 days. If the carpenter could do the framing alone in 16 days, how long would it take the apprentice alone?

_____

25. (Obj. 82) Three fifths of a certain number is two thirds more than half of the same number. Find the number.

_____

CHAPTER 8

PROPORTION, VARIATION, AND PERCENT

8.1   RATIO AND PROPORTION

Car A uses 15 gallons of gas to travel 180 miles. Car B uses 13 gallons of gas to travel 156 miles. Is the ratio of miles to gallons of gas of car A equal to that of car B? In other words, is the following proportion true?

$$\frac{180 \text{ miles}}{15 \text{ gallons}} = \frac{156 \text{ miles}}{13 \text{ gallons}}$$

APPLICATION

85) Write a fraction that shows a ratio comparison of two numbers or two denominate numbers.

86) Determine whether or not a given proportion is true.

OBJECTIVES

A *ratio* is a comparison of a pair of numbers or quantities by indicated division. When two ratios are written with an equal sign between them, the resulting equation is called a *proportion.*

Denominate numbers are represented by number names followed by measurement word names or *units.*

VOCABULARY

Two numbers can be compared by subtraction or by division. Suppose we wish to compare the denominate numbers 12 dollars and 3 dollars. Since $12 - 3 = 9$, we can say that

THE HOW AND WHY OF IT

$12 is nine dollars more than $3.

And since $3\overline{)12}$ or $12 \div 3$ is 4, we can say that

$12 is four times larger than $3.

The indicated division is called a ratio. These are common ways to write the ratio comparison of 12 and 3:

$$12:3 \qquad 12 \div 3 \qquad 12 \text{ to } 3 \qquad \frac{12}{3}$$

Here we shall treat ratios as fractions.

If a car runs 208 miles on 8 gallons of gas, we can compare the unlike denominate numbers "280 miles" and "8 gallons" by writing $\frac{208 \text{ miles}}{8 \text{ gallons}}$. This symbol can be reduced in the same way as a fraction, as long as the measurement word names (or units) are stated:

$$\frac{208 \text{ miles}}{8 \text{ gallons}} = \frac{104 \text{ miles}}{4 \text{ gallons}} = \frac{26 \text{ miles}}{1 \text{ gallon}} = 26 \text{ miles per gallon}$$

It is this process that leads to statements such as "There are 3.1 children to a family," since

$$\frac{31 \text{ children}}{10 \text{ families}} = \frac{3.1 \text{ children}}{1 \text{ family}}$$

The last ratio must be understood as a comparison and not as a fact, since no family has 3.1 children.

If two denominate numbers have the same units, such as $\frac{\$3}{\$100}$, then the unit labels or words may be dropped:

$$\frac{\$3}{\$100} = \frac{3}{100}$$

If two denominate numbers do not have the same units, as in $\frac{26 \text{ miles}}{1 \text{ gallon}}$, the unit labels must be written.

Equations with two ratios set equal, such as

$$\frac{5 \text{ inches}}{2 \text{ feet}} = \frac{20 \text{ inches}}{8 \text{ feet}} \text{ and } \frac{62 \text{ miles}}{1 \text{ hour}} = \frac{100 \text{ kilometers}}{1 \text{ hour}},$$

are called proportions. These are read as "5 inches is to 2 feet as 20 inches is to 8 feet" and "62 miles is to 1 hour as 100 kilometers is to 1 hour." To be true, the ratios must represent equivalent fractions when the units are the same.

The test to determine whether a proportion is true is often called "cross multiplication."

$\frac{14}{8} = \frac{35}{20}$ is a true proportion because

$$\frac{14}{8} \underset{\displaystyle =}{\overset{\displaystyle}{\rightleftharpoons}} \frac{35}{20}$$

$$14 \cdot 20 = 8 \cdot 35$$
$$280 = 280$$

The test is based on the multiplication law of equality. It is equivalent to multiplying both sides of the equation by the product of the denominators.

The proportion $\frac{4}{9} = \frac{2}{3}$ is not true because $4 \cdot 3 \neq 9 \cdot 2$.

To determine if a proportion is true,
1) check that the ratios have the same units;
2) find the cross products;
3) check that the cross products are equal;

that is,

$\frac{a}{b} = \frac{c}{d}$ is true provided $ad = bc$.

EXAMPLES

a)   The ratio of 5 chairs to 6 chairs is

$$\frac{5 \text{ chairs}}{6 \text{ chairs}} = \frac{5}{6}.$$

Since the units are the same they may be dropped.

b)   The ratio of 10 chairs to 8 people is

$$\frac{10 \text{ chairs}}{8 \text{ people}} \text{ or } \frac{5 \text{ chairs}}{4 \text{ people}} \text{ or } \frac{1.25 \text{ chairs}}{1 \text{ person}}.$$

The units must be stated since they are unlike.

c)   The ratio of the length of a room to its width, if it measures 24 feet by 18 feet, is

$$\frac{24 \text{ feet}}{18 \text{ feet}} = \frac{24}{18} = \frac{4}{3}.$$

d)   The ratio of six dimes to fourteen nickels is

$$\frac{6 \text{ dimes}}{14 \text{ nickels}} = \frac{6 \text{ dimes}}{7 \text{ dimes}} = \frac{6}{7},$$

or $\dfrac{6 \text{ dimes}}{14 \text{ nickels}} = \dfrac{12 \text{ nickels}}{14 \text{ nickels}} = \dfrac{12}{14} = \dfrac{6}{7}$,

or $\dfrac{6 \text{ dimes}}{14 \text{ nickels}} = \dfrac{60 \text{ cents}}{70 \text{ cents}} = \dfrac{60}{70} = \dfrac{6}{7}$.

e) The proportion $\dfrac{6}{5} = \dfrac{72}{60}$ is true, since $6 \times 60 = 5 \times 72$.

f) The proportion $\dfrac{2.1}{3.1} = \dfrac{2}{3}$ is false since $(2.1)(3)$ is not equal to $(3.1)(2)$.

g) The proportion $\dfrac{1 \text{ dollar}}{3 \text{ quarters}} = \dfrac{8 \text{ dimes}}{12 \text{ nickels}}$ is true, since we may write it as

$$\dfrac{20 \text{ nickels}}{15 \text{ nickels}} = \dfrac{16 \text{ nickels}}{12 \text{ nickels}} \text{ or } \dfrac{20}{15} = \dfrac{16}{12}$$

and $(20)(12) = (15)(16)$.

EXERCISES Write a ratio to compare the following pairs of numbers. Reduce to lowest terms where possible.

1) 8 people to 11 chairs

2) 3 inches to 12 inches

3) 1 nickel to 1 dime (compare in cents)

4) 4 cm to 8 cm

5) 80 cm to 1 m

Tell whether the following proportions are true or false.

6) $\dfrac{1}{2} = \dfrac{5}{10}$

7) $\dfrac{2}{3} = \dfrac{18}{24}$

8) $\dfrac{\frac{1}{2}}{5} = \dfrac{1}{10}$

9) $\dfrac{7}{8} = \dfrac{11}{12}$

10) $\dfrac{\frac{1}{3}}{\frac{7}{9}} = \dfrac{\frac{1}{2}}{1\frac{1}{6}}$

Write a ratio to compare the following pairs of numbers. Reduce to lowest terms where possible.

11) 6 families to 18 children

12) 1 quarter to 1 dime (compare in cents)

13) 3 dimes to 7 nickels (compare in nickels)

14) 12 meters to 10 meters

15) 600 miles to (per) 8 hours

16) $2.37 to (per) 3 pounds of potatoes

17) 1300 television sets to 1000 houses (reduce to a 1-house comparison)

18) 32 games won to 20 games lost

19) The low gear ratio in a truck's transmission, if the large gear has 189 teeth and the small gear has 14 teeth

Tell whether the following proportions are true or false.

20) $\dfrac{2}{4} = \dfrac{14}{28}$

21) $\dfrac{4\frac{1}{2}}{3} = \dfrac{9}{6}$

22) $\dfrac{8}{34} = \dfrac{9}{32}$

23) $\dfrac{14}{16} = \dfrac{21}{24}$

24) $\dfrac{2.1}{3.2} = \dfrac{1.2}{2.3}$

25) $\dfrac{2.6}{4.8} = \dfrac{3.9}{7.2}$

26) $\dfrac{\frac{1}{4}}{\frac{2}{5}} = \dfrac{8}{5}$

27) $\dfrac{8 \text{ inches}}{2 \text{ feet}} = \dfrac{6 \text{ inches}}{18 \text{ inches}}$

28) $\dfrac{5 \text{ pounds}}{\$1.30} = \dfrac{8 \text{ ounces}}{14 \text{ cents}}$

29) $\dfrac{\$18}{4 \text{ weeks}} = \dfrac{50 \text{ cents}}{1 \text{ day}}$

30) In the application at the beginning of this section, is the proportion true?

8.2 SOLVING PROPORTIONS

APPLICATION

Two gears in mesh have a speed ratio of 2 to 5. If the smaller gear makes 70 rpm, the rpm of the larger gear can be found by solving the following proportion.

$$\frac{2}{5} = \frac{70}{n} \; ; n \text{ is the rpm.}$$

OBJECTIVE

87) Find the missing number that will make a given proportion true.

VOCABULARY

No new vocabulary.

THE HOW AND WHY OF IT

Proportions are used to solve many problems in science and technology which involve ratios in some manner. There are four members in a proportion, and if any three of them are known, it is possible to solve the proportion to find the missing member. The following method of solving a proportion is based upon the fact that the cross products must be equal. Consider the following question: "What number is to 5 as 15 is to 25?" To answer this, we will use a variable (x) to represent the missing number, and we can write the following: x is to 5 as 15 is to 25,

$$\frac{x}{5} = \frac{15}{25}$$

Since the cross products are equal, we have

$$25 \cdot x = 5 \cdot 15$$

or

$$x = 3.$$

Therefore, $\frac{3}{5} = \frac{15}{25}$.

> To solve a proportion,
> 1) find the cross products;
> 2) set the cross products equal to each other;
> 3) solve the equation.

EXAMPLES

Solve the following proportions.

a) $\frac{4}{9} = \frac{8}{x}$

$4x = 72$

$x = 18$

b) $\frac{.6}{x} = \frac{1.2}{.84}$

$1.2x = .504$

$x = .42$

c) $\dfrac{\dfrac{3}{4}}{1\dfrac{2}{3}} = \dfrac{\dfrac{1}{2}}{x}$

$\dfrac{3}{4}x = \dfrac{5}{6}$

$x = 1\dfrac{1}{9}$

EXERCISES

Solve the following proportions.

1) $\dfrac{6}{8} = \dfrac{12}{x}$

2) $\dfrac{x}{42} = \dfrac{5}{7}$

3) $\dfrac{6}{y} = \dfrac{9}{5}$

4) $\dfrac{y}{25} = \dfrac{3}{5}$

5) $\dfrac{5}{2} = \dfrac{w}{9}$

6) $\dfrac{14}{8} = \dfrac{7}{w}$

7) $\dfrac{3}{2} = \dfrac{R}{100}$

8) $\dfrac{5}{4} = \dfrac{R}{100}$

9) $\dfrac{.2}{.3} = \dfrac{8}{x}$

10) $\dfrac{x}{40} = \dfrac{\dfrac{3}{4}}{5}$

11) $\dfrac{\dfrac{1}{2}}{y} = \dfrac{\dfrac{2}{3}}{10}$

12) $\dfrac{8}{9} = \dfrac{\dfrac{1}{3}}{y}$

13) $\dfrac{w}{2.5} = \dfrac{3}{5}$

14) $\dfrac{418}{154} = \dfrac{w}{7}$

15) $\dfrac{6.5}{26} = \dfrac{y}{.04}$

16) $\dfrac{3}{5} = \dfrac{R}{100}$

17) $\dfrac{.05\frac{1}{2}}{1} = \dfrac{R}{100}$

18) $\dfrac{.014}{x} = \dfrac{7}{50}$

19) $\dfrac{3\frac{1}{2}}{10\frac{1}{2}} = \dfrac{8}{x}$

20) $\dfrac{.05}{.9} = \dfrac{y}{4.5}$

21) $\dfrac{y}{3} = \dfrac{15}{16}$

22) $\dfrac{2\frac{1}{2}}{3\frac{1}{3}} = \dfrac{4\frac{1}{4}}{x}$

23) $\dfrac{A}{35} = \dfrac{6.2}{100}$

24) $\dfrac{1.2}{2.7} = \dfrac{3.4}{w}$

25) $\dfrac{2.5}{x} = \dfrac{3.2}{300}$

26) What number is to 36 as 15 is to 18?

27) Five is to one and one-half as fifteen is to what number?

28) Twenty-eight is to what number as seven is to three?

29) In the application at the beginning of this section, what is the rpm of the larger gear?

8.3 WORD PROBLEMS (Proportion)

APPLICATION

Grains and drams are units of weight used in pharmacy. (8 drams = 480 grains.) If a doctor's prescription calls for 24 grains of a drug, what part of a dram is required?

OBJECTIVE

88) Solve word problems using proportions.

VOCABULARY

No new vocabulary.

THE HOW AND WHY OF IT

Many common situations occur in which two quantities are compared by ratios: cost (price per pound), map scale (miles per inch), geometry (length of a shadow to the height of the object), time to do a certain amount of work, and profit to a specific amount of investment, to name a few.

Assuming that the comparison of the two quantities is constant, the given comparison can be used to discover the missing part of a second one. For instance, if 2 pounds of bananas cost $.36, what will 12 pounds of bananas cost?

	Pounds of Bananas	Cost in Dollars
Case 1	2	.36
Case 2	12	

In the preceding table, it is seen that the cost in Case 2 is missing. Assign this missing value the letter y.

	Pounds of Bananas	Cost in Dollars
Case 1	2	.36
Case 2	12	y

Now form the ratios of the like quantities, Case 1 to Case 2. These must be equal if the relationship is constant.

$$\frac{2 \text{ lb of bananas}}{12 \text{ lb of bananas}} = \frac{\$.36}{\$y}$$

Since the units are the same, they may be dropped.

$$\frac{2}{12} = \frac{.36}{y}$$

Solving, we have:
$$2y = 4.32$$
$$y = 2.16$$

The conclusion is that 12 pounds of bananas will cost $2.16.

Although different ratios also remain constant $\left(\frac{\text{dollars}}{\text{pounds}}\right)$ in the relationship, we prefer to form the ones that compare the same units $\left(\frac{\text{pounds}}{\text{pounds}} = \frac{\text{dollars}}{\text{dollars}}\right)$. The denominator of each ratio must come from the same case. The table helps us to keep this in mind because it simulates the ratios we use.

EXAMPLE

a) On a road map of Oregon, $\frac{1}{4}$ inch represents 50 miles. How many miles are represented by $1\frac{1}{2}$ inches?

	Inches	Miles
Case 1	$\frac{1}{4}$	50
Case 2	$1\frac{1}{2}$	N

$$\frac{\frac{1}{4}}{1\frac{1}{2}} = \frac{50}{N}$$

$$\frac{1}{4} \cdot N = 1\frac{1}{2} \cdot 50$$

$$N = 300$$

Therefore, $1\frac{1}{2}$ inches on the map represents 300 miles.

EXERCISES

1) In the application at the beginning of this section, what part of a dram is required?

2) A photograph that measures 8 cm wide and 13 cm high is to be enlarged so that the height will be 26 cm. What will be the width of the enlargement?

3) Merle is knitting a sweater. The knitting gauge is 6 rows to one inch. How many rows must she knit to complete $9\frac{1}{2}$ inches of the sweater?

4) At a certain time of the day, a tree 15 m tall casts a shadow of 12 m, while a second tree casts a shadow of 20 m. How tall is the second tree?

5) If 16 lb of fertilizer will cover 1500 sq ft of lawn, how much fertilizer is needed to cover 2500 sq ft?

6) The Mudville Elementary School expects an enrollment of 980 students. The district assigns teachers at the rate of 3 teachers for every 70 students. The district now employs 32 teachers. How many additional teachers does the district need to hire?

7) The Mitchells pay $810 taxes on their home, which has an assessed value of $28,000. How much will the taxes be on a $42,000 home in the same district?

8) One can of frozen orange juice concentrate mixed with three cans of water makes one liter of juice. At the same rate, how many cans of water are needed to make four liters of juice?

9) A room that contains 24 square yards was carpeted at a cost of $204. If the same kind of carpet is used in a room that contains 16 square yards, what will be the cost?

10) The counter on a tape recorder registers 520 after the recorder has been running for 20 minutes. What would the counter register after half an hour?

11) During the first 320 miles of their vacation trip, the Scaberys used 35 gallons of gasoline. At this rate, to the nearest tenth of a gallon how many gallons of gasoline will be needed to finish the remaining 580 miles of their trip?

12) If it takes 4 men 12 hours to do a certain job, how many of these jobs could they do in 66 hours?

13) In the first 12 games of an 18 game schedule, Dave's basketball team scored a total of 974 points. At this rate, how many points can they expect to score in their remaining games?

14) A 16-ounce can of pears costs $.38 and a 29-ounce can costs $.65. Is the price per ounce the same in both cases? If not, then to the nearest cent, what would the price of the 29-ounce can need to be to equalize the price?

15) A map of the western United States is scaled so that $\frac{3}{4}$ inch represents 100 miles. How many miles is it between San Diego and Seattle if the distance on the map is 9 inches?

16) A doctor requires that the nurse, Ida, give 10 milligrams of a certain drug to his patient. The drug is in a solution that contains 25 milligrams in one cubic centimeter. How many cc's should Ida use for the injection?

17) If Wayne receives $400 for $\frac{2}{3}$ of a ton of strawberries, how much will he receive for $1\frac{3}{5}$ tons?

18) If a 20-foot beam of structural steel contracts .0053 inch for each drop of 5 degrees in temperature, then to the nearest ten-thousandth of an inch, how much would a 13-foot beam of structural steel contract for a drop of 5 degrees in temperature?

19) The ratio of boys to girls taking chemistry is 4 to 3. How many boys are there in a chemistry class of 84 students? (Hint: Fill in the rest of the table.)

	Number of boys	Number of students
Case 1		7
Case 2		84

20) Betty prepares a mixture of nuts that has cashews and peanuts in a ratio of 3 to 5. How many pounds of each will she need to make 48 pounds of the mixture?

21) The estate of the late Mr. John Redgrave is to be divided among his 3 nephews in the ratio of 5 to 2 to 2. How much of the $72,081 in the estate did each nephew receive?

8.4 VARIATION

APPLICATION

The force of attraction between two magnetic poles of opposite polarity varies inversely as the square of the distance between them. What is the force when they are 5 cm apart if it is 36 dynes at a distance of 9 cm?

OBJECTIVES

89) Solve problems involving direct variation.
90) Solve problems involving inverse variation.

VOCABULARY

Two quantities *vary directly* when one is a multiple of the other (their quotient is a constant). In the formula

$$D = 55t \quad \text{(or } \frac{D}{t} = 55\text{)}$$

(distance equals 55 mph times time), *D varies directly* as *t*. Note that as time increases, distance increases, and as time decreases, so does distance.

Two quantities *vary inversely* when their product is constant. In the formula

$$PV = 50$$

(pressure times volume is constant if the temperature remains the same), *P* varies inversely as *V*. Note that pressure increases as volume decreases and that pressure decreases as volume increases.

THE HOW AND WHY OF IT

In our physical world most things are in a state of change or variation. Measurements of changes in temperature, rainfall, light, heat, and fuel supplies are recorded. Changes in a person's height, weight, and blood pressure can be important to health. Many relationships between these can be expressed in formulas. Here we are concerned about only two types of variation, direct and inverse.

Some examples of direct variation are:

Distance traveled at a constant speed varies directly as time: $d = kt$ or $\frac{d}{t} = k$ (quotient is a constant).

The area of a circle varies directly as the square of its radius: $A = \pi R^2$ or $\frac{A}{R^2} = \pi$.

The total cost of a certain number of articles varies directly as the price: $C = np$ or $\dfrac{C}{P} = n$.

In solving a problem involving direct variation, a constant must be identified. Suppose a variable S varies directly as a variable P. If $S = 40$ when $P = 5$, find S when $P = 8$. Since the variables vary directly, we know that $\dfrac{S}{P} = k$. So

$$\frac{S}{P} = k$$

$$\frac{40}{5} = k$$

$$8 = k$$

We now have the formula $\dfrac{S}{P} = 8$ or $S = 8P$ and can solve for S when $P = 8$.

$$S = 8P = 8(8) = 64$$

Some examples of inverse variation are:

The time t it takes to cover a certain distance d varies inversely as the speed R.

$$R \cdot t = d \text{ (product is a constant distance)}$$

The number of vibrations n in a musical string varies inversely as its length ℓ.

$$n \cdot \ell = k$$

When two pulleys are connected by a belt the revolutions per minute each makes varies inversely as their respective diameters.

$$d \cdot r = k \text{ and } D \cdot R = k$$

If the small pulley has a diameter of 10 inches and revolves 80 times per minute, how many rpm will the larger one make if it has a diameter of 16 inches?

smaller pulley	larger pulley
$d \cdot r = k$	$D \cdot R = k$
$(10)(80) = k$	$16 \cdot R = 800$
$800 = k$	$R = 50$

So the larger pulley will make 50 revolutions per minute.

Some formulas show variation between three or more variables. (See Example c.)

EXAMPLES

a) The weight w of a piece of aluminum varies directly as the volume v. If a piece of aluminum containing $1\frac{1}{2}$ cubic feet weighs 254 pounds, find the weight of a piece containing 5 cubic feet.

$$\frac{w}{v} = k \qquad \text{direct variation}$$

$$\frac{254}{1\frac{1}{2}} = k$$

$$k = \frac{508}{3}$$

So $$\frac{w}{v} = \frac{508}{3}$$

$$\frac{w}{5} = \frac{508}{3}$$

$$w = \frac{2540}{3} = 846\frac{2}{3}$$

So the weight of a piece containing 5 cubic feet is $846\frac{2}{3}$ pounds.

b) The base b of a rectangle with constant area k varies inversely with its height. One rectangle has a base of 10 and a height of 6. Find the height of another rectangle whose base is 12.

$$b \cdot h = k \qquad \text{inverse variation}$$

$$(10)(6) = k$$

$$k = 60$$

So $$b \cdot h = 60$$

$$12(h) = 60$$

$$h = 5$$

So the height of the second rectangle is 5.

c)   The interest *i* paid on borrowed money varies directly as the principle *P* and the time *t*. If $80 interest is earned on a loan of $600 in 2 years, how much interest is earned on $1000 borrowed for 3 years?

$$\frac{i}{P \cdot t} = k \qquad \begin{array}{l} i \text{ varies directly as } P \text{ and } t \\ \text{(sometimes read: } i \text{ varies jointly as } P \text{ and } t\text{)} \end{array}$$

$$\frac{80}{(600)\,(2)} = k$$

$$k = \frac{1}{15}$$

So $$\frac{i}{(1000)\,(3)} = \frac{1}{15}$$

$$i = 200$$

So the interest earned is $200.

Solve.   **EXERCISES**

1) In the application at the beginning of this section, what is the force?

2) The weight *w* of a gold brick varies directly as the volume *v*. If a brick containing $\frac{3}{4}$ cubic foot weighs 906 pounds, what will a brick of $1\frac{1}{2}$ cubic feet weigh?

3) The time it takes to make a certain trip varies inversely as the speed. If it takes 5 hours at 50 mph, how long will it take at 60 mph?

4) A car salesman's salary varies directly as his total sales. If he receives $372 for sales of $3100, how much will he receive for sales of $4500?

5) The amount of quarterly income one receives varies directly as the amount of money invested. If Dan earns $40 on a $2000 investment, how much would he earn on a $4300 investment?

6) The number of amperes varies directly as the number of watts. For a reading of 50 watts, the number of amperes is $\frac{5}{11}$. What are the amperes when the watts are 75?

7) The weight of wire varies directly as its length. If 1000 ft of wire weigh 45 lb, what will one mile of wire weigh?

8) The length of a rectangle with a constant area varies inversely as the width. If one rectangle has a length of 12 and a width of 8, what will be the length of a rectangle with a width of 6?

9) The force needed to raise an object with a crowbar varies inversely with the length of the crowbar. If it takes 40 lb of force to lift a certain object with a 2-ft long crowbar, what force will be necessary if you use a 3-ft crowbar?

10) As a rule of thumb, realtors suggest that the price you can afford to pay for a house varies directly with your annual salary. If a person earning $18,500 can purchase a $46,250 home, what price home can a person earning $24,000 annually afford?

11) Assuming that each person works at the same rate, the time it takes to complete a job varies inversely as the number of people assigned to it. If it takes 5 people 12 hours to do a job, how long will it take 3 people?

8.5   WHAT IS PERCENT?

A filbert grower grades the quality of the crop by cracking a sample of 100. He finds that 7 nuts are wormy or blanks. What is the estimated percent of rejects for the entire crop?   **APPLICATION**

91) Write a percent to express a comparison of two numbers or quantities.   **OBJECTIVE**

When using percent to compare a first number (or quantity) to a second number (or quantity), the second number (the number you are comparing to) is called the *base unit*. In comparing 80 to 100, 100 is the base unit. To use percent to compare two numbers or quantities, divide the base unit into 100 equal parts and then find how many of these equal parts it takes to make the first number or quantity. The product of this number of parts and $\frac{1}{100}$ or .01 (each part is $\frac{1}{100}$ of the base unit) is the *percent comparison* or just the *percent*. The percent, $(80)\left(\frac{1}{100}\right)=\frac{80}{100}$, is usually written 80%, where the symbol "%" is read "percent" and $\% = \frac{1}{100} = .01$.   **VOCABULARY**

Consider the comparison of 24 to 100 as a percent. The base unit is 100. Figure 8.1 shows the base unit divided into 100 equal parts, each of size one. We see that 24 of the 100 equal parts are shaded. Since each part represents $\frac{1}{100}$ of the 100, the 24 (shaded part) is $(24)\left(\frac{1}{100}\right)$ or 24% of 100.   **THE HOW AND WHY OF IT**

Figure 8.1 also illustrates that if the first number is smaller than the base unit, then not all of the base unit will be shaded and hence the comparison will be less than 100%. If the first number equals the base unit, the entire unit will be shaded and the comparison will be 100%. If the first number is larger than the base unit, then additional parts will be needed to represent it and the comparison will be more than 100%.

FIGURE 8.1

The percent comparison of two numbers can also be obtained from their ratio. The ratio of 24 to 100 is $\frac{24}{100} = 24 \div 100 = 24 \cdot \frac{1}{100} = 24\%$. This gives rise to the interpretation of percent as "so many per hundred."

The ratio of two numbers can be used to find the percent when the base unit is not 100. Compare 7 to 20. The ratio is $\frac{7}{20}$. Now find the equivalent ratio with denominator 100.

$$\frac{7}{20} = \frac{R}{100} \qquad \text{Write the two ratios as a proportion.}$$

$$20 \cdot R = 7 \cdot 100 \qquad \text{Set the cross products equal.}$$

$$R = 35$$

So, $\frac{7}{20} = \frac{35}{100} = 35 \cdot \frac{1}{100} = 35\%$.

To find the percent comparison of two numbers:
1) write the ratio of the first number to the base number;
2) find the equivalent ratio with denominator 100.
3) $\frac{\text{numerator}}{100} = \text{numerator} \cdot \frac{1}{100} = \text{numerator } \%$.

EXAMPLES

a) The shaded portion represents 55 of the 100 equal parts or $55 \cdot \frac{1}{100} = 55\%$ of the region.

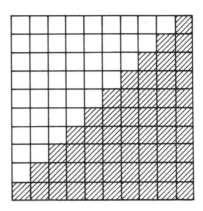

b) At a recent sporting event, there were 30 women among the first 100 people to enter. What percent were women? Each person is $\frac{1}{100}$ of the group, so $30 \cdot \frac{1}{100}$ or 30% were women.

c) The comparison of 42 to 25 as a ratio is $\frac{42}{25}$. As a percent, it is 168%, since

$$\frac{42}{25} = \frac{168}{100} = 168 \cdot \frac{1}{100} \quad .$$

d) Write the comparison of 7 to 13 as a percent.

$$\frac{7}{13} = \frac{R}{100}$$

$$13 \cdot R = 700$$

$$R = 700 \div 13 = 53\frac{11}{13}$$

$$\frac{7}{13} = \frac{53\frac{11}{13}}{100} = 53\frac{11}{13} \cdot \frac{1}{100} = 53\frac{11}{13}\%$$

EXERCISES   What percent of each of the following is shaded?

1)

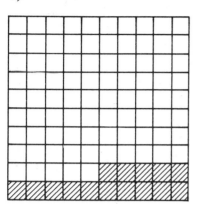

2)

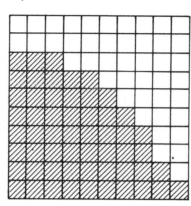

3)

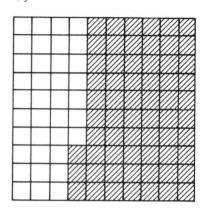

Write a percent to express each of the following comparisons.

4) 27 out of 100   5) 68 parts per 200

6) 16 to 10   7) 215 to 100

8) The ratio of 2 to 25 is the same as_____ to 100 or_____%.

Write each of the following comparisons as a percent.

9) 12 to 25   10) 62 to 62

11) 53 to 500   12) 93 to 80

13) 7 to 9

14) In the application at the beginning of this section, what percent are rejects?

15) The fact that 10% of all women are blond indicates that____?____ out of 100 women are blond.

16) In a recent election, out of every 100 eligible voters, 62 cast their ballots. What percent of the eligible voters exercised the right to vote?

17) Of the people who use Shiny toothpaste, 37 out of 100 report that they have fewer cavities. Out of every 100 people who report, what percent do not report fewer cavities?

18) If the telephone tax rate is 5 cents per dollar, what percent is this?

19) For every $100 spent on gasoline, the state receives $6 tax. What percent of the price of gasoline is the state tax?

20) A bank pays $5.25 interest per year for every $100 in savings. What is the annual interest rate?

21) An electronic calculator originally priced at $100 is on sale for $93. What is the percent of discount? (Discount is the difference between the original price and the sale price.)

8.6 CHANGING DECIMALS TO PERCENTS

APPLICATION

To pass a job entrance examination, an applicant must get a minimum of 136 of the 200 problems correct. This is .68 of the problems correct. Express this as a percent.

OBJECTIVE

92) Write a given decimal or decimal with fraction as a percent.

VOCABULARY

No new vocabulary, but recall:
1) The multiplication property of one, that is, $a \cdot 1 = a$.
2) Fractions in which the numerator and denominator are the same are equal to 1. For example, $\frac{100}{100} = 1$.
3) The reciprocal of the whole number a is $\frac{1}{a}$ (if $a \neq 0$).
4) To divide, multiply by the reciprocal. For instance,

$$\frac{100}{100} = 100 \div 100 = 100 \cdot \frac{1}{100}.$$

THE HOW AND WHY OF IT

In an indicated multiplication, if one of the factors is $\frac{1}{100}$, this indicated multiplication can be read as percent. That is, $75 \left(\frac{1}{100}\right) = 75\%$, $.8 \left(\frac{1}{100}\right) = .8\%$, and $\frac{3}{4}\left(\frac{1}{100}\right) = \frac{3}{4}\%$.

To write .45 as a percent, we pick $100 \cdot \frac{1}{100}$ as a name for one, and we have the following:

$$.45 \ = .45(1) = .45 \ \left(100 \cdot \frac{1}{100}\right) = .45 \ (100) \left(\frac{1}{100}\right)$$

$$= 45 \left(\frac{1}{100}\right) = 45\%$$

To write .2 as a percent, the following is true:

$$.2 = .2(1) = .2 \left(100 \cdot \frac{1}{100}\right) = .2(100)\left(\frac{1}{100}\right)$$

$$= 20 \left(\frac{1}{100}\right) = 20\%$$

5 is written as a percent in the following manner:

$$5 = 5(1) = 5 \left(100 \cdot \frac{1}{100}\right) = 5(100)\left(\frac{1}{100}\right) = 500\left(\frac{1}{100}\right) = 500\%$$

Notice that .45 = 45%, .2 = 20%, and 5 = 500%; in each case the decimal point is moved two places to the right and the percent symbol (%) is inserted.

> To change a decimal or decimal with fraction to a percent, move the decimal point two places to the right (multiply by 100) and insert the percent symbol (%). It might be necessary to insert zeros after the decimal in order to be able to move the decimal point.

EXAMPLES

Change from a decimal to a percent.

a) .35 = 35%

b) .04 = 4%

c) .217 = 21.7%

d) .003 = .3%

e) 9 = 900%

f) .7 = 70%

g) $.25\frac{1}{3} = 25\frac{1}{3}\%$ (recall that $5\frac{1}{3}$ is in the hundredths place)

Write each of the following as a percent.

EXERCISES

1) .36

2) 5.95

3) .08

4) 8.33

5) 1.6

6) .007

7) 20.6

8) 11

9) .531 10) .55 11) .29 12) .74

13) .214 14) .083 15) 7 16) 27

17) 13.21 18) $.27\frac{2}{3}$ 19) .005 20) 1.27

21) .745 22) .0256 23) .4 24) 3.2

25) 1.85 26) $.03\frac{1}{3}$ 27) 5.75 28) $.74\frac{1}{6}$

29) .014 30) .0003

31) In the application at the beginning of this section, what is .68 expressed as a percent?

8.7 CHANGING PERCENTS TO DECIMALS

APPLICATION   An ore sample contains 17% iron. In order to calculate the amount of iron in 327 tons of ore, the decimal equivalent of 17% must be known. What is the decimal representation of 17%?

OBJECTIVE   93) Write a decimal or decimal with fraction which is equivalent to a given percent.

No new vocabulary. Recall:

1) The decimal point for a whole number follows the ones digit;

2) In a decimal with fraction, the fraction part is attached to the digit preceding it and has the same place value as the digit;

3) Any fraction can be thought of as a decimal with fraction, since

$\frac{1}{2} = 0\frac{1}{2}$, $\frac{5}{8} = 0\frac{5}{8}$, and so on.

The definition of percent gives us a clue that we can use to change from a percent to a decimal. The percent symbol indicates multiplication by $\frac{1}{100}$; hence 17% can be written as $17 \cdot \frac{1}{100}$. Since multiplying by the reciprocal of a number is the same as dividing by that number, $17 \cdot \frac{1}{100}$ can be thought of as $17 \div 100$. To divide a number by 100, move the decimal point two places to the left; hence 17% = .17.

> To change a percent to a decimal or a decimal with fraction, move the decimal point two places to the left (multiply by the reciprocal of 100) and drop the percent symbol (%).

EXAMPLES

a) $14.5\% = .145$, since $14.5\% = 14.5 \left(\frac{1}{100}\right) = 14.5 \div 100 = .145$

b) $29\frac{1}{3}\% = 29\frac{1}{3} \left(\frac{1}{100}\right) = 29\frac{1}{3} \div 100 = .29\frac{1}{3}$

c) $35\% = .35$

d) $295\% = 2.95$

e) $83\frac{1}{2}\% = .83\frac{1}{2}$ or $83\frac{1}{2}\% = 83.5\% = .835$

f) $.5\% = .005$

g) $\frac{3}{4}\% = .75\% = .0075$ or $.00\frac{3}{4}$

h) $\frac{1}{3}\% = 0\frac{1}{3}\% = .00\frac{1}{3}$

i) $\frac{1}{3}\% = .3\frac{1}{3}\% = .003\frac{1}{3}$

j) $\frac{1}{3}\% = .33\frac{1}{3}\% = .0033\frac{1}{3}$

same problem

EXERCISES   Write each of the following as a decimal.

1) 14%   2) 27%   3) 92.3%   4) 2.79%

5) 3%   6) .81%   7) 16%   8) 5.9%

9) 82%   10) 36%   11) 2.15%   12) 312%

13) 563%   14) 110.6%   15) 53.7%   16) .04%

17) $\frac{1}{8}$%   18) $\frac{4}{5}$%   19) 135%   20) 112%

21) 314.7%   22) 261.3%   23) .12%   24) .5%

25) .007%   26) $35\frac{1}{6}$%   27) $4.8\frac{1}{3}$%   28) $\frac{1}{4}$%

29) 1%   30) 100%

31) In the application at the beginning of this section, what is the decimal?

8.8 FRACTIONS TO PERCENT

In a survey it was determined that $\frac{5}{8}$ of the residents of Hillsboro, Oregon, own their homes. What percent does this represent?

APPLICATION

94) Change a given fraction or mixed number to percent.

OBJECTIVE

No new vocabulary.

VOCABULARY

Since we already know how to change fractions to decimals and decimals to percent, we can combine the two ideas to change fractions to percent.

THE HOW AND WHY OF IT

> To change a fraction or mixed number to a percent, first change it to a decimal with two decimal places. Then change the decimal to a percent. The decimal may be rounded off or carried out to additional places as directed.

EXAMPLES

a) $\frac{5}{8} = .62\frac{1}{2} = 62\frac{1}{2}\%$ or

$= .625 = 62.5\%$

b) $\frac{5}{6} = .83\frac{1}{3} = 83\frac{1}{3}\%$

c) $2\frac{1}{2} = 2.5 = 250\%$

d) $\frac{3}{7} = .42\frac{6}{7} = 42\frac{6}{7}\%$ or, to the nearest tenth of a percent,

$\approx .429 = 42.9\%$

e) $\frac{1}{40} = .025 = 2.5\%$

f) $\frac{7}{320} = .02\frac{3}{16} = 2\frac{3}{16}\%$ or

$= .0218\frac{3}{4} = 2.18\frac{3}{4}\%$ or, to the nearest tenth of a percent,

$\approx .022 = 2.2\%$

EXERCISES

Change the following fractions to decimals and then change the decimals to percent.

1) $\dfrac{7}{100}$ 2) $\dfrac{50}{100}$ 3) $\dfrac{11}{50}$

4) $\dfrac{3}{10}$ 5) $\dfrac{17}{20}$ 6) $\dfrac{4}{25}$

7) $\dfrac{1}{2}$ 8) $\dfrac{3}{5}$ 9) $\dfrac{3}{12}$

10) $2\dfrac{1}{4}$ 11) $1\dfrac{3}{20}$ 12) $\dfrac{21}{20}$

Change each fraction to a two-digit decimal with fraction. Then change the decimal with fraction to percent.

13) $\dfrac{2}{3}$ 14) $\dfrac{1}{6}$ 15) $1\dfrac{7}{8}$

16) $3\dfrac{7}{9}$ 17) $\dfrac{21}{400}$ 18) $\dfrac{14}{15}$

19) $\dfrac{17}{200}$ 20) $2\dfrac{5}{7}$

Change each fraction to a decimal rounded to the nearest thousandth. Then change the decimal to a percent. (The result is the fraction expressed as a percent to the nearest tenth.)

21) $\dfrac{1}{3}$ 22) $\dfrac{7}{500}$ 23) $\dfrac{1}{80}$

24) $2\dfrac{5}{9}$ 25) $\dfrac{8}{21}$ 26) $\dfrac{5}{6}$

27) $3\dfrac{11}{16}$ 28) $\dfrac{17}{10,000}$

29) In the application at the beginning of this section, what percent of the residents own their homes?

30) Steve applied for a loan to buy a motorcycle. The annual interest charged was 23¢ for each $2 of the loan $\left(\dfrac{23}{200}\right)$. What was the annual rate (percent) of interest?

31) Four-sevenths of the eligible voters turned out for the recent Hillsboro city elections. What percent of the voters turned out, to the nearest tenth of a percent?

8.9 PERCENTS TO FRACTIONS

APPLICATION

A census determined that $37\frac{1}{2}\%$ of the residents of a certain city were of age 40 or over and 50% were of age 25 or under. What fraction of the residents are between the ages of 25 and 40?

OBJECTIVE

95) Change percents to fractions or mixed numbers.

VOCABULARY

No new vocabulary.

THE HOW AND WHY OF IT

Recall that $6.5\% = 6.5 \times \frac{1}{100}$. This gives a very efficient method for changing a percent to a fraction. Proceed by changing 6.5 to a fraction and then performing the indicated multiplication.

> To change a percent to a fraction or a mixed number, replace the percent symbol (%) with an indicated multiplication by $\frac{1}{100}$. Before multiplying, rewrite the other factor (if necessary) as a fraction.

EXAMPLES

a) $6.5\% = 6.5 \times \frac{1}{100} = 6\frac{5}{10} \times \frac{1}{100} = \frac{65}{10} \times \frac{1}{100} = \frac{65}{1000} = \frac{13}{200}$

b) $312\% = 312 \times \frac{1}{100} = \frac{312}{100} = 3\frac{12}{100} = 3\frac{3}{25}$

c) $12\frac{2}{3}\% = 12\frac{2}{3} \times \frac{1}{100} = \frac{38}{3} \times \frac{1}{100} = \frac{38}{300} = \frac{19}{150}$

d) $.05\frac{1}{3}\% = .05\frac{1}{3} \times \frac{1}{100} = 5\frac{1}{3} \times \frac{1}{100} \times \frac{1}{100} = \frac{16}{3} \times \frac{1}{100} \times \frac{1}{100}$

   $= \frac{4 \times 4 \times 1 \times 1}{3 \times 4 \times 25 \times 4 \times 25} = \frac{1}{1875}$

Change each of the following to a fraction or mixed number.

1) 5%   2) 35%   3) 125%

4) 20%   5) 70%   6) 400%

7) 56%   8) 82%   9) 112%

10) 6.5%   11) $\frac{3}{4}$%   12) $16\frac{1}{4}$%

13) $9\frac{1}{5}$%   14) 32.6%   15) $16\frac{1}{2}$%

16) 285%   17) $\frac{1}{3}$%   18) .0016%

19) $11\frac{1}{9}$%   20) .25%   21) 2.5%

22) 20.5%

23) In the application at the beginning of this section, what fraction of the residents are between the ages of 25 and 40?

24) George and Ethel paid 16% of their annual income in taxes last year. What fractional part of their income went to taxes?

25) The enrollment at City Community College this year is 118% of last year's enrollment. What fractional increase in enrollment took place this year?

8.10 FRACTIONS, PERCENTS, AND DECIMALS

APPLICATION   See the previous five sections.

OBJECTIVES   96) Write a percent as a decimal and as a fraction.
                  97) Write a fraction as a percent and as a decimal.
                  98) Write a decimal as a percent and as a fraction.

VOCABULARY   No new vocabulary.

THE HOW AND   Decimals, fractions, and percents can each be expressed in terms of
WHY OF IT   the others. This can be shown as follows:

$$50\% = 50 \cdot \frac{1}{100} = \frac{50}{100} = \frac{1}{2} \text{ and } 50\% = 50\,(.01) = .50$$

$$\frac{3}{4} = 3 \div 4 = .75 \text{ and } \frac{3}{4} = .75 = 75\%$$

$$.65 = 65\,(.01) = 65\% \text{ and } .65 = \frac{65}{100} = \frac{13}{20}$$

EXAMPLES

Fill in the empty spaces with the related fraction, decimal, or percent.

Percent	Fraction	Decimal
30%	_____	_____
_____	$\frac{7}{8}$	_____
_____	_____	.62

We find $30\% = .30$, and $30\% = \frac{3}{10}$;

$\frac{7}{8} = .87\frac{1}{2}$ (also .875), and $\frac{7}{8} = 87\frac{1}{2}\%$ (also 87.5%);

$.62 = 62\%$, and $.62 = \frac{31}{50}$. The filled-in chart is:

Percent	Fraction	Decimal
30%	$\frac{3}{10}$	.30 or .3
$.87\frac{1}{2}\%$ or 87.5%	$\frac{7}{8}$	$.87\frac{1}{2}$ or .875
62%	$\frac{31}{50}$	.62

Fill in the empty spaces with the related percent, fraction, or decimal.

EXERCISES

Percent	Fraction	Decimal
	$\frac{1}{10}$	
30%		
		.75
	$\frac{9}{10}$	
145%		
	$\frac{3}{8}$	
		.001
		1
	$2\frac{1}{4}$	

	Percent	Fraction	Decimal
*	$5\frac{1}{2}\%$		.8
	$5\frac{1}{2}\%$		
*			.875
	$\frac{1}{2}\%$		
*			.6
*	$62\frac{1}{2}\%$		
	$\frac{3}{4}\%$		
*		$\frac{2}{3}$	
*	25%		
*			.20
*	40%		
*	$33\frac{1}{3}\%$		
*			.125
*		$\frac{7}{10}$	

*If you are going to be working with problems that will involve percent in your job or your personal finances (loans, savings, insurance, and so on), it is advisable to know (memorize) these special relationships.

8.11 SOLVING PERCENT PROBLEMS

A six-sheave pulley block is only 80% efficient. Find the actual load L that can be raised if the theoretical load is 1,320 lb. (L is 80% of 1,320 lb.) **APPLICATION**

99) Solve problems that are written in the form "A is $R\%$ of B" or "$R\%$ of B is A." **OBJECTIVE**

To *solve* a percent problem means to do one of the following: **VOCABULARY**
1) Find A, given R and B.
2) Find B, given R and A.
3) Find R, given A and B.

In the statement "$R\%$ of B is A,"
R is the *rate* of percent and is followed by the "%" symbol.
B is the *base* unit and follows the word "of."
A is the *amount* that is compared to B.

We show two methods for solving percent problems. It is recommended that you look over both methods, pick your favorite, and use it for doing percent problems. **THE HOW AND WHY OF IT**

Method One: Formula	*Method Two: Proportion*
The word "of" in the statement above and in other places in mathematics indicates multiplication. The word "is" describes the relationship "is equal to" or "=". Thus we may write $R\%$ of B is A in the form:	Since $R\%$ is a comparison of A to B and we have seen that this comparison can be thought of as a ratio, the following proportion can be formed:
$$R\% \cdot B = A.$$	$$\frac{R}{100} = \frac{A}{B}$$
When any two of R, B, and A are known, we can find the other value. 25% of 60 is what number? $25\% \cdot 60 = A$ Since $25\% = .25$,	Now if any one of A, B, or R is missing, it can be found by solving the proportion. 25% of 60 is what number? Here $R = 25$, $B = 60$, and $A = ?$ So $$\frac{25}{100} = \frac{A}{60}$$

$$.25 \, (60) = A$$
$$15.00 = A$$

So 25% of 60 is 15.

$$(25) \, (60) = 100 \, (A)$$
$$1500 = 100 \, (A)$$
$$15 = A$$

So 25% of 60 is 15.

EXAMPLES

a)   135% of_____?_____is 54.

Method One	*Method Two*
$(135\%) \, (B) = 54$	$R = 135, B = ?, A = 54$
$(1.35) \, (B) = 54$	$\dfrac{135}{100} = \dfrac{54}{B}$
$B = 40$	$(135) \, (B) = 5400$
So 135% of 40 is 54.	$B = 40$
	So 135% of 40 is 54.

b)   50 is_____?_____% of 180. (To the nearest tenth of one percent.)

Method One	*Method Two*
$50 = (R\%) \, (180)$	$A = 50, R = ?, B = 180$
$50 = (R) \, (.01) \, (180)$	$\dfrac{R}{100} = \dfrac{50}{180}$
$50 = (R) \, (1.80)$	$(180) \, (R) = 5000$
$27.777 \approx R$	$R \approx 27.777$
So 50 is 27.8% of 180 to the nearest tenth of one percent.	So 50 is 27.8% of 180 to the nearest tenth of one percent.

c) $27\frac{2}{3}\%$ of 60 is ___?___.

Method One	*Method Two*
$\left(27\frac{2}{3}\%\right)(60) = A$	$\dfrac{27\frac{2}{3}}{100} = \dfrac{A}{60}$
$\left(27\frac{2}{3}\right)\left(\dfrac{1}{100}\right)(60) = A$	$1660 = (100)(A)$
	$16.6 = A$
$\dfrac{83}{5} = A$	
$16\frac{3}{5} = A$	

Since $16\frac{3}{5} = 16.6$, $27\frac{2}{3}\%$ of 60 is $16\frac{3}{5}$ or 16.6.

EXERCISES

1) 80% of 45 is___?___.

2) 64 is ___?___% of 80.

3) ___?___% of 65 is 13.

4) 74% of 38 is___?___.

5) 9.3% of 60 is___?___.

6) $\frac{3}{4}$% of _____?_____ is 21.

7) _____?_____% of 56 is 14.

8) 6 is _____?_____% of 4.

9) 175% of 30 is _____?_____ .

10) 1 is _____?_____% of 1000.

11) $36\frac{3}{4}$% of 28 is_____?_____ .

12) 19% of_____?_____ is 19.

13) 16 is_____?_____% of 16.

14) 3.2% of .7 is_____?_____.

15) _____?_____ is $\frac{1}{2}$% of .5.

16) 28% of_____?_____ is 36. (To the nearest tenth.)

17) 47 is_____?_____% of 30. (To the nearest tenth of one percent.)

18) _____?_____ is 31.6% of 57.8. (To the nearest tenth.)

19) 82.5 is_____?_____% of 37.6. (To the nearest whole number percent.)

20) $5\frac{1}{3}$% of $6\frac{1}{2}$ is_____?_____. (As a fraction.)

21) 219 is 12% of_____?_____.

22) _____?_____ is 53% of $15\frac{2}{3}$. (As a two-place decimal with fraction.)

23) 1.25% of 1250 is ____?____ .

24) _____?_____ % of 82 is 105.5. (To the nearest tenth of one percent.)

25) $5\frac{1}{2}$% of_____?_____is 34.5. (To the nearest hundredth.)

26) In the application at the beginning of this section, what is the actual load?

8.12   WORD PROBLEMS (Percent)

APPLICATION   During a six-year period, the cost of maintaining a diesel engine averages $2650 and the cost of maintaining a gasoline engine averages $4600. What is the percent of saving of the diesel compared to the gasoline engine? (To the nearest whole percent.)

OBJECTIVE   100) Solve word problems involving percent.

No new vocabulary.

When a word problem is translated into the simpler word form,

   "What percent of what is what?"

the answers can be found by the procedures of the last section. Either
Method One or Method Two can be used. The two methods are used
alternately in the examples.

EXAMPLES

a)   Rod bought a car with a 12% annual loan. If the interest payment was $54 per
     year, how much was his loan?

     The $54 interest is 12% of the loan, so we can translate the problem into the
     simpler word form is

$$12\% \text{ of what is } 54?$$

     Using the formula $R\% \cdot B = A$, we write the equation

$$12\% \, (B) = 54$$

$$.12\,B = 54$$

$$B = 450$$

so Rod's loan was for $450.

b)   The population of Century County is now 130% of its population ten years ago.
     The population ten years ago was 117,000. What is the present population?

     The simpler word form is

$$130\% \text{ of } 117{,}000 \text{ is what number?}$$

and, using a proportion,

$$\frac{130}{100} = \frac{A}{117{,}000}$$

$$15{,}210{,}000 = (100)\,(A)$$

$$152{,}100 = A$$

The population is now 152,100.

c)   A car has depreciated to 65% of its original cost. If the value of the car is
     presently $1469, what did it cost originally? (Method One.)

$$65\% \text{ of what is } 1469? \text{ (simpler word form)}$$

$$.65B = 1469 \qquad \text{(equation)}$$

$$B = 2260$$

The original cost was $2260.

d) The Goliath Bakery has 500 loaves of day-old bread they want to sell. If the price was originally 52¢ a loaf and they sell it for 36¢ a loaf, what percent discount, based on the original price, should they advertise? (Method Two.)

The discount is 16 cents so:

What percent of 52 is 16? (simpler word form)

$$\frac{R}{100} = \frac{16}{52} \text{ (equation)}$$

$$(52)(R) = 1600$$

$$R \approx 30.8 \text{ (about 30.8\%)}$$

The bakery will probably advertise "over 30% off."

e) In a poll taken among a group of students, 2 said they walked to their school, 7 said they rode the bus, 10 drove in car pools, and 3 drove their own cars. What percent of the group rode the bus? (Method One.)

There are 22 students in the group, so the problem can be translated to

What percent of 22 is 7? (simpler word form)

$$R(.01)(22) = 7 \text{ (equation)}$$

$$.22R = 7$$

$$R = 31\frac{9}{11} \approx 32$$

$31\frac{9}{11}\%$ or approximately 32% of the students take the bus.

EXERCISES

1) In the application at the beginning of this section, what is the percent of savings for the diesel engine?

2) Dan bought a used motorcycle for $955. He made a down payment of 18%. How much cash did he pay as a down payment?

3) Adams High School's basketball team finished the season with a record of 15 wins and 9 losses. What percent of the games played were won?

4) Carol pays $225, or 32% of her monthly income, for rent. To the nearest dollar, what is her monthly income?

5) Vera's house is valued at $38,500 and rents for $3,960 per year. What percent of the value of the house is the annual income from rent? (To the nearest tenth of a percent.)

6) A furniture store has a sale on sofas. Every sofa is marked 20% off. What is the sale price of a sofa that was priced at $249.95?

7) The population of Port City increased 15% since the last census. If the former population was 124,000 what is the present population?

8) For customers who use their charge card, a bank charges $1\frac{1}{4}\%$ finance charge on monthly accounts that have a balance of $400 or less. Merle's finance charge for August was $2.80. What was the amount of her account for that month?

9) Eddie and his family went to a restaurant for dinner. The dinner check was $13.20. He left the waiter a tip of $2. What percent of the check was the tip? (To the nearest whole number percent.)

10) Floyd earns a monthly salary of $805. He spends $130 a month at the supermarket. What percent of the salary is spent at the supermarket? (To the nearest whole number percent.)

11) In a certain state there is a 20% gasoline tax. The federal tax is 4¢ per gallon. If gasoline is 30¢ per gallon before taxes, what is the total price per gallon including tax?

12) In preparing a mixture of concrete Jerry uses 300 pounds of gravel, 100 pounds of cement, and 200 pounds of sand. What percent of the mixture is gravel?

13) Last year Jack had 14% of his salary withheld for taxes. If the total amount withheld was $1254.40, what is Jack's yearly salary?

14) George has two savings accounts. One of them pays $4\frac{1}{2}$% interest, and the other pays $5\frac{1}{4}$% interest. If he has $340 in the first account and $621.30 in the second, how much interest did his savings earn last year? (To the nearest cent.)

15) The cost of dairy items increased an average of 2.3% during the month of February. If the price of eggs on February 1 was 68¢ per dozen, what was the price on March 1? (Assume that the rate of increase for eggs was close to the average increase.)

16) Gene bought 125 shares of IWW Inc. for $21.75 a share and sold them for $29.33 a share. If the brokerage fee on both transactions together totaled $39.50, what percent profit did he make on his investment? (To the nearest whole percent.)

17) St. Joseph's Hospital has eight three-bed wards, twenty four-bed wards, twelve two-bed wards, and ten private rooms. What percent of the capacity of St. Joseph's Hospital is in private rooms? (To the nearest tenth of a percent.)

18) The town of Verboort has a population of 15,560, which is 45% male. Of the men, 32% are of age 40 or older. How many men are there in Verboort who are younger than 40?

19) A discount house advertises that they sell all appliances at cost plus 10%. If Kathy buys a TV set for $150, what is the profit for the store?

20) It is claimed that in 15,000 hours, or six years, a gasoline engine will be down 32 days for rountine maintenance while a diesel engine will be down only 13 days. What percent less time is the diesel engine down compared to the gasoline engine? (To the nearest whole percent.)

1. (Obj. 86) Is the following proportion true or false?

$$\frac{3 \text{ feet}}{2 \text{ yards}} = \frac{12 \text{ inches}}{2 \text{ feet}}$$

_____

2. (Obj. 92) Write $.13\frac{1}{8}$ as a percent.

_____

3. (Obj. 93) Write 119% as a decimal.

_____

4. (Obj. 95) Change 145% to a mixed number.

_____

5. (Obj. 99) 63% of what number is 75.6?

_____

6. (Obj. 99) What percent of 64 is 72?

_____

7. (Obj. 94) Change $\frac{7}{8}$ to a percent.

_____

8. (Obj. 93) Write 1.8% as a decimal.

_____

9. (Obj. 99) What number is 11.6% of 210?

_____

10. (Obj. 87) Solve the following proportion.

$$\frac{8}{27} = \frac{y}{81}$$

_____

11. (Obj. 86) Is the following proportion true or false?

$$\frac{15}{36} = \frac{20}{52}$$

_____

12. (Obj. 94) Change $1\frac{3}{11}$ to a percent. (Give answer to the nearest tenth of a percent.)

_____

13. (Obj. 85) Write a ratio to compare 8 hours to 2 days and reduce. (Compare in hours)

_____

14. (Obj. 95) Change $26\frac{1}{3}\%$ to a fraction.

_____

15. (Obj. 87) Solve the following proportion.

$$\frac{\frac{1}{3}}{2} = \frac{2\frac{3}{4}}{x}$$

_____

16. (Obj. 92) Write 1.6 as a percent. _____

17. (Obj. 88) On the assembly line at the local electronics firm John can make 45 welds in 3 hours. At this rate, how many welds will he make in an 8-hour day? _____

18. (Obj. 88) If Kathy is paid $24.85 for 7 hours of work, how much would she expect to earn for 16 hours work? _____

19. (Obj. 89) The weight of a piece of metal varies directly as the volume. If a piece of the metal containing $2\frac{1}{3}$ cubic feet weighs 310 lb, find the weight of a piece containing 4 cubic feet? _____

20. (Obj. 90) The time it takes to travel a certain distance varies inversely with the speed. If it takes 4 hours to cover the distance at 45 mph, how long will it take at 60 mph? _____

21. (Obj.100) If 44 out of every 50 people are right-handed, what percent of the population is right-handed? _____

22. (Obj.100) The local nurseryman sells an average of 112 dozen flowering plants a day during the spring planting season. If 38% of what he sells are petunias, how many dozen petunias does he sell in a day? (Nearest dozen.) _____

23. (Obj.100) Carol's baby weighed $6\frac{3}{4}$ lb when he was born. On his first birthday he weighed 22 lb. What was the percent of increase during the year to the nearest tenth of a percent? _____

24. (Obj.100) A discount store advertises 30% off the suggested retail price. What does the store sell an item for that has a suggested price of $38.50? _____

25. (Obj. 88) In preparing a mixture of cement, Charlie uses 300 lb of gravel and 100 lb of cement. How many pounds of cement should he use for a mixture containing 720 lb of gravel? _____

ROOTS AND EXPONENTS

9.1 SQUARE ROOTS AND CUBE ROOTS

What is the length of the side of a square whose area is 64 cm²? (The formula for the length of the side of a square is $S = \sqrt{A}$.)

APPLICATION

101) Find the square roots of perfect squares.
102) Find the approximate positive square root of a positive real number.
103) Find the cube root of perfect cubes.

OBJECTIVES

The words *perfect square* or *perfect cube* used in this section refer to whole numbers and fractions that are the squares or cubes of other whole numbers or fractions. $16 = 4^2$ and $\frac{1}{9} = \left(\frac{1}{3}\right)^2$, so 16 and $\frac{1}{9}$ are perfect *squares*. $7, 8, \frac{1}{5}$, and $\frac{4}{11}$ are not perfect squares.

VOCABULARY

The *square roots* of a given number are the positive number and the negative number that when squared (used as a factor two times) yield that given number as a product. The symbol "$\sqrt{a}$" means the *positive square root of a*, and the symbol "$-\sqrt{a}$" means the *negative square root of a*.

The *cube root* of a given number is the number (positive or negative) that when cubed (used as a factor three times) yields that given number as a product. "$\sqrt[3]{a}$" means the *cube root of a.*

The symbols $\sqrt{}$ and $\sqrt[3]{}$ are called *radicals*.

The square roots of non-negative numbers can be located on the number line. All numbers on the number line are called *real numbers*. ($\sqrt{9}, \sqrt{14}, \sqrt{3.14}$, and $\sqrt{\frac{7}{8}}$ are real numbers.) Square roots of negative numbers cannot be located on the number line and are not real numbers. ($\sqrt{-4}$ is not a real number.)

THE HOW AND
WHY OF IT

Some whole numbers or fractions are squares of other whole numbers or fractions. For example, the whole number 4 is the square of 2 or −2 since 2 · 2 = 4 and (−2)(−2) = 4. Therefore, we can say

$$\sqrt{4} = 2 \quad \text{The positive square root of 4 is 2.}$$
$$-\sqrt{4} = -2 \quad \text{The negative square root of 4 is } -2$$
$$\sqrt{9} = 3 \quad \text{The positive square root of 9 is 3.}$$
$$-\sqrt{9} = -3 \quad \text{The negative square root of 9 is } -3.$$
$$\sqrt{\frac{4}{9}} = \frac{2}{3} \quad \text{The positive square root of } \frac{4}{9} \text{ is } \frac{2}{3}.$$
$$\sqrt[3]{8} = 2 \quad \text{The cube root of 8 is 2.}$$

Some numbers (such as 2, 3, and 5) are not perfect squares, but their square roots can be approximated. Also, some squares of whole numbers are large and it is difficult to determine the square root by observation. These roots and approximations can be found by using a calculator. It will be assumed from this point on that the reader has access to a calculator. The following roots were found by calculator.

$$\sqrt{2} \approx 1.41 \qquad \sqrt{35} \approx 5.92 \qquad \sqrt{571536} = 756$$

EXAMPLES

a) $\sqrt{16} = 4$ since $4^2 = 16$

b) $\sqrt{25} = 5$ since $5^2 = 25$

c) $-\sqrt{64} = -8$ since $(-8)^2 = 64$

d) $\sqrt{\frac{9}{25}} = \frac{3}{5}$ since $\left(\frac{3}{5}\right)^2 = \frac{9}{25}$

e) $\sqrt[3]{27} = 3$ since $3^3 = 27$

f) $\sqrt[3]{64} = 4$ since $4^3 = 64$

g) $\sqrt[3]{\frac{8}{27}} = \frac{2}{3}$ since $\left(\frac{2}{3}\right)^3 = \frac{8}{27}$

h) $\sqrt{88} \approx 9.381$ (approximation by calculator, rounded to three decimal places)

i) $\sqrt{412} \approx 20.30$ (approximation by calculator, rounded to two decimal places)

j) $\sqrt{3025} = 55$ (root found by calculator)

Compute the following roots.

1) $\sqrt{1}$

2) $-\sqrt{1}$

3) $\sqrt[3]{-1}$

4) $\sqrt{9}$

5) $\sqrt[3]{27}$

6) $\sqrt{81}$

7) $\sqrt{64}$

8) $-\sqrt{25}$

9) $\sqrt{\dfrac{64}{81}}$

10) $-\sqrt[3]{\dfrac{64}{125}}$

11) $\sqrt[3]{-8}$

12) $-\sqrt{-27}$

13) $\sqrt{196}$

14) $\sqrt{\dfrac{144}{169}}$

15) $\sqrt[3]{\dfrac{1}{216}}$

Compute the following to two decimal places of accuracy when the answer is not a whole number. Use a calculator.

16) $\sqrt{8}$

17) $\sqrt{196}$

18) $\sqrt{961}$

19) $\sqrt{24}$

20) $\sqrt{1215}$

21) $\sqrt{892}$

22) $\sqrt{596}$

23) $\sqrt{10.6}$

24) $\sqrt{3.48}$

25) $\sqrt{.041}$

26) In the application at the beginning of this section, what is the length of the side of the square?

27) Evaluate the formula $t = 2\pi\sqrt{\dfrac{\ell}{g}}$ for t if $\ell = 36$, $g = 49$, and $\pi \approx \dfrac{22}{7}$.

28) A formula for computing the area of a triangle is
$A = \sqrt{s(s-a)(s-b)(s-c)}$, where the sides of the triangle are a, b, and c, and $s = \frac{1}{2}(a + b + c)$. What is the area of a triangle whose sides are 6′, 8′, and 10′?

29) The lens setting of a camera varies inversely as the square root of the shutter speed. If the speed is $\frac{1}{100}$ sec for a setting of 20, what should the lens opening be for a speed of $\frac{1}{25}$ sec?

30) The number of vibrations of a pendulum varies inversely as the square root of the length. If the number of vibrations is 28 when the length is 9, what will be the number of vibrations when the length is 16?

31) To meet the city code the attic of a new house needs a vent with a minimum area of 706 square inches. What is the radius of a circular vent that will meet the code requirement? The radius of a circle in terms of the area is given by $r = \sqrt{.32A}$. Compute the radius to one decimal place.

9.2   THE PYTHAGOREAN THEOREM

APPLICATION   What is the length of a rafter that has a rise of 6′ and a run of 12′?

OBJECTIVE   104) Solve right triangles using the Pythagorean theorem.

VOCABULARY   The *Pythagorean theorem* (theorem of Pythagoras) describes a relationship between the sides of a right triangle.

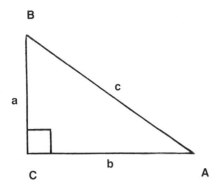

In the above right triangle the legs are a and b and the hypotenuse is c. It is always true for a right triangle that $a^2 + b^2 = c^2$. That is, the square of the hypotenuse is equal to the sum of the squares of the legs.

It is also true that if the square of one side is equal to the sum of the squares of the other two sides, the triangle is a right triangle.

Using the laws for solving equations, we can write

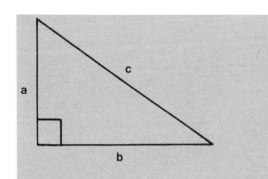

$$a^2 + b^2 = c^2$$
or
$$b^2 = c^2 - a^2$$
or
$$a^2 = c^2 - b^2$$

If any two sides of a right triangle are known, the third can be found by substituting the known values into one of the preceding formulas.

Consider a right triangle with one leg 6″ and the hypotenuse 10″. To determine leg b we use the formula $b^2 = c^2 - a^2$ and we have

$$b^2 = 10^2 - 6^2$$

$$b^2 = 100 - 36$$

$$b^2 = 64$$

$$b = \sqrt{64}$$

$$b = 8$$

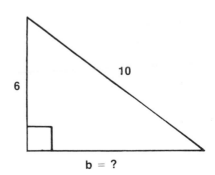

EXAMPLES

a) Find the length of the leg of a right triangle whose second leg is 12 cm and hypotenuse is 15 cm.

$$a^2 = c^2 - b^2$$

$$a^2 = 15^2 - 12^2$$

$$a^2 = 225 - 144$$

$$a^2 = 81$$

$$a = 9$$

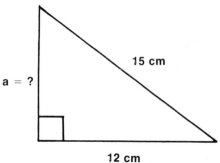

a = ?

15 cm

12 cm

b) Given a right triangle whose legs are 5′ and 6′, respectively, what is the length of the hypotenuse?

$$a^2 + b^2 = c^2$$

$$5^2 + 6^2 = c^2$$

$$25 + 36 = c^2$$

$$61 = c^2$$

$$\sqrt{61} = c \text{ (exact value)}$$

$$7.81 \approx c \text{ (approximate value)}$$

c = ?

5′

6′

c) Is the triangle whose sides are 12′, 18′, and 24′ a right triangle?

Does $12^2 + 18^2 = 24^2$?

$$12^2 + 18^2 \overset{?}{=} 24^2$$

$$144 + 324 \overset{?}{=} 576$$

$$468 \neq 576$$

Therefore the triangle is not a right triangle.

EXERCISES

Find the missing side of the following right triangles. Round decimal answers to the nearest hundredth.

1) $a = 3$, $b = 4$, $c = ?$ 2) $a = 6$, $b = ?$, $c = 10$

3) $a = ?, b = 12, c = 13$

4) $a = 8, b = 15, c = ?$

5) $a = 9, b = 12, c = ?$

6) $a = 12, b = ?, c = 13$

7) $a = ?, b = 8, c = 17$

8) $a = 9, b = 7, c = ?$

9) $a = 27, b = ?, c = 45$

10) $a = 1, b = 2, c = ?$

11) $a = ?, b = 10, c = 15$

Are the following right triangles?

12) $a = 16, b = 30, c = 34$

13) $a = 6, b = 8, c = 9$

14) In the application at the beginning of this section, what is the length of the rafter? (To the nearest tenth of a foot.)

15) What is the length of cable needed to replace a brace on a 50′ power pole that is attached to a ground level anchor that is 35′ from the base of the pole? (To the nearest tenth of a foot.)

16) What is the rise of a rafter that is 20 feet in length and has a run of 16 feet?

17) A plane is flying south at a speed of 200 miles per hour. A wind is blowing from the west at a rate of 50 miles per hour. To the nearest tenth of a mile, how many miles does the plane actually fly in a southeasterly direction? (The following figure illustrates the problem.)

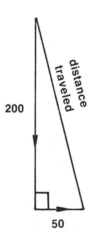

18) A baseball "diamond" is actually a square which is 90 feet on each side (between the bases). To the nearest tenth of a foot, what is the distance the catcher must throw when attempting to put out a runner who is stealing second base? (The following figure illustrates the problem.)

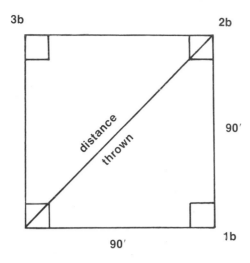

3b   2b

distance thrown

90′

90′   1b

9.3   MULTIPLYING RADICAL EXPRESSIONS (SQUARE ROOTS)

The height of an equilateral triangle is given by the formula $h = \frac{s}{2}\sqrt{3}$, where s is the length of a side. Find the height when $s = \sqrt{6}$.

APPLICATION

105) Multiply two or more radical expressions (square roots).

OBJECTIVE

No new vocabulary.

VOCABULARY

Two square roots can be multiplied using the procedures for multiplying whole numbers and fractions together with the following rule.

THE HOW AND WHY OF IT

The product of two square root expressions is the square root of the product of the two numbers. In other words, for all positive numbers a and b,

$$\sqrt{a} \cdot \sqrt{b} = \sqrt{a \cdot b}$$

EXAMPLES

a) $\sqrt{5} \cdot \sqrt{6} = \sqrt{5 \cdot 6} = \sqrt{30}$

b) $\sqrt{2} \cdot \sqrt{8} = \sqrt{2 \cdot 8} = \sqrt{16} = 4$

c) $\sqrt{3} \cdot \sqrt{5} \cdot \sqrt{2} = \sqrt{30}$

d) $\sqrt{3}\,(2 - \sqrt{2}) = 2\sqrt{3} - \sqrt{2} \cdot \sqrt{3} = 2\sqrt{3} - \sqrt{6}$

e) $(3 + \sqrt{2})\,(3 - \sqrt{2}) = 3^2 - (\sqrt{2})^2 = 9 - 2 = 7$

EXERCISES   Multiply and simplify. Assume all variables represent positive numbers.

1) $\sqrt{2} \cdot \sqrt{3}$

2) $\sqrt{3} \cdot \sqrt{5}$

3) $\sqrt{a} \cdot \sqrt{b}$

4) $\sqrt{2} \cdot \sqrt{5}$

5) $\sqrt{3} \cdot \sqrt{7}$

6) $\sqrt{2}\,(\sqrt{3} + \sqrt{5}\,)$

7) $\sqrt{2} \cdot \sqrt{7}$

8) $\sqrt{5} \cdot \sqrt{7}$

9) $\sqrt{3}\,(\sqrt{7} + \sqrt{10}\,)$

10) $\sqrt{10} \cdot \sqrt{7}$

11) $\sqrt{37} \cdot \sqrt{2}$

12) $\sqrt{5} \cdot \sqrt{20}$

13) $\sqrt{7} \cdot \sqrt{14} \cdot \sqrt{2}$

14) $\sqrt{3} \cdot \sqrt{5} \cdot \sqrt{7}$

15) $\sqrt{x} \cdot \sqrt{y}$

16) $\sqrt{a} \cdot \sqrt{b} \cdot \sqrt{c}$

17) $\sqrt{a} \cdot \sqrt{ab} \cdot \sqrt{b}$

18) $\sqrt{5} \cdot \sqrt{3}$

19) $\sqrt{45} \cdot \sqrt{5}$

20) $\sqrt{2}\,(3 + \sqrt{3}\,)$

21) $\sqrt{xy} \cdot \sqrt{wz}$

22) $\sqrt{6}\,(\sqrt{6} - \sqrt{5}\,)$

23) $\sqrt{6}\,(3 + \sqrt{5}\,)$

24) $2\,(\sqrt{3} - \sqrt{7}\,)$

25) $\sqrt{15}(\sqrt{2}+\sqrt{7})$

26) $(4-\sqrt{2})(4+\sqrt{2})$

27) $(\sqrt{5}-3)(\sqrt{5}+3)$

28) $(\sqrt{5}-\sqrt{3})(\sqrt{5}+\sqrt{3})$

29) $\sqrt{9}(\sqrt{4}+\sqrt{25}-\sqrt{49})$

30) In the application at the beginning of this section, what is the height? (To nearest tenth.)

9.4 SIMPLIFYING RADICAL EXPRESSIONS

A city lot has an area of 5600 square feet. If the lot is a square, what is the length of one side? $S=\sqrt{A}$, where S is the length of a side and A is the area. Express S in terms of a simplified radical.   APPLICATION

106) Simplify a radical expression.   OBJECTIVE

No new vocabulary.   VOCABULARY

A square root (or cube root) can be simplified when the number under the radical can be factored using a perfect square (or perfect cube) as one of the factors.   THE HOW AND WHY OF IT

Consider $\sqrt{24}$. $\sqrt{24}$ can be expressed as $\sqrt{4\cdot6}$, where 4 is the perfect square. ($\sqrt{2\cdot12}$ and $\sqrt{3\cdot8}$ are not used since none of these numbers are perfect squares.) So,

$$\sqrt{24}=\sqrt{4}\cdot\sqrt{6}\text{ (product of radicals)}$$
$$\text{or }\sqrt{24}=2\sqrt{6}$$

A radical expression is in simplified form when the two steps below have been completed.
1) Write the number under the radical as the product of two factors. One of the factors is the largest possible perfect square (or perfect cube).
2) Replace the factor involving the perfect square (or perfect cube) with the square (or cube) root.

We use $\sqrt{32} = \sqrt{16} \cdot \sqrt{2}$ not $\sqrt{4} \cdot \sqrt{8}$ because 16 is a larger perfect square than 4. So,

$$\sqrt{32} = \sqrt{16} \cdot \sqrt{2}$$
$$= 4\sqrt{2}$$

EXAMPLES

a) $\sqrt{8} = \sqrt{4} \cdot \sqrt{2} = 2\sqrt{2}$

b) $\sqrt{x^3} = \sqrt{x^2} \cdot \sqrt{x} = x\sqrt{x}$, where x represents a positive number

c) $\sqrt[3]{54} = \sqrt[3]{27} \cdot \sqrt[3]{2} = 3\sqrt[3]{2}$

d) $\sqrt{128x^5} = \sqrt{64x^4} \cdot \sqrt{2x} = 8x^2\sqrt{2x}$, where x represents a positive number

EXERCISES

Simplify the following. Assume all variables represent positive numbers.

1) $\sqrt{8}$ 2) $\sqrt{12}$ 3) $\sqrt[3]{16}$

4) $\sqrt{a^2}$ 5) $\sqrt{20}$ 6) $\sqrt[3]{24}$

7) $\sqrt{x^2 y}$ 8) $\sqrt{28}$ 9) $\sqrt{32}$

10) $\sqrt[3]{32}$ 11) $\sqrt{27}$ 12) $\sqrt[3]{40}$

13) $\sqrt[3]{56}$ 14) $\sqrt{12x^3}$ 15) $\sqrt{25a^4}$

16) $\sqrt{18a^3b^2}$ 17) $\sqrt[3]{x^5y^8z}$ 18) $\sqrt{250}$

19) $\sqrt{128}$ 20) $\sqrt[3]{128}$ 21) $\sqrt{12x^3y^5}$

22) $\sqrt{16a^5b^2c^9}$ 23) $\sqrt{75x}$ 24) $\sqrt{288x}$

25) In the application at the beginning of this section, what is the length of a side?

9.5 DIVIDING RADICAL EXPRESSIONS

The grandfather clock in the Barker residence has a pendulum length   APPLICATION
of 30 inches. The time it takes for one cycle of the pendulum is given by

$$T = 2\pi \frac{\sqrt{L}}{\sqrt{32}}$$

where L is the length of the pendulum. Simplify the right side of the formula by rationalizing the denominator and reducing.

107) Divide two radical expressions.   OBJECTIVE

Rewriting a radical expression containing a radical in the denom-   VOCABULARY
inator so that the denominator is a whole number is called *rationalizing*
the denominator. $\sqrt{\frac{3}{2}} = \frac{\sqrt{6}}{2}$

THE HOW AND
WHY OF IT
The quotient of two radicals (square root) is equal to the square root of the quotient of the two numbers.

$$\frac{\sqrt{a}}{\sqrt{b}} = \sqrt{\frac{a}{b}}$$

This quotient can be approximated by dividing the numbers and approximating the square root on a calculator.

$$\frac{\sqrt{3}}{\sqrt{2}} = \sqrt{\frac{3}{2}} = \sqrt{1.5} \approx 1.225$$

The common way of stating the exact value of the quotient is to write an indicated division with a whole number for the denominator. This is called rationalizing the denominator.

> To rationalize the denominator,
> 1) Write the quotient as a fraction, if necessary.
> 2) Multiply by 1 ($a \cdot 1 = a$). The name for 1 that is chosen will eliminate the radical in the denominator.
> 3) Simplify.

For instance, to divide $\sqrt{6}$ by $\sqrt{5}$ or $\frac{\sqrt{6}}{\sqrt{5}}$, we multiply by $\frac{\sqrt{5}}{\sqrt{5}}$ (name for 1) so the denominator will be $\sqrt{25}$ or 5.

$$\sqrt{6} \div \sqrt{5} = \frac{\sqrt{6}}{\sqrt{5}} = \frac{\sqrt{6}}{\sqrt{5}} \cdot \frac{\sqrt{5}}{\sqrt{5}} = \frac{\sqrt{30}}{5}$$

EXAMPLES

a) $\sqrt{\frac{5}{3}} = \frac{\sqrt{5}}{\sqrt{3}} = \frac{\sqrt{5}}{\sqrt{3}} \cdot \frac{\sqrt{3}}{\sqrt{3}} = \frac{\sqrt{15}}{3}$

b) $\sqrt{15} \div \sqrt{18} = \frac{\sqrt{15}}{\sqrt{18}} = \frac{\sqrt{15}}{\sqrt{18}} \cdot \frac{\sqrt{2}}{\sqrt{2}} = \frac{\sqrt{30}}{\sqrt{36}} = \frac{\sqrt{30}}{6}$

c) $4\sqrt{3} \div 2\sqrt{8} = \frac{4\sqrt{3}}{2\sqrt{8}} = \frac{4\sqrt{3}}{2\sqrt{8}} \cdot \frac{\sqrt{2}}{\sqrt{2}} = \frac{4\sqrt{6}}{2\sqrt{16}}$

$$= \frac{4\sqrt{6}}{2 \cdot 4} = \frac{4\sqrt{6}}{8} = \frac{\sqrt{6}}{2}$$

d) $\sqrt{\frac{y}{x}} = \frac{\sqrt{y}}{\sqrt{x}} \cdot \frac{\sqrt{x}}{\sqrt{x}} = \frac{\sqrt{xy}}{\sqrt{x^2}} = \frac{\sqrt{xy}}{x}$

x and *y* represent positive numbers

e) $\frac{x}{\sqrt[3]{y^2}} = \frac{x}{\sqrt[3]{y^2}} \cdot \frac{\sqrt[3]{y}}{\sqrt[3]{y}} = \frac{x\sqrt[3]{y}}{\sqrt[3]{y^3}} = \frac{x\sqrt[3]{y}}{y}$

x and *y* represent positive numbers

f) $\dfrac{3}{4+\sqrt{3}} = \dfrac{3}{4+\sqrt{3}} \cdot \dfrac{4-\sqrt{3}}{4-\sqrt{3}} = \dfrac{3(4-\sqrt{3})}{(4+\sqrt{3})(4-\sqrt{3})}$

$$= \dfrac{12-3\sqrt{3}}{16-3}$$

$$= \dfrac{12-3\sqrt{3}}{13}$$

Divide by rationalizing the denominator. Assume all variables represent positive numbers.

EXERCISES

1) $\dfrac{1}{\sqrt{2}}$

2) $\dfrac{1}{\sqrt{3}}$

3) $\sqrt{\dfrac{2}{3}}$

4) $\dfrac{2}{\sqrt{5}}$

5) $\dfrac{1}{\sqrt{6}}$

6) $\sqrt{2} \div \sqrt{3}$

7) $\sqrt{2} \div \sqrt{5}$

8) $\sqrt{2} \div \sqrt{7}$

9) $\dfrac{1}{3+\sqrt{2}}$

10) $\dfrac{\sqrt{4}}{\sqrt{2}}$

11) $\sqrt{7} \div \sqrt{6}$

12) $\sqrt{13} \div \sqrt{5}$

13) $3 \div \sqrt{8}$

14) $\sqrt[3]{2} \div \sqrt[3]{9}$

15) $\sqrt{\dfrac{2}{5}}$

16) $\dfrac{3\sqrt{7}}{4\sqrt{6}}$

17) $\dfrac{2}{6\sqrt{3}}$

18) $\dfrac{\sqrt[3]{5}}{\sqrt[3]{4}}$

19) $\sqrt{\dfrac{1}{x}}$ 20) $\dfrac{3}{\sqrt[3]{x^2}}$ 21) $\sqrt{\dfrac{x}{y}}$

22) $\dfrac{1}{3+\sqrt{2}}$ 23) $\dfrac{2}{4-\sqrt{3}}$ 24) $\dfrac{9}{1-\sqrt{3}}$

25) In the application at the beginning of this section, what is the time for one cycle?

9.6   COMBINING RADICAL EXPRESSIONS

APPLICATION   A plot of ground is surveyed as two equilateral triangles. One has for the length of its sides 25 yd and the other 20 yd. What is the area of the entire plot of ground? The area of an equilateral triangle is given by the formula

$$A = \frac{1}{4}S^2 \sqrt{3} \text{ (S is the length of a side.)}$$

OBJECTIVE   108) Combine radical expressions.

VOCABULARY   No new vocabulary.

THE HOW AND WHY OF IT   Radical expressions are combined using the same laws of algebra that are used to combine algebraic terms.

If two radicals are the same and have the same number under the radical ($\sqrt{7}$ and $\sqrt{7}$, but not $\sqrt{7}$ and $\sqrt[3]{7}$ and not $\sqrt{7}$ and $\sqrt{17}$), we can call them *like radical numbers* (similar to *like terms*).

> To combine like radical numbers, add (or subtract) the numerical coefficients and annex the like radical number. It is sometimes necessary to simplify the radicals in the radical expression before it is possible to combine the expressions.

EXAMPLES
(Assume all variables represent positive numbers.)

a) $\sqrt{2} + \sqrt{2} = (1+1)\sqrt{2} = 2\sqrt{2}$

b) $3\sqrt{5} + 4\sqrt{5} = (3+4)\sqrt{5} = 7\sqrt{5}$

c) $8\sqrt[3]{2} - 4\sqrt[3]{2} = (8-4)\sqrt[3]{2} = 4\sqrt[3]{2}$

d) $\begin{aligned} 4\sqrt{x} + 3\sqrt{y} + 7\sqrt{x} &= 4\sqrt{x} + 7\sqrt{x} + 3\sqrt{y} \\ &= (4+7)\sqrt{x} + 3\sqrt{y} \\ &= 11\sqrt{x} + 3\sqrt{y} \end{aligned}$

e) $\begin{aligned} \sqrt{75} + \sqrt{20} - \sqrt{\frac{1}{3}} &= \sqrt{25} \cdot \sqrt{3} + \sqrt{4} \cdot \sqrt{5} - \frac{\sqrt{1}}{\sqrt{3}} \\ &= 5\sqrt{3} + 2\sqrt{5} - \frac{\sqrt{3}}{3} \\ &= (5 - \frac{1}{3})\sqrt{3} + 2\sqrt{5} \\ &= \frac{14}{3}\sqrt{3} + 2\sqrt{5} \end{aligned}$

f) $\begin{aligned} \sqrt{8x^3} - \sqrt{50x^3} + \sqrt[3]{27x^3} &= \sqrt{4x^2} \cdot \sqrt{2x} - \sqrt{25x^2} \cdot \sqrt{2x} + 3x \\ &= 2x\sqrt{2x} - 5x\sqrt{2x} + 3x \\ &= -3x\sqrt{2x} + 3x \end{aligned}$

g) $\begin{aligned} \sqrt{24} - 6\sqrt{6} + \sqrt{50} \cdot \sqrt{3} &= \sqrt{4} \cdot \sqrt{6} - 6\sqrt{6} + \sqrt{25} \cdot \sqrt{2} \cdot \sqrt{3} \\ &= 2\sqrt{6} - 6\sqrt{6} + 5\sqrt{2} \cdot \sqrt{3} \\ &= 2\sqrt{6} - 6\sqrt{6} + 5\sqrt{6} \\ &= \sqrt{6} \end{aligned}$

Combine the following radical expressions. Assume all variables represent positive numbers.

EXERCISES

1) $\sqrt{2} + \sqrt{2}$

2) $2\sqrt{2} + \sqrt{2}$

3) $\sqrt[3]{2} + \sqrt[3]{2}$

4) $2\sqrt{5} + \sqrt{5}$

5) $\sqrt{8} + \sqrt{2}$

6) $\sqrt{9} + \sqrt{16}$

7) $4\sqrt{7} - 3\sqrt{7}$

8) $2\sqrt{2} + \sqrt{18}$

9) $5\sqrt{2} + \sqrt{2} - 3\sqrt{2}$

10) $\sqrt{6} + 2\sqrt{6}$

11) $\sqrt[3]{5} - 4\sqrt[3]{5}$

12) $3\sqrt{2} - 6\sqrt{2} + 8\sqrt{2}$

13) $6\sqrt{x} + 5\sqrt{x} - 3\sqrt{x}$

14) $\sqrt{8} + 3\sqrt{2}$

15) $3\sqrt{7} - \sqrt{28}$

16) $4\sqrt[3]{2} + 5\sqrt[3]{16} - 6\sqrt[3]{2}$

17) $3a\sqrt{ab} + 7\sqrt{a^3b} - 5a\sqrt[3]{a}$

18) $\sqrt[3]{x^4} - 6x\sqrt[3]{x} + \sqrt[3]{8x^4}$

19) $4\sqrt{2} + 3\sqrt[3]{5} - \sqrt{8} + \sqrt[3]{5}$

20) $\sqrt{72} - \sqrt{75} + \sqrt{98} - \sqrt{27}$

21) $\sqrt{16x} - \sqrt[3]{8x}$

22) $\sqrt{6x} - \sqrt{24x} - \sqrt{54x}$

23) $\sqrt{24} + \sqrt{32} - \sqrt{128}$

24) $x\sqrt{xy} + 3\sqrt{x^3y} + x\sqrt{16xy}$

25) $\sqrt{2} + \sqrt{\dfrac{1}{2}}$

26) $\sqrt{12} + \sqrt{\dfrac{1}{3}}$

27) $\sqrt{\dfrac{2}{3}} + \sqrt{\dfrac{2}{27}}$

28) $6\sqrt{32} + 8\sqrt{50} - \dfrac{3}{\sqrt{2}}$

29) In the application at the beginning of this section, what is the area of the plot of ground?

9.7 FRACTIONAL EXPONENTS

The formula for the radius of a circle in terms of the area is $r \approx \sqrt{.32A}$. Express the right side of the equation using fractional exponents.

APPLICATION

109) Write radical expressions in exponential form.

110) Write expressions written in exponential form as radical expressions.

OBJECTIVES

$(a + b)^3$ is said to be written in *exponential form* since rather than writing it as $(a + b)(a + b)(a + b)$, the three factors of $a + b$ are indicated with the exponent 3.

VOCABULARY

It is possible to write radical expressions in exponential form. If the laws of exponents for squaring a number are accepted, we can consider the following:

THE HOW AND WHY OF IT

$$(a^{1/2})^2 = a^{1/2} \cdot a^{1/2} = a^{1/2 \, + \, 1/2} = a^1 = a \text{ or}$$
$$(a^{1/2})^2 = a^{1/2 \, \cdot \, 2} = a^1 = a$$

We also know that the following is true:

$$(\sqrt{a})^2 = \sqrt{a} \cdot \sqrt{a} = a$$

Since $(a^{1/2})^2$ and $(\sqrt{a})^2$ both yield an answer of a, we make the following definition:

$$\sqrt{a} = a^{1/2}, \text{ where } a \text{ represents a positive number}$$

A similar argument with cube root leads to the following definition:

$$\sqrt[3]{a} = a^{1/3}, \sqrt[3]{a^2} = a^{2/3}$$

Using the laws of exponents adopted earlier we can see that

$$x^{2/3} = x^{2 \cdot 1/3} = (x^2)^{1/3} = \sqrt[3]{x^2} \text{ and}$$
$$\sqrt{a^3} = (a^3)^{1/2} = a^{3 \cdot 1/2} = a^{3/2}$$

EXAMPLES

Write the following in exponential form.

a) $\sqrt{5} = 5^{1/2}$

b) $\sqrt[3]{x} = x^{1/3}$

c) $\sqrt{xy^3} = (xy^3)^{1/2} = x^{1/2} \cdot y^{3 \cdot 1/2} = x^{1/2} y^{3/2}$

Write the following as radical expressions.

d) $7^{1/2} = \sqrt{7}$

e) $9^{2/3} = \sqrt[3]{9^2}$

f) $x^{1/3} y^{1/3} = (xy)^{1/3} = \sqrt[3]{xy}$

EXERCISES

Write the following in exponential form.

1) $\sqrt{2}$

2) $\sqrt[3]{4}$

3) $\sqrt{6}$

4) $\sqrt[3]{a^2}$

Write the following as radical expressions.

5) $a^{1/3}$

6) $9^{1/3}$

7) $a^{1/2}$

8) $5^{1/2}$

9) $a^{1/2} b^{1/2}$

Write the following in exponential form.

10) $\sqrt{7}$

11) $\sqrt[3]{2}$

12) $\sqrt{x}$

13) $\sqrt[3]{y}$

14) $\sqrt{57}$

15) $\sqrt{ab}$

16) $\sqrt[3]{xy}$

17) $\sqrt[3]{x^2}$

Write the following as radical expressions.

18) $6^{1/2}$

19) $12^{1/3}$

20) $x^{1/3}$

21) $y^{1/2}$

22) $(7b)^{1/2}$

23) $xy^{1/3}$

24) $(rs)^{1/3}$

25) $ab^{1/2}$

26) $a^{2/3}$

27) $(xy)^{2/3}$

28) $(xy^2)^{1/3}$

29) In the application at the beginning of this section, write the right-hand member using fractional exponents.

9.8 NEGATIVE EXPONENTS

In photography the time for taking a picture (exposure time) is given   APPLICATION
by the formula

$$t = kd^{-2}$$

where d is the diameter of the camera lens and k is a constant. Write the formula for t without the negative exponent.

111) Write expressions in an equivalent form using negative ex-   OBJECTIVES
ponents.
112) Write an expression written with negative exponents as an expression without negative exponents.

VOCABULARY An expression such as $x^3 + 3x + x^{-1}$ has a *negative exponent.* In this case it is -1. The expression is read "x cubed plus three x plus x to the negative one."

THE HOW AND WHY OF IT

Using the laws of exponents for multiplication, we have the following:

$$x^2 \cdot x^{-2} = x^{2+(-2)} = x^0 = 1$$

$$x^{-3} \cdot x^3 = x^{-3+3} = x^0 = 1$$

We also know from multiplication of rational expressions that the following is true:

$$x^2 \cdot \frac{1}{x^2} = 1 \text{ and } \frac{1}{x^3} \cdot x^3 = 1$$

Therefore, x^{-2} and $\frac{1}{x^2}$ must name the same number, and x^{-3} and $\frac{1}{x^3}$ must name the same number.

In general we can state that

$$x^{-n} = \frac{1}{x^n}, x \neq 0.$$

The negative sign in the exponent indicates "the reciprocal of."

EXAMPLES
(Assume all variables represent positive numbers.)

a) $3^{-1} = \dfrac{1}{3}$

b) $x^{-\frac{1}{2}} = \dfrac{1}{x^{\frac{1}{2}}}$

c) $(a + b)^{-1} = \dfrac{1}{a + b}$

d) $(x + y)^{-4} = \dfrac{1}{(x + y)^4}$

e) $\dfrac{x}{a^4} = xa^{-4}$

f) $\dfrac{1}{x} + \dfrac{1}{y} = x^{-1} + y^{-1}$

g) $\dfrac{1}{(x + y)^{\frac{1}{3}}} = (x + y)^{-\frac{1}{3}}$

h) $\dfrac{1}{(a + 2b + c)^3} = (a + 2b + c)^{-3}$

Write the following without fractions using negative exponents.

1) $\dfrac{1}{a}$

2) $\dfrac{2}{a^2}$

3) $\dfrac{5}{a^3}$

4) $\dfrac{2}{x^3}$

Write the following using positive exponents.

5) 4^{-1}

6) a^{-1}

7) x^{-2}

8) $4a^{-1}$

9) $(4a)^{-1}$

Write the following expressions without fractions using negative exponents. Assume all variables represent positive numbers.

10) $\dfrac{1}{4}$

11) $\dfrac{1}{x}$

12) $\dfrac{7}{x^2}$

13) $\dfrac{1}{x+y}$

14) $\dfrac{1}{a}+\dfrac{1}{b}$

15) $\dfrac{2}{x+y}$

16) $\dfrac{5}{x+y+z}$

17) $\dfrac{x}{y}$

18) $\dfrac{a^{1/2}b}{c^2}$

19) $\dfrac{1}{(a+b)^{1/2}}$

Write the following expressions using positive exponents.

20) x^{-1}

21) a^{-2}

22) $(a)^{-3}$

23) $(x+y)^{-2/3}$

24) $2x^{-2}$

25) $\dfrac{1}{y^{-1}}$

26) $\dfrac{b^{-1/2}}{a^{-1/2}}$

27) $\dfrac{x}{y^{-1}}$

28) $\dfrac{x^{-1}}{(x+y)^{-1}}$

29) $\dfrac{x}{x^{-1}}$

30) In the application at the beginning of this section, write the formula for t without the negative exponent.

31) An equation of a probability curve is $y = e^{-x^2}$. Write the right-hand member with all exponents positive.

9.9 SCIENTIFIC NOTATION

APPLICATION

The distance from the earth to the sun is approximately 93,000,000 miles. Write this number in scientific notation.

OBJECTIVES

113) Write a number in scientific notation.
114) Change scientific notation to place value notation.

VOCABULARY

To write a number in *scientific notation* means to write the number as a product of two numbers. The first number is between one and ten and the second number is a power of ten. The power of ten is written in exponential form, i.e.,

Number = (number between one and ten) × (a power of ten)
.023 = 2.3×10^{-2}

THE HOW AND WHY OF IT

Many numbers used in science and industry are very large or very small. Scientific notation is a means of writing these numbers as a product of two numbers—mathematical shorthand. For example:

The wave length of x-rays range from .000001 cm to .000000001 cm, or 1×10^{-6} cm to 1×10^{-9} cm.

One coulomb is equal to about 6,280,000,000,000,000,000 electrons, or 6.28×10^{18} electrons.

Calculators are limited in the number of digits they display. As a result, the answers are approximated using scientific notation. For example:

$$(92,835)(101,000) = 9.38 \times 10^9$$
(rounded to two decimal places)

> To write a number in scientific notation, write the given number as a number with one digit to the left of the decimal point and multiply by the power of ten that will yield the original number as the product.

If a number is multiplied by a power of ten, e.g., 10^1, 10^2, or 10^3, this has the effect of moving the decimal point to the right the same number of places as the exponent of ten. If a number is multiplied by a negative power of ten, e.g., 10^{-1}, 10^{-2}, or 10^{-3}, this has the effect of moving the decimal point to the left the same number of places as the absolute value of the exponent. This is the same as dividing by powers of ten, since $10^{-1} = \frac{1}{10}$, $10^{-2} = \frac{1}{10^2}$, $10^{-3} = \frac{1}{10^3}$, and so on.

A number between one and ten has only one digit to the left of the decimal when written in place value notation.

Consider 265. This can be written in scientific notation in the following way:

2.65 has one digit to the left of the decimal.

The decimal point has been moved two places to the left; to move the decimal point two places to the right, multiply by 10^2. We now have

$$265 = 2.65 \times 10^2$$

Consider .0123. This can be written in scientific notation in the following way:

1.23 has one digit to the left of the decimal point.

The decimal point has moved two places to the right; to move two places to the left, multiply by 10^{-2}. We now have

$$.0123 = 1.23 \times 10^{-2}$$

> To change from scientific notation to place value notation, perform the indicated multiplication.
>
> $6.27 \times 10^3 = 6.27 \times 1000 = 6270.$
>
> $1.76 \times 10^{-3} = 1.76 \times \frac{1}{10^3} = 1.76 \times \frac{1}{1000} = 1.76 \div 1000 = .00176$

EXAMPLES

Write in scientific notation.

a) 27,600 = 2.7600 × 10⁴ or 2.76 × 10⁴

b) .00265 = 2.65 × 10⁻³

Write in place value notation.

c) 6.74 × 10⁵ = 674,000

d) 5.86 × 10⁻⁶ = .00000586

EXERCISES   Write the following in scientific notation.

1) 211,000   2) 3,250,000   3) .00121

4) 38,940,000   5) .0000682

Write the following in place value notation.

6) 2.1×10^2 7) 3.21×10^{-1} 8) 4.31×10^3

9) 2.14×10^{-2}

Write in scientific notation.

10) 2765   11) 182,000   12) 4,860,000

13) .00276   14) .00000135   15) .0000124

16) .000369

17) .7623

18) 1,235,000,000

19) 387,600,000

Write in place value notation.

20) 6.85×10^3

21) 2.76×10^{-3}

22) 7.962×10^{-7}

23) 3.82×10^5

24) 9.863×10^{-8}

25) In the application at the beginning of this section, write the number in scientific notation.

**CHAPTER 9
TEST**

1. (Obj. 105)   Multiply. $\sqrt{7} \cdot \sqrt{6}$ _____

2. (Obj. 106)   Simplify. $\sqrt{12}$ _____

3. (Obj. 101)   Evaluate. $\sqrt{49}$ _____

4. (Obj. 107)   Rationalize the denominator.

$$\frac{\sqrt{2}}{\sqrt{7}}$$
_____

5. (Obj. 112)   Write $\dfrac{4x^{-2}}{y^{-2}}$ with positive exponents. _____

6. (Obj. 107)   Rationalize the denominator and simplify.

$$\sqrt{16} \div \sqrt{6}$$
_____

7. (Obj. 106)   Simplify. $\sqrt{128x^3 y^2}$ _____

8. (Obj. 107)   Rationalize the denominator.

$$\frac{2}{5 - \sqrt{3}}$$
_____

9. (Obj. 108)   Add. $\sqrt{8} + \sqrt{2}$ _____

10. (Obj. 110)   Write $3x^{1/2} y^{1/2}$ in radical form. _____

11. (Obj. 114)   Write 2.31×10^{-5} in place value notation. _____

12. (Obj. 105)   Multiply. $(\sqrt{6} + \sqrt{3})(\sqrt{5} - \sqrt{7})$ _____

13. (Obj. 109)   Write $\sqrt{ab^3}$ in exponential form. _____

14. (Obj. 103)   Evaluate. $\sqrt[3]{64}$ _____

15. (Obj. 108)   Subtract. $\sqrt[3]{16} - \sqrt[3]{2}$ _____

16. (Obj. 111)   Write $\dfrac{1}{a + b}$ without fractions using negative exponents. _____

17. (Obj. 105)   Multiply. $\sqrt{5}(\sqrt{2} + \sqrt{3})$ _____

18. (Obj. 108)   Add. $7\sqrt{6} + \sqrt{54} + 3\sqrt{24}$ _____

19. (Obj. 106)   Simplify. $\sqrt[3]{8x^4}$ _____

20. (Obj. 113)   Write 6,230,000 in scientific notation.   _____

21. (Obj. 104)   In a right triangle, if side $b = 15$ and side $c = 17$, what is the length of side a?   _____

22. (Obj. 104)   A fresh air duct will have a cross section area of 256.25 square inches. If the duct will be a square, what is the length of a side? (To the nearest tenth of an inch.)   _____

23. (Obj. 104)   What is the length of the hypotenuse of a right triangle whose legs are 6 and 12, respectively? (Leave answer in simplified radical form.)   _____

24. (Obj. 104)   What is the length of the side of a square food crib that has a floor area of 1089 sq ft?   _____

25. (Obj. 104)   What is the length of the diagonal of a rectangle whose length is 15 m and whose width is 8 m?   _____

QUADRATIC EQUATIONS

10.1 QUADRATIC EQUATIONS—SOLVED BY SQUARE ROOTS

A fluid moving quickly through a pipe can actually erode the inner surface of the pipe and reduce its life. Therefore, it is important to determine the velocity through a pipe of a given size for a required volumetric flow rate. The flow rate through the pipe is

APPLICATION

$$F = \frac{\pi}{4} D^2 V *$$

where F is the volumetric flow rate, ft^3/sec
D is the diameter of the pipe, ft
V is the velocity of the fluid, ft/sec
Find the diameter of the pipe when F = 3.5 ft^3/sec and V = 40 ft/sec.

115) Solve a quadratic equation that can be written in the form $x^2 = a$, where a is not negative.

OBJECTIVE

Recall that every positive number has two square roots, one positive and one negative, and that the square of any real number is not negative.

VOCABULARY

The symbol ± indicates a positive or a negative number. So $x = \pm 6$ means $x = +6$ or $x = -6$.

Given a quadratic equation written in the form $x^2 = a$, where a is not negative, the roots of the equation are found by taking the square root of both sides. (The ± is omitted in front of the x since it does not affect the solution.)

THE HOW AND WHY OF IT

So if $x^2 = a$
$x = \pm\sqrt{a}$

_____
*Formula courtesy of Exxon Co.

Consider the equation

$$x^2 - 25 = 0$$
$$x^2 = 25 \qquad \text{(form of } x^2 = a\text{)}$$
$$x = \pm\sqrt{25} \text{ (square root of both sides)}$$
$$x = \pm 5$$

$x^2 = -16$ has no real solution, since the square of any positive number is positive and the square of any negative number is positive.

Therefore, $x^2 = a$, when a is negative, has no real solutions. If $x^2 = 0$, then $x = 0$ is the only solution.

EXAMPLES

a) $x^2 = 16$
$\quad x = \pm\sqrt{16}$
$\quad x = \pm 4$
$\quad$ Check: $(+4)^2 = 16 \qquad (-4)^2 = 16$
$\qquad\qquad\quad 16 = 16 \qquad\quad 16 = 16$

b) $4x^2 = 25$
$\quad x^2 = \dfrac{25}{4}$

$\quad x = \pm\sqrt{\dfrac{25}{4}}$

$\quad x = \pm\dfrac{5}{2}$

$\quad$ Check: $4\left(+\dfrac{5}{2}\right)^2 = 25 \qquad 4\left(-\dfrac{5}{2}\right)^2 = 25$

$\qquad\qquad\quad 4\left(\dfrac{25}{4}\right) = 25 \qquad\quad 4\left(\dfrac{25}{4}\right) = 25$

$\qquad\qquad\qquad\quad 25 = 25 \qquad\qquad\quad 25 = 25$

c) $6 + 2x^2 = x^2 + 9$
$\quad\quad\ x^2 = 3$
$\quad\quad\ x = \pm\sqrt{3}$

d) $3x^2 + 15 = x^2 + 9$
$\quad\quad\ 2x^2 = -6$
$\quad\quad\ x^2 = -3$

No real solution because x^2 cannot be negative.

EXERCISES   Solve.

1) $x^2 = 25$

2) $x^2 - 81 = 0$

3) $121 = x^2$

4) $0 = 1600 - y^2$

5) $x^2 = 17$

6) $x^2 = -36$

7) $x^2 - 10 = 49$

8) $x^2 = 7$

9) $2x^2 - 35 = x^2 - 19$

10) $3 - 2x - x^2 = 5 - 2x - 2x^2$

11) $4(x^2 - 5) = 3(x^2 + 6)$

12) $3x^2 - 5x^2 = 2x^2 - 7x^2$

13) $x^4 = 256$ (Hint: Take the square root twice.)

14) In the application at the beginning of this section, what is the diameter of the pipe? (To the nearest hundredth.)

15) A square plot of land has an area of 2116 sq ft. What is the length of a side?

16) A lawn sprinkler covers a circular area of 1809 sq ft. How far must it be placed from a sidewalk so that the walk will not be sprinkled? (Find to the nearest tenth, use $\pi \approx 3.14$.)

17) For railroad cars equipped with Type E or F couplers and coupled to the base car, the minimum radius curve that can be negotiated is given by:

$$R = \frac{B^2 - D^2 - E^2}{2E} *$$

where

R = minimum radius of curve (ft)
B = one-half of distance over coupling lines (ft)
D = one-half of distance over truck centers (ft)
E = offset at coupling line (ft)

Find B when $E = 1.3$, $D = 23.1$, and $R = 188$. (Find to the nearest tenth.)

*Formula courtesy of FMC Corp.

10.2 QUADRATIC EQUATIONS—SOLVED BY QUADRATIC FORMULA

APPLICATION

An arrow is shot directly upward from a height of 6 feet. If its distance from the ground is given by the formula $S = 6 + 150t - 16t^2$, how long will it take to reach a height of 300 feet? (S in feet, t in seconds.)

OBJECTIVE

116) Find the solutions of a quadratic equation by using the quadratic formula.

VOCABULARY

A quadratic equation is an equation that can be written in the form $ax^2 + bx + c = 0$, $a \neq 0$. The quadratic formula, $x = \dfrac{-b \pm \sqrt{b^2 - 4ac}}{2a}$, gives the roots of the quadratic equation.

THE HOW AND WHY OF IT

To solve a quadratic by using the formula, first write the equation in the form $ax^2 + bx + c = 0$. Then identify the coefficients a, b, and c. Substitute these in the formula and evaluate.

Consider

$x^2 - 3x - 10 = 0$	form $ax^2 + bx + c = 0$
$a = 1, b = -3, c = -10$	identify a, b, and c
$x = \dfrac{-b \pm \sqrt{b^2 - 4ac}}{2a}$	quadratic formula
$x = \dfrac{-(-3) \pm \sqrt{(-3)^2 - 4(1)(-10)}}{2(1)}$	substitute
$x = \dfrac{3 \pm \sqrt{49}}{2}$	evaluate
$x = \dfrac{10}{2}$ or $x = \dfrac{-4}{2}$	
$x = 5$ or -2	solutions

The radical part of the formula, $\sqrt{b^2 - 4ac}$ (called the discriminant), identifies whether the equation has real roots. Recall that the square root of a negative number is not a real number. So if $b^2 - 4ac$ is negative, the equation has no real roots.

EXAMPLES

a) $4x^2 + 7x - 1 = 0$
$ax^2 + bx + c = 0$
$a = 4, b = 7, c = -1$

$$x = \frac{-b \pm \sqrt{b^2 - 4ac}}{2a}$$

$$x = \frac{-(7) \pm \sqrt{(7)^2 - 4(4)(-1)}}{2(4)}$$

$$x = \frac{-7 \pm \sqrt{49 + 16}}{8}$$

$$x = \frac{-7 \pm \sqrt{65}}{8}$$

b) $x^2 = 2x + 11$
$x^2 - 2x - 11 = 0$
$ax^2 + bx + c = 0$
$a = 1, b = -2, c = -11$

$$x = \frac{-b \pm \sqrt{b^2 - 4ac}}{2a}$$

$$x = \frac{-(-2) \pm \sqrt{(-2)^2 - 4(1)(-11)}}{2(1)}$$

$$x = \frac{2 \pm \sqrt{4 + 44}}{2}$$

$$x = \frac{2 \pm \sqrt{48}}{2}$$

$$x = \frac{2 \pm 4\sqrt{3}}{2}$$

$$x = 1 \pm 2\sqrt{3}$$

c) $x^2 - 3x + 5 = 0$
$a = 1, b = -3, c = 5$

$$x = \frac{-(-3) \pm \sqrt{(-3)^2 - 4(1)(5)}}{2(1)}$$

$$x = \frac{3 \pm \sqrt{9 - 20}}{2}$$

$$x = \frac{3 \pm \sqrt{-11}}{2}$$

Since $\sqrt{-11}$ is not real, the equation has no real roots.

EXERCISES   Solve.

1) $x^2 - 3x - 18 = 0$ 2) $x^2 + 6x - 40 = 0$

3) $y^2 + 12y + 35 = 0$ 4) $x^2 + 11x + 18 = 0$

5) $6x^2 - 5x - 6 = 0$ 6) $4x^2 - 27x - 7 = 0$

7) $20x^2 - 23x - 21 = 0$ 8) $2x^2 + 11x - 6 = 0$

9) $12x^2 - 7x + 1 = 0$ 10) $4x^2 - 4x - 5 = 0$

11) $4x^2 - 4x + 5 = 0$ 12) $y^2 - 2y - 16 = 0$

13) $x^2 - 5x + 10 = 0$ 14) $x^2 - 5x - 10 = 0$

15) $y^2 - y - 14 = 0$

16) $x(2x - 4) - 3 = 0$

17) $5x^2 - 3x - 1 = 0$

18) $-x^2 + 6x - 10 = 0$

19) $17x^2 + 2x - 13 = 0$

20) $26x + 45 = 7x^2$

21) In the application at the beginning of this section, how long did it take the arrow to reach the height of 300 ft? (To the nearest hundredth.)

22) The reciprocal of a number is $\dfrac{16}{15}$ less than the number itself. What is the number?

23) A rectangular plot of ground is 12 ft longer than it is wide. What are its dimensions if it has an area of 640 sq ft?

24) A rectangular piece of cardboard is 3 inches longer than it is wide. A 2-inch square is cut out of each corner and the edges turned up to form a container. If the volume of the container is 496 in^3, what are the dimensions of the piece of cardboard? (To the nearest hundredth.)

25) A variable electrical current is given by $i = t^2 - 8t + 15$. If t is in seconds, at what time is the current (i) equal to 16 amperes? (To the nearest hundredth.)

10.3 QUADRATIC EQUATIONS: A REVIEW

APPLICATION   A 4 × 6 photo is mounted in a frame of uniform width. What are the outside dimensions of the frame if the area of the frame and the area of the photo are equal?

OBJECTIVE   117) Solve quadratic equations by factoring, taking the square root of each member, or by using the quadratic formula.

VOCABULARY   No new vocabulary.

THE HOW AND WHY OF IT   Although all quadratic equations can be solved by using the quadratic formula, the other two methods are sometimes easier.

When given a quadratic equation use the easiest of the following:
1.   Check to see if it can be written in the form $x^2 = a$. If so, solve by taking the square root of each member.
2.   Write the quadratic equation in the form $ax^2 + bx + c = 0$ and attempt to factor the left member. If you succeed, solve by setting each factor equal to zero.
3.   Solve by using the quadratic formula.
$$x = \frac{-b \pm \sqrt{b^2 - 4ac}}{2a}$$

EXAMPLES

a)
$$3x^2 - 4x + 17 = 25 - 4x$$
$$3x^2 = 8$$
$$x = \pm\sqrt{\frac{8}{3}} = \pm\frac{2}{3}\sqrt{6}$$

b)
$$\frac{x}{x+1} + \frac{10x}{(x+1)(x+3)} - \frac{15}{x+3} = 0$$
$$x(x+3) + 10x - 15(x+1) = 0$$
$$x^2 + 3x + 10x - 15x - 15 = 0$$
$$x^2 - 2x - 15 = 0$$
$$(x-5)(x+3) = 0$$

(Note that $x \neq -1$ or -3 because division by 0 is not defined.)

$x - 5 = 0$ or $x + 3 = 0$
$x = 5$ or $x = -3$
Since $x \neq -3$ or -1, the solution is $x = 5$.

c)
$$(x-5)(x+4) = x(3x-5)+3$$
$$x^2 - x - 20 = 3x^2 - 5x + 3$$
$$2x^2 - 4x + 23 = 0$$
$$x = \frac{-(-4) \pm \sqrt{(-4)^2 - 4(2)(23)}}{2(2)}$$
$$x = \frac{4 \pm \sqrt{-168}}{4}$$

No real roots.

Solve.

1) $x^2 = 4x + 5$

2) $16 = 5x - 2x^2$

3) $3x^2 - 8x - 60 = 0$

4) $x(x-5) = 5(4-x)$

5) $16x^2 + 4x = 8x^2$

6) $7x = 5 - 6x^2$

7) $3x^2 - x - 9 = 0$

8) $3x^2 + 8 = -12x$

9) $(x - 4)(2x + 3) = x^2 - 4x + 8$

10) $3x(x - 6) + 2 = (2x - 4)(x - 7) + 3$

11) $2(x - 1)^2 = (x + 3)^2 - 4$

12) $(x + 5)(x + 6) = (2x - 3)(x + 7)$

13) $(2x + 6)(3x - 5) + 32 = (2x - 3)(x + 7)$

14) $x(2x - 1) + 3x(1 - 4x) + 8 = 0$

15) $3 + \dfrac{2}{x} - \dfrac{5}{x^2} = 0$

16) $1 - \dfrac{3}{x} - \dfrac{40}{x^2} = 0$

17) $\dfrac{3}{x-2} + 2 = \dfrac{2}{x+5}$

18) $\dfrac{x}{x-9} + \dfrac{3}{x-4} = \dfrac{3}{(x-9)(x-4)}$

19) $\dfrac{6}{x-10} = \dfrac{x+2}{8}$

20) $(x+3)^2 - 39 = (2x-5)^2$

21) $\dfrac{x}{x-7} - \dfrac{2x}{(x-7)(x-5)} + \dfrac{6}{x-5} = 0$

22) $x - \dfrac{5x}{x+2} = \dfrac{10}{x+2}$

23) $\dfrac{x^2}{x-4} = 9 + \dfrac{56}{x-4}$

24) $\dfrac{10}{x+3} = x + 6$

25) In the application at the beginning of this section, what are the dimensions of the frame?

26) A metal worker must cut a rectangle with an area of 180 sq cm out of a flat piece of iron. If the rectangle is to be 3 cm longer than it is wide, what are the desired dimensions?

27) The radius of a circular arch of a certain height (h) and span (b) is given by

$$r = \frac{b^2 + 4h^2}{8h}$$

Find h when $r = 30$ ft and $b = 24$ ft.

28) A sheet metal worker needs to construct a rectangular duct with a cross section area of 340 sq cm. If the width is 3 cm less than the length, what are the needed dimensions?

29) A variable electrical current is given by $i = t^2 - 4t + 16$. If t is in seconds, at what time is the current (i) equal to 61 amperes?

10.4 EQUATIONS INVOLVING RADICALS

APPLICATION   At what height above sea level must a citizens band microwave antenna be placed if it is to transmit a signal 30 miles? Distance traveled D is given by $D = \sqrt{h\,(h + 7800)}$ (h is height). h and D are in miles.

118) Find the solution of an equation that contains the variable within a radical expression.

An *extraneous root* is a root derived from an equation by multiply- ing each number by zero or raising each member to a power. The extraneous root is not a root of the original equation and can be eliminated by checking. For instance, an extraneous root is obtained from $x = 3$ by squaring both members: $(x)^2 = (3)^2$ or $x^2 = 9$. The roots of $x^2 = 9$ are $x = \pm 3$. -3 does not check in the original equation, $x = 3$, because $-3 \neq 3$.

In order to solve an equation containing the variable in a radical expression, we eliminate the radical. Squaring a radical expression elimin- ates the radical. Assume all expressions under the radical represent nonnegative numbers. $(\sqrt{3})^2 = 3$, $(\sqrt{x})^2 = x$, $(\sqrt{x - 2})^2 = x - 2$, and $(\sqrt{x^2 - 5})^2 = x^2 - 5$.

> When a radical is in an equation, square each member of the equation and solve. Check possible solutions for extraneous roots.

Consider

$$\sqrt{x} = 9$$
$$(\sqrt{x})^2 = (9)^2 \qquad \text{square both members}$$
$$x = 81 \qquad \text{the solution is } possibly \text{ 81}$$

Check:

$$\sqrt{81} = 9$$
$$9 = 9$$

Therefore, $x = 81$ is the solution.

If the radical is not isolated, it will not be eliminated by squaring.

$$\sqrt{x - 1} + 1 = 7$$
$$(\sqrt{x - 1} + 1)^2 = (7)^2 \qquad \text{square both members}$$
$$x - 1 + 2\sqrt{x - 1} + 1 = 49 \qquad \text{radical not eliminated}$$

If the radical is not isolated, it is necessary to isolate it prior to squaring.

$$\sqrt{x - 1} + 1 = 7$$
$$\sqrt{x - 1} = 6 \qquad \text{isolate the radical}$$
$$(\sqrt{x - 1})^2 = (6)^2 \qquad \text{square both members}$$
$$x - 1 = 36$$
$$x = 37 \qquad \text{the solution is } possibly \text{ 37}$$

Check:

$$\sqrt{37-1}+1 = 7$$
$$\sqrt{36}+1 = 7$$
$$6+1 = 7$$
$$7 = 7$$

Therefore, $x = 37$ is the solution.

If more than one radical is involved in the equation, isolate them one at a time. It may take several squarings to eliminate all radicals.
Consider

$$\sqrt{x-5}+\sqrt{x} = 5$$
$$\sqrt{x-5} = 5-\sqrt{x} \quad \text{isolate one radical}$$
$$(\sqrt{x-5})^2 = (5-\sqrt{x})^2 \text{ square both members}$$
$$x-5 = 25-10\sqrt{x}+x$$
$$-30 = -10\sqrt{x}$$
$$3 = \sqrt{x} \quad \text{isolate second radical}$$
$$(3)^2 = (\sqrt{x})^2 \quad \text{square both members}$$
$$9 = x \quad \text{the solution is } \textit{possibly } 9$$

Check:

$$\sqrt{9-5}+\sqrt{9} = 5$$
$$\sqrt{4}+\sqrt{9} = 5$$
$$2+3 = 5$$
$$5 = 5$$

Therefore, $x = 9$ is the solution.

Consider

$$\sqrt{x+4} = 2x-7$$
$$x+4 = 4x^2-28x+49 \qquad \text{square both sides}$$
$$4x^2-29x+45 = 0$$
$$x = 5 \text{ or } x = \frac{9}{4} \qquad \text{possible solutions found by quadratic formula}$$

Check:

$$\sqrt{5+4} = 2(5)-7 \qquad \sqrt{\frac{9}{4}+4} = 2\left(\frac{9}{4}\right)-7$$
$$3 = 3$$
$$\frac{5}{2} = -\frac{5}{2} \text{ false}$$

Therefore, the solution is $x = 5$.

EXAMPLES

a) $\sqrt{5-x} = -3$
No solution because $\sqrt{5-x}$ cannot be negative.

b) $\sqrt{3-x} + x = 6$
$\sqrt{3-x} = 6 - x$
$(\sqrt{3-x})^2 = (6-x)^2$
$3 - x = 36 - 12x + x^2$
$x^2 - 11x + 33 = 0$
$a = 1, b = -11, c = 33$
$x = \dfrac{-(-11) \pm \sqrt{(-11)^2 - 4(1)(33)}}{2(1)}$
$x = \dfrac{11 \pm \sqrt{121 - 132}}{2}$
$x = \dfrac{11 \pm \sqrt{-11}}{2}$
No real solution because $\sqrt{-11}$ is not real.

c) $\sqrt{x+3} - \sqrt{x-5} = 2$
$\sqrt{x+3} = 2 + \sqrt{x-5}$
$(\sqrt{x+3})^2 = (2 + \sqrt{x-5})^2$
$x + 3 = 4 + 4\sqrt{x-5} + x - 5$
$4 = 4\sqrt{x-5}$
$1 = \sqrt{x-5}$
$(1)^2 = (\sqrt{x-5})^2$
$1 = x - 5$
$x = 6$
Check:

$$\begin{aligned}
\sqrt{x+3} - \sqrt{x-5} &= 2 \\
\sqrt{6+3} - \sqrt{6-5} &= 2 \\
\sqrt{9} - \sqrt{1} &= 2 \\
3 - 1 &= 2 \\
2 &= 2
\end{aligned}$$

Therefore, $x = 6$ is the solution.

EXERCISES

Solve.

1) $\sqrt{x} = 8$

2) $\sqrt{x+9} = 3$

3) $\sqrt{x-8} = -5$

4) $\sqrt{9-x} = 0$

5) $\dfrac{1}{\sqrt{x}} = 3$

6) $\sqrt{x} - 6 = -2$

7) $\dfrac{6}{\sqrt{x}} = 1$

8) $\sqrt{2x + 1} = 5$

9) $\sqrt{x + 6} + 10 = 3$

10) $\sqrt{x + 8} = \sqrt{-x}$

11) $\sqrt{x - 5} = x - 5$

12) $\sqrt{x + 8} + 4 = x$

13) $13 + \sqrt{x + 7} = x$

14) $6 - \sqrt{x - 9} = 3$

15) $\sqrt{x} - \sqrt{x + 2} = 6$

16) $\sqrt{x + 3} + \sqrt{x - 2} = 10$

17) $\sqrt{2x - 2} = x - 5$

18) $\dfrac{2}{\sqrt{x + 3}} + 5 = 0$

19) $\dfrac{1}{\sqrt{x - 5}} - \dfrac{2}{\sqrt{x + 10}} = 0$

20) $\sqrt{x^2 - 5x} = 2\sqrt{6}$

21) $\sqrt{x} - \dfrac{3}{\sqrt{x}} = 2$

22) In the application at the beginning of this section, what height is necessary?

23) Find the stroke of a 4-cylinder, 180-in^3 displacement engine if the bore of each cylinder is 3.5 inches. (To the nearest hundredth.)

$$\text{Bore} = \sqrt{\frac{1}{.7854} \cdot \left(\frac{\text{Displacement}}{\text{\# of cylinders}}\right) \cdot \frac{1}{\text{Stroke}}}$$

CHAPTER 10 TEST

Solve.

1. (Obj. 116) $x^2 - 3x - 5 = 0$ _____

2. (Obj. 118) $\sqrt{x + 39} + \sqrt{x + 7} = 8$ _____

3. (Obj. 115) $4x^2 - 8 = 16$ _____

4. (Obj. 117) $x^2 - 6x - 40 = 0$ _____

5. (Obj. 118) $\dfrac{1}{\sqrt{x - 1}} + 3 = 2$ _____

6. (Obj. 117) $\dfrac{x}{x + 3} + \dfrac{2}{x - 6} = \dfrac{3x}{(x + 3)(x - 6)}$ _____

7. (Obj. 118) $\sqrt{x + 5} - 6 = 13$ _____

8. (Obj. 115) $x^2 - 100 = 0$ _____

9. (Obj. 116) $2x^2 + 7x = 1$ _____

10. (Obj. 116) $x^2 - 5x + 11 = 0$ _____

11. (Obj. 116)   A rectangular plot of ground has its length equal to 6 less than twice its width. If the area is 108 sq m, what are the dimensions of the plot?   _____

12. (Obj. 117)   The height (S) of a rocket fired from ground level with an initial velocity of 320 ft/sec at any time (t) is given by the formula

$$S = 320t - 16t^2$$

(S in feet and t in seconds)
In how many seconds will the rocket reach a height of 1600 ft?   _____

CHAPTER 11

GRAPHING

11.1 RECTANGULAR COORDINATE SYSTEM

Locate the town on the map of the state of Washington with co-ordinates (5,B). (See p. 382 for map of Washington.)

APPLICATION

119) Given the coordinates for a point, locate that point on a rectangular coordinate system.

120) Find the coordinates of a point, given its graph.

OBJECTIVES

When two number lines are set at right angles to each other with the intersection at 0, the resulting figure is called a *rectangular coordinate system*. The lines (*axes*) divide the plane into four parts called *quadrants*. The horizontal axis is called the *x-axis* and the vertical axis the *y-axis*. Points on the plane are identified by *ordered pairs* of numbers (x, y), called *coordinates* of the point. The *x-value*, the first member of the ordered pair, is called the *abscissa*. The *y-value*, the second member, is called the *ordinate*. A *graph* is a picture of a set of points.

VOCABULARY

To find the graph of (6, 3) we will first need to construct the rectangular coordinate system. Draw a horizontal number line and label.

THE HOW AND WHY OF IT

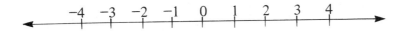

Now draw a vertical number line intersecting the first one at 0. Label the horizontal line x and the vertical one y.

Map courtesy of Best Western Motels.

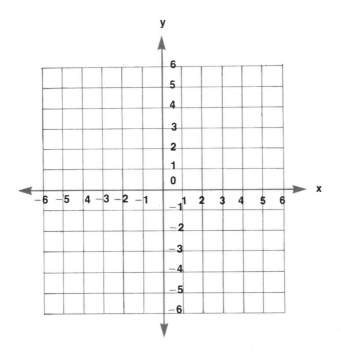

To plot the point (6, 3), proceed 6 units to the right (positive) on the x-axis and 3 units up (positive) on the y-axis. Complete the rectangle and the corner opposite the origin is the graph of (6, 3).

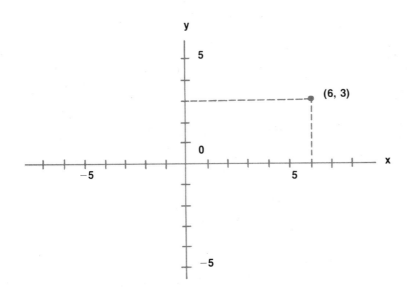

Note that the point could have been found by proceeding right 6 units and up 3 units.

To find the coordinates of a point when its graph is given, read the x and y values off the axes.

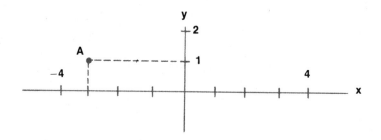

The point A has -3 for the x-value and 1 for the y-value, so the coordinates are $(-3, 1)$.

EXAMPLES

a) Plot the following points A $(3, -2)$, B $(-5, 3)$, C $(-4, -1)$, D $(1.5, 5)$ and E $(3, -4.5)$.

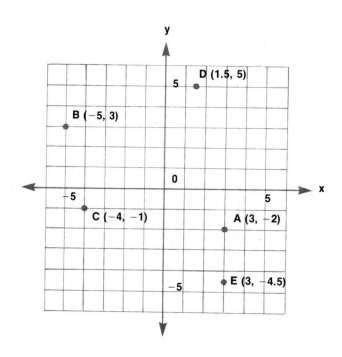

b) Identify the coordinates of the points on the following graph.

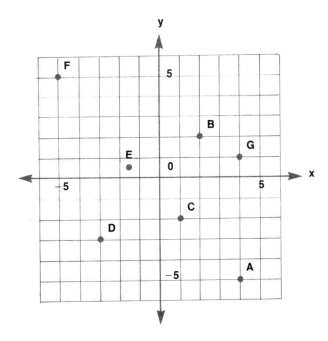

The coordinates are A (4, −5), B (2, 2), C (1, −2), D (−3, −3), E (−1.5, .5), F (−5, 5), and G (4, 1).

Construct a rectangular coordinate system and plot the following points. **EXERCISES**

1) (2, 3)

2) (−6, 7)

3) (−2, −1)

4) (−5, 6)

5) $(4, -3\frac{1}{2})$

6) $(1\frac{2}{3}, -3\frac{1}{3})$

7) $(\frac{4}{5}, -1)$

8) (5, −9)

9) $(-3, -2\frac{1}{2})$

10) $(4\frac{1}{2}, 6\frac{2}{3})$

11) Name each of the following points using ordered pairs.

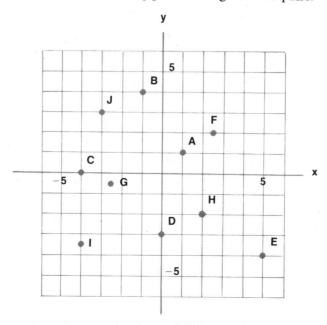

12) In the application at the beginning of this section, what is the name of the town?

11.2 GRAPHS OF LINEAR EQUATIONS

APPLICATION   The cost of renting a circular saw is $5 for the first day and $3 for each additional day the saw is kept. This can be expressed by the equation $y = 3x + 2$, where y is the total rental fee and x is the number of days the saw is kept. Draw the graph of this equation.

OBJECTIVE   121) Draw the graph of a linear equation by plotting and connecting several points.

VOCABULARY   A *linear equation* is one that can be written in the form $ax + by + c = 0$ with a and b not both zero. The graph of a linear equation is a straight line.

Consider the linear equation $3x + 2y - 6 = 0$. Its graph can be found by setting up a table of values. To make a table of values, assign any value for x and solve for y.

THE HOW AND WHY OF IT

$$x = 0$$
$$3(0) + 2y - 6 = 0$$
$$2y = 6$$
$$y = 3$$

$$x = 3$$
$$3(3) + 2y - 6 = 0$$
$$9 + 2y - 6 = 0$$
$$2y = -3$$
$$y = -1.5$$

$$x = -2$$
$$3(-2) + 2y - 6 = 0$$
$$-6 + 2y - 6 = 0$$
$$2y = 12$$
$$y = 6$$

$$x = 4$$
$$3(4) + 2y - 6 = 0$$
$$2y = -6$$
$$y = -3$$

x	y	(x, y)
0	3	$(0, 3)$
3	-1.5	$(3, -1.5)$
-2	6	$(-2, 6)$
4	-3	$(4, -3)$

Now construct a rectangular coordinate system, plot the points in the table and draw the line joining the points.

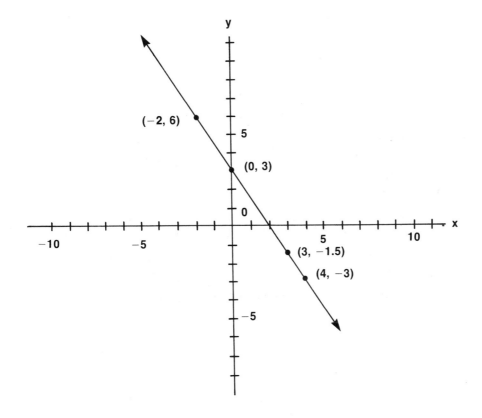

The significance of the graph lies in the fact that every ordered pair of numbers that makes the equation true represents a point that lies on the line. Every point on the line has coordinates that make the equation true.

> To draw a graph of a line, given its equation,
> 1) determine three or more points by assigning values to x and solving for y;
> 2) plot the points on a rectangular coordinate system;
> 3) draw the line joining the points.

EXAMPLE

Draw the graph of $x = 5y + 10$.

x	y
10	0
0	−2
15	1
5	−1

Two of the easiest coordinates to find are the ones with one value equal to zero.

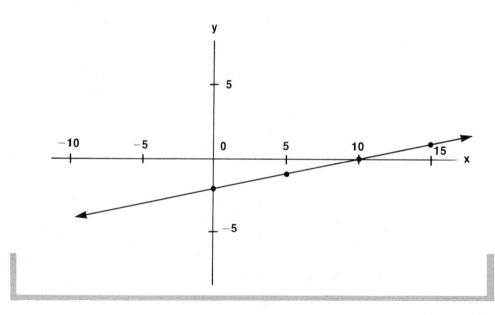

EXERCISES

Make a table of values and graph each of the following.

1) $y = \dfrac{1}{2}x$

2) $y = 2x - 5$

3) $y = -x$

4) $y = -3x + 2$

5) $y + x = 6$

6) $2y + 2x = 5$

7) $5x - 7y = 10$

8) $6x + 5y = 15$

9) $3x = 4y + 12$

10) $7y - x = 14$

11) $9x - 5y = 45$

12) $\dfrac{x}{4} + \dfrac{y}{7} = \dfrac{1}{28}$

13) In the application at the beginning of this section, sketch the graph from $x = 1$ to $x = 10$. Note that the graph has meaning only when x represents a whole number of days.

14) If twice the length x of a rectangle is added to three times the width of y, the sum is 37 (i.e., $2x + 3y = 37$). The graph of the equation will give the length and width of all such rectangles. (Disregard coordinates with negative values.) Draw the graph.

15) The sum of two numbers is 14. This can be stated as an equation with x and y representing the numbers: $x + y = 14$. Draw the graph that illustrates all such numbers.

11.3 PROPERTIES OF LINES

The roof of a house rises 3 ft for every 5 ft of run. $5y = 3x$ is the equation that will determine the amount of rise y that corresponds to a run x. The slope of the line determined by the equation is called the pitch of the roof.

APPLICATION

122) Find the x-intercept given the equation of a line.
123) Find the y-intercept given the equation of a line.
124) Find the slope of a line given two points on the line.
125) Find the slope of a line given its equation.

OBJECTIVES

The *x and y intercepts* are the points where a line crosses (intercepts) the respective axes. The description of the slant of the line is called the *slope*. The slope is expressed as a ratio of the vertical change to the horizontal change.

VOCABULARY

A look at the coordinate system will verify the fact that every point on the y-axis has its abscissa (x-value) zero and every point on the x-axis has its ordinate (y-value) zero.

THE HOW AND WHY OF IT

To find the y-intercept, substitute 0 for x in the equation and solve for y.

To find the x-intercept, substitute 0 for y in the equation and solve for x.

Consider

$$3x + 4y = 12$$
$$\text{If } x = 0, \quad 3(0) + 4y = 12$$
$$4y = 12$$
$$y = 3$$

Therefore, the y-intercept is at $(0, 3)$.

$$\text{If } y = 0, \quad 3x + 4(0) = 12$$
$$3x = 12$$
$$x = 4$$

Therefore the x-intercept is at $(4, 0)$.

The slope of a line is the ratio of the change in the vertical to the change in the horizontal between any two points on the line. The slope is constant regardless of the two points chosen.

Consider

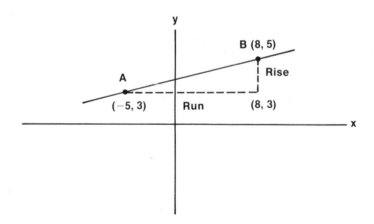

Notice that if you start at point A and proceed to point B, you can proceed horizontally (run) until below point B and then vertically (rise) until you reach B itself. The slope is then the ratio $\dfrac{\text{rise}}{\text{run}}$. Slope $= m = \dfrac{\text{rise}}{\text{run}}$.

The rise is from $y = 3$ to $y = 5$ and is given by $5 - 3 = 2$. The run is from $x = -5$ to $x = 8$ and is given by $8 - (-5) = 13$. Therefore,

$$m = \frac{5-3}{8-(-5)} = \frac{2}{13}$$

In general the slope of a line given two points, $A\,(x_1, y_1)$ and $B\,(x_2, y_2)$, is

$$\text{Slope} = m = \frac{y_2 - y_1}{x_2 - x_1} \text{ or } \frac{y_1 - y_2}{x_1 - x_2}$$

The slope of a line can also be found by solving the equation for y.

When written in the form $y = mx + b$, the coefficient of x is the slope and $(0, b)$ is the y-intercept.

Consider

$$3y + 2x = 8$$
$$3y = -2x + 8$$
$$y = -\frac{2}{3}x + \frac{8}{3}$$

So $m = -\dfrac{2}{3}$ and the y-intercept is $(0, \dfrac{8}{3})$.

EXAMPLES

a) Find the slope of the line passing through the points $(6, 4)$ and $(-5, 13)$.

$$m = \frac{y_2 - y_1}{x_2 - x_1} = \frac{13 - 4}{-5 - 6} = \frac{9}{-11} = -\frac{9}{11}$$

b) Find the slope of the line represented by the equation $x - y = 6$. Identify two points

x	y
0	-6
6	0

$$m = \frac{0 - (-6)}{6 - 0} = \frac{6}{6} = 1$$

c)
$$3x - 2y = 18$$
$$-2y = -3x + 18$$
$$y = \frac{3}{2}x - 9$$

Therefore, the slope is $\frac{3}{2}$ and the y-intercept is $(0, -9)$

d)
$$6x + 5y = 11$$
$$5y = -6x + 11$$
$$y = -\frac{6}{5}x + \frac{11}{5}$$

Therefore, the slope is $-\frac{6}{5}$ and the y-intercept is $(0, \frac{11}{5})$.

EXERCISES

Find the x-intercept and the y-intercept of the graph of each of the following equations.

1) $5x + 2y = 10$

2) $14x - 3y = 42$

3) $6x - 5y = 15$

4) $y - 6x = 0$

5) $2x - 9y = 1$

6) $11x = 2y - 22$

Find the slope of the line that joins the following pairs of points.

7) $(7, 5), (3, 1)$

8) $(-3, 2), (5, -1)$

9) $(-1, -1), (-6, -6)$

10) $(7, -5), (-3, -7)$

11) $(3, 6), (-5, 11)$

12) $(-13, 12), (-1, 0)$

Find the slope of the line and its y-intercept.

13) $3x - 8y = 16$

14) $3y + 7x = 15$

15) $8y = x - 7$

16) $2x = 11y$

17) $3y = 15x - 2$

18) $x + y = 12$

19) In the application at the beginning of this section, what is the pitch of the roof?

11.4 DISTANCE BETWEEN TWO POINTS

A triangle has vertices at A (0, 0), B (10, 0), and C (6, 8). Find the lengths of its sides.   APPLICATION

126) Find the distance between two points on a graph, given their coordinates.   OBJECTIVE

No new vocabulary.   VOCABULARY

THE HOW AND WHY OF IT

The distance between two points on a graph is given by

$$d = \sqrt{(x_2 - x_1)^2 + (y_2 - y_1)^2},$$

where (x_1, y_1) and (x_2, y_2) are the coordinates of the points A and B.

This formula can be developed using the theorem of Pythagoras and the following illustration.

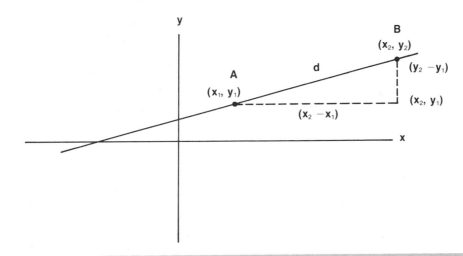

EXAMPLES

a)   Find the distance between (3, 2) and (5, −1). Let (x_1, y_1) = (3, 2) and (x_2, y_2) = (5, −1).

$$d = \sqrt{(5 - 3)^2 + (-1 - 2)^2}$$
$$d = \sqrt{(2)^2 + (-3)^2} = \sqrt{4 + 9} = \sqrt{13}$$

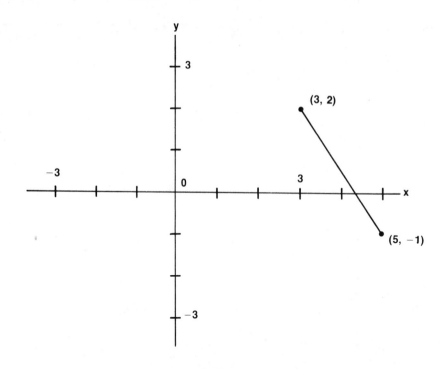

b) Find the distance between $(-3, -7)$ and $(-4, 3)$. Let $(x_1, y_1) = (-3, -7)$ and $(x_2, y_2) = (-4, 3)$.

$$d = \sqrt{[-4 - (-3)]^2 + [3 - (-7)]^2}$$
$$d = \sqrt{(-1)^2 + (10)^2}$$
$$d = \sqrt{1 + 100}$$
$$d = \sqrt{101} \approx 10.05$$

EXERCISES   Find the distance between each of the following pairs of points.

1) $(6, 5), (9, 9)$ 2) $(-6, 3), (3, 9)$

3) $(3, 4), (2, -1)$ 4) $(-2, -3), (-5, 1)$

5) $(2.1, 3.5), (1.5, -2.5)$ 6) $(4, -7), (3.5, -7.5)$

7) $(-11, -13), (0, 0)$

8) $(-5, 7), (-1, -3)$

9) In the application at the beginning of this section, what are the lengths of the sides of the triangle?

10) A bicycle rider follows a path that is the graph of the equation $y = 3x$. If he starts at the point $(0, 0)$, how far has he traveled when he reaches the point $(4, 12)$?

11) What is the length of a line segment whose end points are $(3, 7)$ and $(-1, 5)$?

12) Find the area of the rectangle whose vertices are at the points $(1, 7)$, $(5, 3)$, $(10, 8)$, and $(6, 12)$.

11.5 GRAPHING LINES USING SLOPE AND INTERCEPT

A truck driver averages 80 km/hr. The formula for the distance driven is $d = 80t$, where d represents distance (in km) and t represents time (in hr). Draw a graph showing the relationship between the distance and time. (Let each unit on the x-axis represent 1 hour and each unit on the y-axis represent 100 km.)   APPLICATION

127) Draw the graph of a line, given its equation, using the slope and the y-intercept.   OBJECTIVE

VOCABULARY   No new vocabulary.

THE HOW AND   Consider the equation $2x - 3y = 6$. First write it in the slope-
WHY OF IT   intercept form $y = mx + b$.

$$2x - 3y = 6$$
$$-3y = -2x + 6$$
$$y = \frac{2}{3}x - 2$$

Therefore, $m = \frac{2}{3}$ and the y-intercept is at $(0, -2)$. We know the line crosses the y-axis at $(0, -2)$, and for every 3 units of run there is a corresponding 2 units of rise. We utilize this to draw the graph.

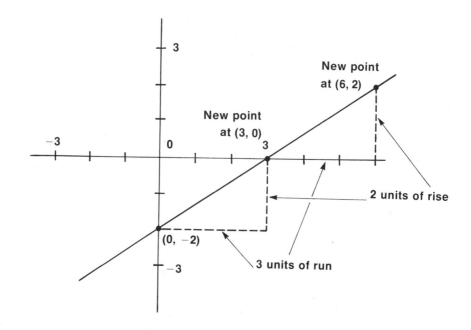

By sketching the run and rise we are able to obtain as many new points as we wish.

> To draw the graph of a line using the slope and the y-intercept,
> 1)   write the equation in the form $y = mx + b$;
> 2)   plot the y-intercept $(0, b)$;
> 3)   using the slope m, starting at the y-intercept, locate an additional point;
> 4)   draw the line joining the points.

EXAMPLE

Draw the graph of $4x + 5y = 10$.

$$4x + 5y \quad 10$$
$$5y \quad -4x + 10$$
$$y \quad -\frac{4}{5}x + 2$$

$$m = -\frac{4}{5} \quad y\text{-intercept at } (0, 2).$$

Note: Since the slope is negative, the rise will be down 4 (drop).

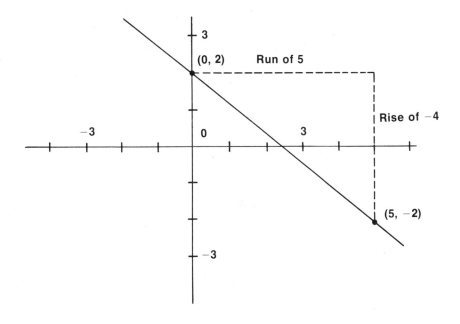

When using a negative slope, we will consider the rise to be negative and the run positive.

Draw the graph of the line given its slope and y-intercept.   **EXERCISES**

1) $m = \frac{2}{3}, (0, 5)$

2) $m = -\frac{3}{4}, (0, -1)$

3) $m = 3, (0, -5)$

4) $m = -2, (0, 2)$

5) $m = \dfrac{5}{7}, (0, 0)$

6) $m = -\dfrac{7}{9}, (0, 8)$

Draw the graph of each of the following equations using the slope and the y-intercept.

7) $4x + 5y = 20$

8) $7x - 3y = 21$

9) $2y = x - 4$

10) $3x = 4y + 5$

11) $5x - 6y = 3$

12) In the application at the beginning of this section, draw the graph.

11.6 SOLVING SYSTEMS OF LINEAR EQUATIONS BY GRAPHING

Two cross country runners leave the same starting point one hour apart. The first to depart runs at an average of 5 mph and the second at an average of 7 mph. How far will they have run when the second runner overtakes the first one? $d = rt$, where d represents distance, r represents rate, and t represents time. If t represents the time of the first runner, the time of the second runner is one hour less or $t - 1$. The system of equations is

APPLICATION

$$d = 5t \qquad \text{first runner}$$
$$d = 7(t - 1) \qquad \text{second runner}$$

OBJECTIVE   128) Solve a system of linear equations by graphing.

VOCABULARY   A system of equations is a set of two or more equations. If the graphs of two linear equations have no points in common, the equations are called an *independent and inconsistent system.* If the lines have one point in common, the equations are called an *independent and consistent system.* If the lines have all points in common, the equations are called *a dependent system.*

THE HOW AND   If two equations are graphed on the same set of axes, only three
WHY OF IT   possibilities exist. These are shown below.

The graphs of both equations are the same. A pair of such equations are

$3x - y = 2$, equation of ℓ_1
$9x - 3y = 6$, equation of ℓ_2

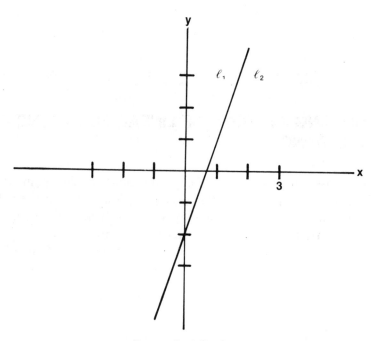

Dependent System

The graphs of the equations are parallel lines. Two such equations are

$3x - 7y = 9$, equation of ℓ_1
$3x - 7y = 2$, equation of ℓ_2

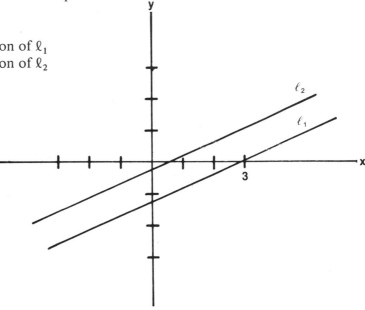

Independent-Inconsistent System

The graphs of the equations are lines that intersect in exactly one point. The coordinates of this common point are the solution of the system. Two such equations are

$3x - 2y = 6$, equation of ℓ_1
$3x + 2y = 6$, equation of ℓ_2

The solution is $(2, 0)$.

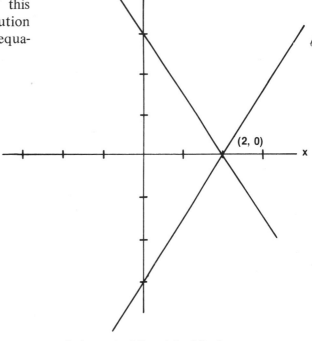

Independent-Consistent System

To solve a system of equations, graph each of the equations. Identify which of the three preceding cases is involved, and if the system is consistent and independent, write the common solution.

EXAMPLES

a) Solve the system.

$$2x - y - 1 = 0$$
$$8x - 4y - 3 = 0$$

ℓ_1

$$2x - y - 1 = 0$$
$$-y = -2x + 1$$
$$y = 2x - 1$$
$$m = \frac{2}{1}$$

y-intercept

$$(0, -1)$$

ℓ_2

$$8x - 4y - 3 = 0$$
$$-4y = -8x + 3$$
$$y = 2x - \frac{3}{4}$$
$$m = \frac{2}{1}$$

y-intercept

$$(0, -\frac{3}{4})$$

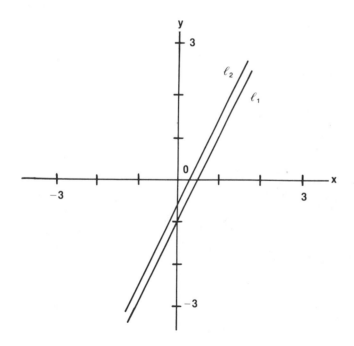

The lines do not intersect and therefore there is no common solution. The system is independent and inconsistent.

b) $4x - 5y - 10 = 0$
     $x + 3y + 9 = 0$

ℓ_1
$4x - 5y - 10 = 0$
$-5y = -4x + 10$
$y = \dfrac{4}{5}x - 2$
$m = \dfrac{4}{5}$
y-intercept
$(0, -2)$

ℓ_2
$x + 3y + 9 = 0$
$3y = -x - 9$
$y = -\dfrac{1}{3}x - 3$
$m = -\dfrac{1}{3}$
y-intercept
$(0, -3)$

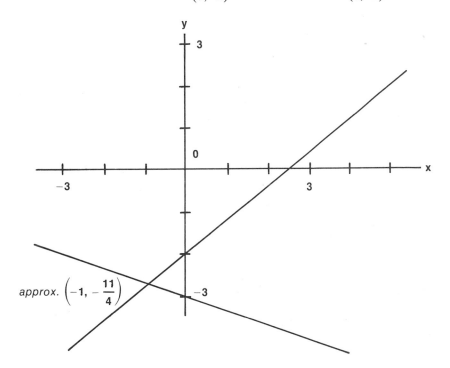

$approx. \left(-1, -\dfrac{11}{4}\right)$

The equations have one common solution, which is estimated to be $(-1, -\dfrac{11}{4})$. [The exact solution is $(-\dfrac{15}{17}, -\dfrac{46}{17})$.] A more precise method of finding the exact solution will be given in the next section.

Solve the following systems by graphing.

EXERCISES

1) $7x - y = 7$
     $x - y = 1$

2) $2x + y = 8$
     $2x - y = -4$

3) $y = 4x + 1$
$y = 3x - 4$

4) $3x + 4y = 6$
$x - 2y = 2$

5) $x - 2y = 1$
$2x - y = 1$

6) $2x + y = 4$
$4x + 2y = 16$

7) In the application at the beginning of this section, how far did they run?

8) The sum of two numbers is 12. The difference of the same numbers is 2. Write the system of equations and solve by graphing.

11.7 SOLVING SYSTEMS OF LINEAR EQUATIONS BY USING A LINEAR COMBINATION

Ida's mother requires a daily diabetic diet. The doctor's prescription for her diet includes 99 g of carbohydrate, 91 g of protein, and 30 g of fat. Her breakfast and lunch menus for Saturday have already provided for 66 g of carbohydrate, 61 g of protein, and 18 g of fat. The bread, vegetables, and dessert at dinner will account for an additional 33 g of carbohydrate and 9 g of protein. The (approximate) food values of liver and bacon are (given in grams per 100 grams)

APPLICATION

	carbohydrate	protein	fat
liver	0	25	8
bacon	0	20	40

How many grams of bacon and how many grams of liver should Ida prepare for her mother's dinner to balance her diet for the day? (Round all weights to the nearest gram.)

129) Solve a system of linear equations by using a linear combination.

OBJECTIVE

The equation $M(2x - y - 6) + N(3x - y + 4) = 0$ is called a *linear combination* of $2x - y + 6 = 0$ and $3x - y + 4 = 0$.

VOCABULARY

To solve a system of equations by a linear combination, we choose M and N so that the coefficients of one of the variables will be opposites. The variable will drop out of the combination and the resulting equation can be solved for the remaining variable.

THE HOW AND WHY OF IT

Consider the system

$$\begin{cases} 2x - y - 6 = 0 \\ 3x - y + 4 = 0 \end{cases}$$

We choose $M = -3$ and $N = 2$ so that the coefficients of x will be opposites. Forming the linear combination we have

$$-3(2x - y - 6) + 2(3x - y + 4) = 0$$
$$-6x + 3y + 18 + 6x - 2y + 8 = 0$$
$$y + 26 = 0$$
$$y = -26$$

Now, substitute for y in one of the equations.

$$2x - (-26) - 6 = 0$$
$$2x + 20 = 0$$
$$x = -10$$

The solution is $(-10, -26)$.

If the choice of M and N results in both variables being eliminated, the graphs of the equations are either parallel or the same line. They are parallel (the system is inconsistent and independent) if the equation is false (such as $8 = 0$) and the system has no solution. They are the same line (the system is consistent and dependent) if the equation is true (such as $0 = 0$) and the system has an infinite number of solutions.

> To solve a system of linear equations by a linear combination,
> 1) write each equation in the form $Ax + By + C = 0$;
> 2) choose factors M and N so that the linear combination $M(A_1 x + B_1 y + C_1) + N(A_2 x + B_2 y + C_2) = 0$ will make the coefficients of x (or y) opposites;
> 3) simplify the equation so that x (or y) is eliminated and solve for y (or x);
> 4) use this value of y (or x) to find the other unknown;
> 5) write the solution as an ordered pair.

EXAMPLES

a) **Solve the system.**

$$\begin{cases} 3x + 2y + 8 = 0 \\ 2x - 3y + 6 = 0 \end{cases}$$

If $M = 2$ and $N = -3$, the coefficients of x will be opposites. The linear combination is:

$$2(3x + 2y + 8) - 3(2x - 3y + 6) = 0$$
$$6x + 4y + 16 - 6x + 9y - 18 = 0$$
$$13y - 2 = 0$$
$$13y = 2$$
$$y = \frac{2}{13}$$

Now substitute $y = \frac{2}{13}$ into one of the original equations and solve for x

$$3x + 2y + 8 = 0$$

$$3x + 2\left(\frac{2}{13}\right) + 8 = 0$$
$$39x + 4 + 104 = 0$$
$$39x = -108$$
$$x = -\frac{108}{39} = -\frac{36}{13}$$

The solution is $\left(-\frac{36}{13}, \frac{2}{13}\right)$.

b) Solve the system.

$$\begin{cases} 9x - 4y - 36 = 0 \\ 18x - 8y - 11 = 0 \end{cases}$$

If $M = -2$ and $N = 1$, the coefficients of x will be opposites.

$$-2(9x - 4y - 36) + 1(18x - 8y - 11) = 0$$
$$-18x + 8y + 72 + 18x - 8y - 11 = 0$$
$$61 = 0$$

Both variables were eliminated and the resulting equation is false. The system has no solutions.

Solve the following systems by using linear combinations.

EXERCISES

1) $x + 3y = 25$
    $2x + y = 15$

2) $5c - 3d = -23$
    $4c + d = 2$

3) $2h + 3k = 10$
    $3h + 2k = 10$

4) $x + 5y + 1 = 0$
    $2x + 7y - 1 = 0$

5) $3x + 4y = 24$
$2x + y = 11$

6) $3a - 5b = 10$
$7a - 3b = 14$

7) $7m - 8n = 9$
$4m - 5n = 7$

8) $7w - 3v = -12$
$5w - 2v = -8$

9) $t + s = 35$
$t - 5s = 17$

10) $4h + 5k = -2$
$5h - 2k = 3$

11) $3x - 4y = 12$
$6x - 8y = 15$

12) $5w + 2v = -2$
$w - v = 8$

13) $3x + y = 7$

$2x - 5y = -1$

14) $x - \dfrac{5y}{2} = \dfrac{19}{2}$

$\dfrac{3x}{4} + y = -\dfrac{3}{2}$

15) In the application at the beginning of this section, how many grams of liver and how many grams of bacon should Ida prepare?

16) The sum of two numbers is 7. If the second number is subtracted from twice the first the result is 2. What are the numbers? Write a system of equations and solve.

17) Greg has 80 coins, all nickels and dimes. The value of the coins is $6.75. Write a system of equations and solve for the number of each kind of coin. Hint:

Let x = number of nickels

Let y = number of dimes

$x + y = 80$ (number of coins)

$.05x + .10y = 6.75$ (value of coins)

18) How many ounces each of gold 85% pure and gold 70% pure are needed to make up 15 ounces of gold 75% pure?

19) Given a tank of solution A which is 10% salt by weight and a tank of solution B which is 50% salt by weight, how many gallons of each of the solutions A and B must be mixed to make 100 gallons of a 20% by weight salt solution? Assume that the solutions A and B have the same densities (weigh the same per gallon). (Courtesy of Exxon Co.)

20) How many kilograms each of a brass alloy containing 30% zinc and pure zinc are needed to make 12 kilograms of a brazing alloy that is 50% zinc? (Compute to the nearest hundredth of a kilogram.)

11.8 SOLVING SYSTEMS OF LINEAR EQUATIONS BY SUBSTITUTION

A landscaper needs 600 pounds of grass seed that is 45% blue grass. He has two mixtures available, one contains 40% blue grass and the other 70% blue grass. How much of each will he use?

APPLICATION

130) Solve a system of linear equations by substitution.

OBJECTIVE

No new vocabulary.

VOCABULARY

In the system of equations

THE HOW AND WHY OF IT

$$\begin{cases} 2x - y - 6 = 0 \\ 3x - y + 4 = 0 \end{cases}$$

we can readily solve for y in either equation by adding y to both sides of the equation. For instance, in the first equation,

$$2x - 6 = y \quad \text{or} \quad y = 2x - 6$$

If this expression for y (that is, $2x - 6$) is substituted for y in the second equation, the variable y is eliminated and we can solve for x.

$$3x - (2x - 6) + 4 = 0$$
$$3x - 2x + 6 + 4 = 0$$
$$x = -10$$

As before, we can find y by using -10 for x.

$$y = 2x - 6$$
$$y = 2(-10) - 6$$
$$y = -26$$

The solution is $(-10, -26)$.

To solve a system of linear equations by substitution,
   1) solve *either* equation for x (or y);
   2) substitute this expression for x (or y) in the other equation and solve for the remaining unknown;
   3) use this value of y (or x) to find the second unknown;
   4) write the solution as an ordered pair.

EXAMPLES

a) Solve the system.

$$\begin{cases} x + 3y = 25 \\ 2x + y = 15 \end{cases}$$

Step 1: Solve the first equation for x.

$$x = 25 - 3y$$

Step 2: Substitute $25 - 3y$ for x in the second equation and solve for y.

$$2(25 - 3y) + y = 15$$
$$50 - 6y + y = 15$$
$$-5y = -35$$
$$y = 7$$

Step 3: $x = 25 - 3y = 25 - 3(7) = 25 - 21 = 4$
Step 4: The solution is $(4, 7)$.

b) Solve the system.

$$\begin{cases} x = 2y + 3 \\ 2x - 4y = 7 \end{cases}$$

$$2(2y + 3) - 4y = 7 \qquad \text{substitute } 2y + 3 \text{ for } x \text{ in the second equation}$$

$$4y + 6 - 4y = 7 \qquad \text{solve for } y$$
$$6 = 7$$

Since this equation is false, there is no solution.

EXERCISES   Solve.

1) $4x + y = 2$
    $5x + 3y = 20$

2) $x - 5y - 14 = 0$
    $2x + y + 5 = 0$

3) $2r + 3s = 1$
$\quad\ r - 4s = 6$

4) $2p - q = -15$
$\quad\ p + 2q = 5$

5) $7w - 3x = -8$
$\quad\ w - 5x = 13$

6) $a + b = 44$
$\quad 3a - 2b = 37$

7) $y = \dfrac{1}{2}x + 4$
$\quad 2x + 5y = 11$

8) $x = \dfrac{2}{3}y - 8$
$\quad 8x - y = 1$

9) $y = x - 3$
$\quad 2x = y + 7$

10) $x + 2y = \dfrac{3}{4}$
$\quad 2x + y = -\dfrac{3}{8}$

11) $6x - y = 7$
$2x + 3y = 5$

12) $2x + 3y = -4$
$3x + 2y = 6$

13) $2x - 4y = 6$
$3x - 6y = 9$

14) $y = 2x + 5$
$6x - 3y = 1$

15) In the application at the beginning of this section, how many pounds of each kind of seed did he use?

16) The sum of two numbers is 84 and the difference is 22. What are the numbers?

17) The length of the smaller of two rectangles is 12 cm and the length of the larger rectangle is 18 cm. The sum of their areas is 246 sq cm and the difference between the areas is 78 sq cm. What is the width of each rectangle?

18) Mr. Mitchell has $3000 in two savings accounts. One account pays 6% simple interest and the other pays 7.5% simple interest. If the annual interest earned is $207, how much money is invested in each account?

19) Flying with the wind an airplane can fly a distance of 1080 miles in 6 hours. If the plane was flying against the wind, it would take 9 hours to fly the same distance. What is the speed of the airplane in still air and what is the speed of the wind?

20) A cereal company intends to add enough dried fruit to each box of cereal so that each box will contain 21 grams of protein and 338 grams of carbohydrate. The approximate food values are

	protein	carbohydrate
dried fruit	2%	75%
cereal	10%	85%

How much cereal and how much dried fruit should each box contain?

21) Ellie needs some change for her weekend garage sale. She has enough nickels and wants three times as many quarters as dimes. If she takes $10.20 to the bank, how many quarters and how many dimes can she get?

1. (Obj. 121)   Draw the graph of $2y + 3x - 12 = 0$. _____

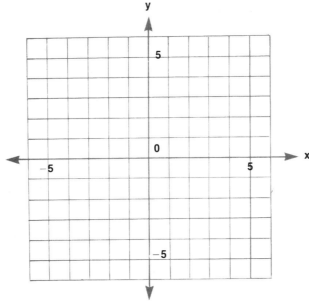

2. (Obj. 126)   Find the distance between the points $(-3, -2)$ and $(-1, 3)$. _____

3) (Obj. 123)   Write the coordinates of the y-intercept of the line represented by $x = 2y + 8$. _____

4) (Obj. 124)   Write the slope of the line that passes through the points $(-3, -2)$ and $(-1, 3)$. _____

5) (Obj. 119)   Locate the following points on the coordinate system: $A(-6, 4)$, $B(0, -3)$, $C(-1, -5)$, $D(-5, 0)$, and $E(5, 1)$. _____

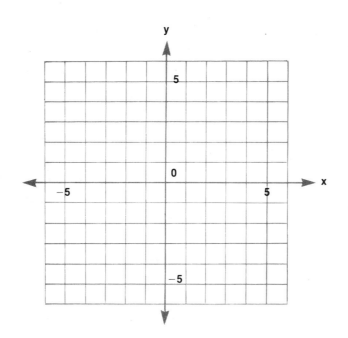

6) (Obj. 128) Solve the following system of equations by graphing. _____

$$4x + 3y = 6$$
$$2x - y = -2$$

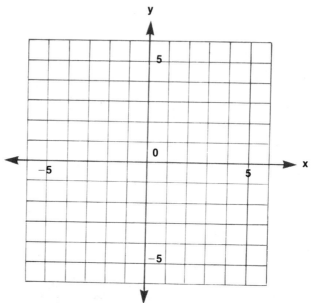

7) (Obj. 122) Write the coordinates of the x-intercept of the line represented by the equation $3x - 4y - 12 = 0$. _____

8) (Obj. 129) Solve the following system of equations using a linear combination.

$$5s + 4t = -2$$
$$-2s + 5t = 3$$
_____

9) (Obj. 127) Draw the graph of the line through (4, 0) that has slope $m = -\dfrac{1}{2}$. _____

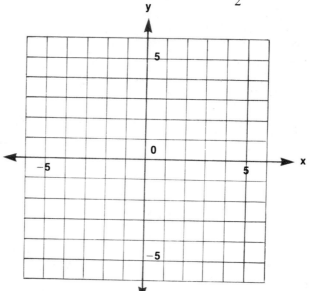

10) (Obj. 125) Write the slope of the line represented by the equation $-2y = 3x + 6$. _____

11) (Obj. 120) Identify the coordinates of the points on the graph below.

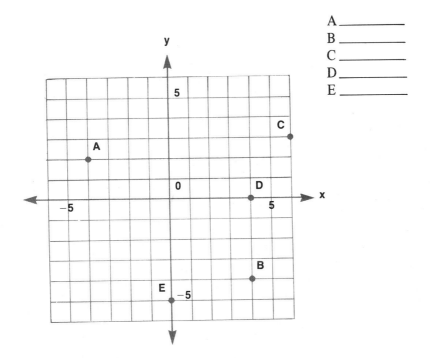

A _____
B _____
C _____
D _____
E _____

12) (Obj. 130) The voltage across a resistor is equal to the product of the current times the resistance. When a current of 8 amperes passes through a resistor and 5 amperes passes through another resistor, the sum of the voltages is 180 volts. If the sum of the two resistors is 30 ohms, what is the value of each resistor? _____

TRIGONOMETRY

12.1 MEASUREMENT OF ANGLES

A stairway makes an angle of 42 degrees with the downstairs floor. **APPLICATION**
What angle does it make with the wall?

131) Find the degree measure of the third angle in a triangle given **OBJECTIVE**
two of the angles.

The word *line* is used here to mean a straight line which has no end **VOCABULARY**
in either direction. Any two points on the line can be used to name the
line.

Line *MN* or line *MP* or line *NP* are possible names.
A *half line* contains all the points on either side of a point on a line.
A *ray* is a half line plus its endpoint.

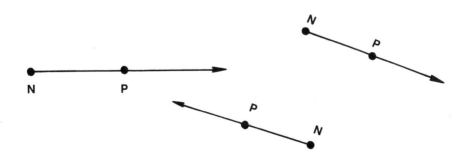

The above rays are read "ray *NP*" (not "ray *PN*").

An *angle (L)* is formed when two rays start from a common point.

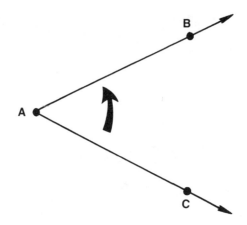

Ray *AB* and ray *AC* form angle *A* (or angle *CAB*). *A* is used to designate the angle or the measure of the angle.

A *degree* is a commonly used measure of angles.

Angles are classified according to their measures. *Obtuse angles* have measures between 90° and 180°; a *right angle* measures 90°; and *acute angles* have measures between 0° and 90°.

Minutes and *seconds* are smaller units of measure for an angle.

$$360 \text{ degrees} = 1 \text{ revolution}$$
$$60 \text{ minutes} = 1 \text{ degree}$$
$$60 \text{ seconds} = 1 \text{ minute}$$

If *C* = 36 degrees, 20 minutes, 30 seconds, we can write *C* = 36° 20′ 30″.

In this text, angle measures of less than a degree will be expressed to the nearest tenth of a degree. So, $C \approx 36.3°$.

The angles in a triangle always have measures whose sum is 180°.

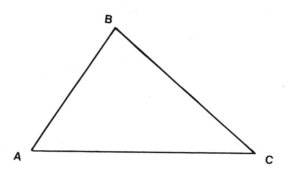

In triangle *ABC* (△ *ABC*) we have

$$A + B + C = 180°$$

> If the measure of two angles of a triangle is known, the measure of the third can be found by subtracting the first two measures from 180°.

THE HOW AND
WHY OF IT

Consider $\triangle RST$.

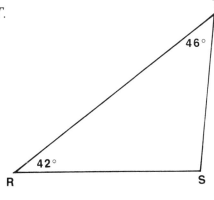

$$R = 42°$$
$$T = 46°$$
$$S = ?$$

$$R + S + T = 180°$$
$$42° + S + 46° = 180°$$
$$S = 180° - 88°$$
$$S = 92°$$

EXAMPLES

a)

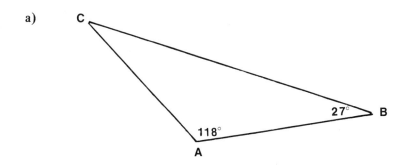

In $\triangle ABC$
$A = 118°$
$B = 27°$
$C = ?$

$$A + B + C = 180°$$
$$118° + 27° + C = 180°$$
$$C = 180° - 145°$$
$$C = 35°$$

b)

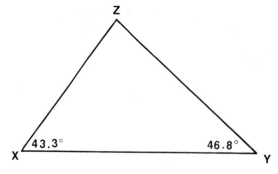

In △ XYZ
$X = 43.3°$
$Y = 46.8°$
$Z = ?$

$$X + Y + Z = 180°$$
$$43.3° + 46.8° + Z = 180°$$
$$Z = 180° - 90.1°$$
$$Z = 89.9°$$

EXERCISES

In each triangle find the measure of the third angle.

1) In △ ABC, $A = 30°$, $B = 60°$

2) In △ DEF, $D = 45°$, $E = 45°$

3) In △ MNP, $P = 60°$, $M = 40°$

4) In △ LMN, $L = 90°$, $M = 80°$

5) In △ PDQ, $D = 40°$, $Q = 50°$

6) In △ ABC, $A = 23°$, $B = 88°$

7) In $\triangle DEF$, $D = 41°$, $F = 41°$

8) In $\triangle XYZ$, $Y = 50°$, $Z = 101°$

9) In $\triangle RST$, $R = 110°$, $S = 65°$

10) In $\triangle PQR$, $P = 35.6°$, $R = 37.3°$

11) In $\triangle GHJ$, $H = 57.5°$, $J = 62.3°$

12) In $\triangle LMN$, $M = 87.4°$, $N = 86.2°$

13) In $\triangle ACE$, $A = 75.9°$, $E = 75.3°$

14) In $\triangle TUV$, $V = 114.1°$, $U = 27.8°$

15) In $\triangle WXY$, $X = 38.2°$, $Y = 47.1°$

16) In $\triangle ADG$, $D = 55.5°$, $G = 64°$

17) In the application at the beginning of this section, what angle does the stairway make with the wall?

12.2 TRIGONOMETRIC RATIOS: PART I

APPLICATION A road rises 9 feet for every 100 feet of horizontal distance. The grade of the road is the tangent of the angle expressed as a percent.

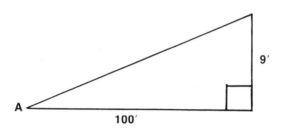

OBJECTIVE 132) Given any two sides of a right triangle find the sine, cosine, and/or tangent of either acute angle.

VOCABULARY A *right triangle* is any triangle that has one angle whose measure is 90°.

In the figure below, triangle *ABC* is a right triangle with right angle *C* (*C* = 90°). Triangle *ADE* is also a right triangle with right angle *D*.

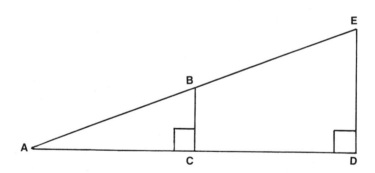

These right triangles are *similar triangles*. From geometry we know that similar triangles have equal corresponding angles and proportional corresponding sides. Therefore,

$$B = E \qquad \text{corresponding angles}$$
$$C = D \qquad \text{both are } 90° \text{ angles}$$

$$\frac{\overline{BC}}{\overline{AC}} = \frac{\overline{ED}}{\overline{AD}}$$

$\overline{BC}$ represents the measure of the side of the triangle opposite angle *A*.

$$\frac{\overline{BC}}{\overline{AB}} = \frac{\overline{ED}}{\overline{AE}}$$

$$\frac{\overline{AC}}{\overline{AB}} = \frac{\overline{AD}}{\overline{AE}}$$

Since these ratios are always equal (and very useful), they have special names. The *sine* of angle A = $\sin A = \dfrac{\overline{BC}}{\overline{AB}}$ $\left(\text{or}\ \dfrac{\overline{ED}}{\overline{AE}}\right)$.

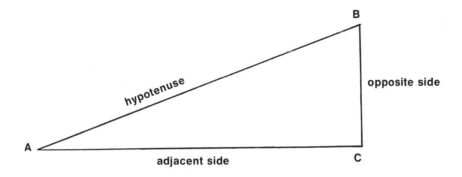

We can say

$$\sin A = \frac{\text{opposite}}{\text{hypotenuse}} = \frac{\overline{BC}}{\overline{AB}}$$

$$\text{cosine}\ A = \cos A = \frac{\text{adjacent}}{\text{hypotenuse}} = \frac{\overline{AC}}{\overline{AB}}$$

$$\text{tangent}\ A = \tan A = \frac{\text{opposite}}{\text{adjacent}} = \frac{\overline{BC}}{\overline{AC}}$$

These three ratios are called *trigonometric ratios.*

If we had chosen angle B, the opposite side would have been $\overline{AC}$ and the adjacent side would have been $\overline{BC}$, so that

$$\sin B = \frac{\text{opp}}{\text{hyp}} = \frac{\overline{AC}}{\overline{AB}}$$

$$\cos B = \frac{\text{adj}}{\text{hyp}} = \frac{\overline{BC}}{\overline{AB}}$$

$$\tan B = \frac{\text{opp}}{\text{adj}} = \frac{\overline{AC}}{\overline{BC}}$$

THE HOW AND
WHY OF IT

In this triangle

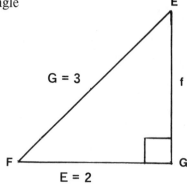

$g = \overline{EF} = 3$ and $e = \overline{FG} = 2$; therefore,

$$
\begin{aligned}
f^2 + e^2 &= g^2 \\
f^2 + 4 &= 9 \\
f^2 &= 5 \\
f &= \sqrt{5}
\end{aligned}
$$

To find the trigonometric ratios of an acute angle of a right triangle, substitute into the following formulas.

$$\sin A = \frac{\text{opp}}{\text{hyp}}, \quad \cos A = \frac{\text{adj}}{\text{hyp}}, \quad \tan A = \frac{\text{opp}}{\text{adj}}$$

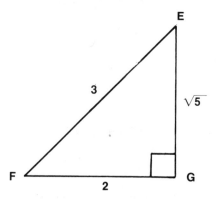

$$\sin F = \frac{\text{opp}}{\text{hyp}} = \frac{\sqrt{5}}{3}$$

$$\cos F = \frac{\text{adj}}{\text{hyp}} = \frac{2}{3}$$

$$\tan F = \frac{\text{opp}}{\text{adj}} = \frac{\sqrt{5}}{2}$$

$$\sin E = \frac{\text{opp}}{\text{hyp}} = \frac{2}{3}$$

$$\cos E = \frac{\text{adj}}{\text{hyp}} = \frac{\sqrt{5}}{3}$$

$$\tan E = \frac{\text{opp}}{\text{adj}} = \frac{2}{\sqrt{5}} = \frac{2\sqrt{5}}{5}$$

EXAMPLES

a)   In $\triangle RST$ find sin S, cos S, and tan S.

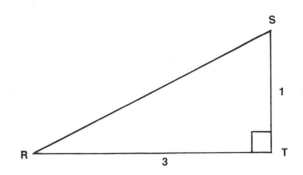

$s^2 + r^2 = t^2$ $\sin S = \dfrac{3}{\sqrt{10}} = \dfrac{3\sqrt{10}}{10}$

$9 + 1 = t^2$ $\cos S = \dfrac{1}{\sqrt{10}} = \dfrac{\sqrt{10}}{10}$

$10 = t^2$ $\tan S = \dfrac{3}{1} = 3$

$\sqrt{10} = t$

b)   In $\triangle UVW$ find sin V, cos V, and tan V.

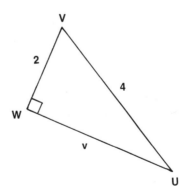

$$v^2 + u^2 = w^2$$
$$v^2 + 4 = 16$$
$$v^2 = 12$$
$$v = \sqrt{12} = 2\sqrt{3}$$

$$\sin V = \frac{2\sqrt{3}}{4} = \frac{\sqrt{3}}{2}$$

$$\cos V = \frac{2}{4} = \frac{1}{2}$$

$$\tan V = \frac{2\sqrt{3}}{2} = \sqrt{3}$$

EXERCISES

Find $\sin A$, $\cos A$, and $\tan A$.

1) $\triangle ABC$ has $b = 6$, $c = 10$, and $C = 90°$

2) $\triangle ABC$ has $a = 15$, $c = 39$, and $C = 90°$

3) $\triangle ABC$ has $b = 3$, $a = 3$, and $C = 90°$

4) $\triangle ABC$ has $b = 13$, $c = 12$, and $B = 90°$

Find $\sin T$, $\cos T$, and $\tan T$.

5) $\triangle RST$ has $s = 15$, $r = 9$, and $S = 90°$

6) $\triangle RST$ has $s = 10$, $t = 9$, and $S = 90°$

7) $\triangle RST$ has $t = 4$, $s = 6$, and $R = 90°$

8) $\triangle RST$ has $t = 7$, $s = 7$, and $R = 90°$

Find $\sin D$, $\cos D$, and $\tan D$.

9) $\triangle DEF$ has $f = 3$, $e = 1$, and $F = 90°$

10) $\triangle DEF$ has $e = 3$, $d = 1$, and $F = 90°$

11) $\triangle DEF$ has $d = 3$, $e = 1$, and $F = 90°$

12) $\triangle DEF$ has $f = 2$, $e = 3$, and $E = 90°$

13) In the application at the beginning of this section, what is the grade of the road, (a) as a percent, and (b) as a decimal?

14) If the tread of a stairway is 20 cm and the rise is 16.5 cm, what is the tangent of the angle the stairway makes with the floor?

12.3 TRIGONOMETRIC RATIOS: PART II

APPLICATION   A tuned circuit in Greg's CB receiver has a reactance of 153.29 ohms. Find the impedance of the circuit if the phase angle is 52.7°. The formula for impedance is

$$Z_T = \frac{X}{\sin \theta}$$

where X is the reactance measured in ohms, Z_T is the impedance measured in ohms, and θ is the phase angle between the resistance and reactance of the tuned circuit.

OBJECTIVES   133) Find the exact value of the sine, cosine, and/or tangent of an angle of 30°, 45°, and 60°.
        134) Find the approximate value of the sine, cosine and/or tangent of any acute angle from a table (see table at end of chapter) or a calculator with keys for trigonometric ratios.
        135) Find the acute angle, to the nearest tenth of a degree, given the value of its sine, cosine, and/or tangent. Use a table or calculator.

VOCABULARY   A *30° - 60° right triangle* is shown below.

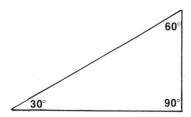

        A *45° right triangle* has two angles and two sides equal (isosceles) as shown.

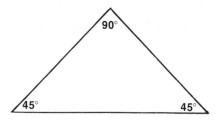

        From geometry we know that the lengths of the sides are multiples of the following lengths.

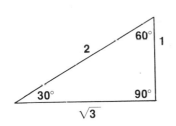
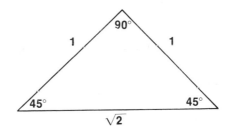

The two special triangles shown above are used in many applications, so it is useful to memorize the exact values of the trigonometric ratios of these angles.

**THE HOW AND
WHY OF IT**

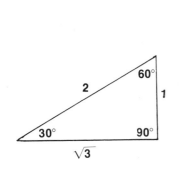
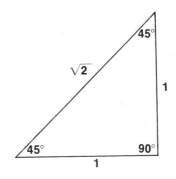

sin 30°	$\dfrac{\text{opp}}{\text{hyp}}$	$\dfrac{1}{2}$
sin 45°	$\dfrac{\text{opp}}{\text{hyp}}$	$\dfrac{1}{\sqrt{2}}$ or $\dfrac{\sqrt{2}}{2}$
sin 60°	$\dfrac{\text{opp}}{\text{hyp}}$	$\dfrac{\sqrt{3}}{2}$

cos 30°	$\dfrac{\text{adj}}{\text{hyp}}$	$\dfrac{\sqrt{3}}{2}$
cos 45°	$\dfrac{\text{adj}}{\text{hyp}}$	$\dfrac{1}{\sqrt{2}}$ or $\dfrac{\sqrt{2}}{2}$
cos 60°	$\dfrac{\text{adj}}{\text{hyp}}$	$\dfrac{1}{2}$

tan 30°	$\dfrac{\text{opp}}{\text{adj}}$	$\dfrac{1}{\sqrt{3}}$ or $\dfrac{\sqrt{3}}{3}$
tan 45°	$\dfrac{\text{opp}}{\text{adj}}$	$\dfrac{1}{1}$ or 1
tan 60°	$\dfrac{\text{opp}}{\text{adj}}$	$\dfrac{\sqrt{3}}{1}$ or $\sqrt{3}$

To find the three ratios of any other acute angle, unless the sides of the triangle are given, we must use a table of values or a calculator. Tables usually show 4 or 5 decimal places and calculators may display 7, 8, or 9 decimal places. (A four-place table is available at the end of this chapter.) In Examples *a* through *d* the four-place table is used; in examples *e* through *h* a calculator was used and the value rounded to five places.

When the sine, cosine, or tangent of an angle is known, the table is read in reverse to find the angle. In Examples *i* through *k* the table was used to find the angles to the nearest tenth of a degree, and a calculator was used to find the value and round it to the nearest thousandth of a degree.

EXAMPLES

a) $\sin 74°$ ≈ 0.9613 (four-place table)

b) $\cos 45°$ ≈ 0.7071 (four-place table)

c) $\tan 17.5°$ ≈ 0.3153 (four-place table)

d) $\cos 66.4°$ ≈ 0.4003 (four-place table)

e) $\sin 37°$ ≈ 0.60182 (calculator)

f) $\cos 22.8°$ ≈ 0.92186 (calculator)

g) $\tan 80.6°$ ≈ 6.04051 (calculator)

h) $\sin 43.2°$ ≈ 0.68455 (calculator)

i) $\sin\ A = 0.3435$
   $A \approx 20.1°$ (four-place table)
   $A \approx 20.090°$ (calculator)

j) $\cos\ B = 0.8321$
   $B \approx 33.7°$ (four-place table)
   $B \approx 33.685°$ (calculator)

k) $\tan\ C = 0.7444$
   $C \approx 36.7°$ (four-place table)
   $C \approx 36.664°$ (calculator)

EXERCISES Find value correct to four decimal places of each trigonometric ratio.

1) $\sin 21°$ 2) $\cos 74°$ 3) $\tan 45°$

4) $\cos 23°$ 5) $\tan 7°$ 6) $\sin 54.8°$

7) cos 26.6°

8) tan 23.4°

9) cos 10.2°

10) sin 82.6°

11) cos 46.2°

12) sin 20.4°

13) tan 8.4°

Find the measure of the following acute angles to the nearest tenth of a degree.

14) $\cos B = 0.6211$

15) $\tan Y = 0.6297$

16) $\sin B = 0.3338$

17) $\tan A = 3.7214$

18) $\cos C = 0.4217$

19) $\sin M = 0.9146$

20) $\sin A = 0.9061$

21) $\cos B = 0.9061$

22) $\tan C = 0.9061$

23) $\cos M = 0.6000$

24) $\tan N = 1.7000$

25) $\sin P = 0.3444$

26) sin T = 0.9849

27) In the application at the beginning of this section, what is the impedance of the circuit?

28) If the tread of a stairway is 20 cm and the rise is 16.5 cm, what angle does the stairway make with the floor? (To the nearest tenth of a degree.)

12.4   RIGHT TRIANGLE SOLUTIONS

APPLICATION   What angle does a rafter make with the horizontal if it has a rise of 4.5 feet in a horizontal run of 13.2 feet?

OBJECTIVE   136) Solve a right triangle.

VOCABULARY   To *solve* a right triangle is to find the measures of the sides and angles not already given.

THE HOW AND   If three parts of a right triangle are known, including the right angle
WHY OF IT   and at least one side, the measures of the remaining angles and sides can be found. To find them we can use any of the following formulas.

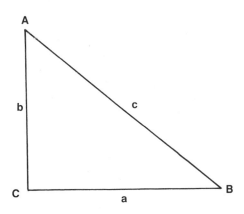

$$a^2 + b^2 = c^2 \qquad\qquad A + B + C = 180°$$

$$\sin A = \frac{a}{c} \qquad\qquad \cos A = \frac{b}{c} \qquad\qquad \tan A = \frac{a}{b}$$

$$\sin B = \frac{b}{c} \qquad\qquad \cos B = \frac{a}{c} \qquad\qquad \tan B = \frac{b}{a}$$

In the use of tables or calculators for trigonometric ratios, it is under-stood that the values are approximations. The Examples illustrate how the formulas may be applied.

EXAMPLES

a)

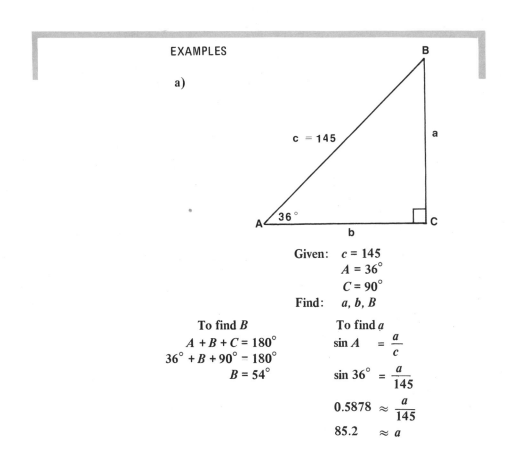

Given: $c = 145$
         $A = 36°$
         $C = 90°$

Find: a, b, B

To find B	To find a
$A + B + C = 180°$	$\sin A = \dfrac{a}{c}$
$36° + B + 90° - 180°$	
$B = 54°$	$\sin 36° = \dfrac{a}{145}$
	$0.5878 \approx \dfrac{a}{145}$
	$85.2 \approx a$

To find b

$$\cos A \ = \frac{b}{c}$$

$$\cos 36° = \frac{b}{145}$$

$$0.8090 \approx \frac{b}{145}$$

$$117.3 \ \approx b$$

b)

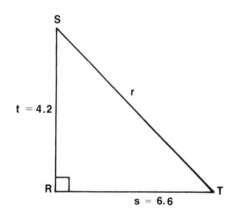

Given: $R = 90°$
$t = 4.2$
$s = 6.6$

Find: r, S, T

To find r

$$t^2 + s^2 = r^2$$
$$17.64 + 43.56 = r^2$$
$$7.8 \approx r$$

To find S

$$\tan S \ = \frac{s}{t}$$

$$\tan S \ = \frac{6.6}{4.2} \approx 1.5714$$

$$S \approx 57.5°$$

$$S + R + T = 180°$$
$$57.5° + 90° + T \approx 180°$$
$$T \approx 32.5°$$

c) A tunnel is to be dug down at an angle of 11.3° from a level surface. What is the vertical distance (to the nearest tenth of a meter) between two points that are 500 meters apart along the tunnel? First sketch a figure and label it.

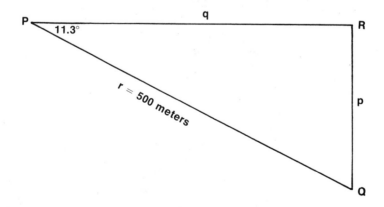

$$\overline{PQ} = r = 500 \text{ meters}$$
$$P = 11.3°$$

Find p

$$\sin P = \frac{p}{r} \quad \text{So, } \sin 11.3° = \frac{p}{500}$$

$$0.1959 \approx \frac{p}{500}$$

$$98.0 \approx p$$

So the vertical distance is approximately 98 meters.

Find lengths to the nearest tenth of a unit and angles to the nearest tenth of a degree.

1) Given: $e = 6$
   $E = 30°$
   $D = 90°$

   Find: d, f, F

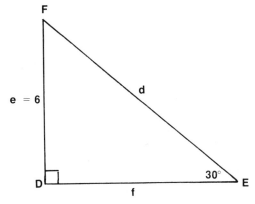

2) Given: $r = 10.1$
   $Q = 45°$
   $P = 90°$

   Find: p, q, R

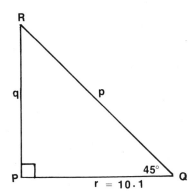

3) Given: $b = 7$
   $a = 10$
   $A = 90°$

   Find: c, B, C

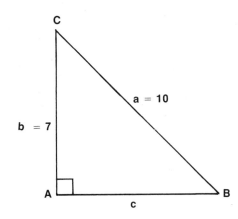

4) Given: $a = 37.4$
           $b = 25$
           $C = 90°$

    Find: c, A, B

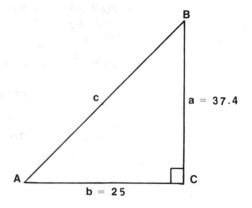

5) Given: $s = 77.3$
           $R = 21.1°$
           $S = 90°$

    Find: r, t, T

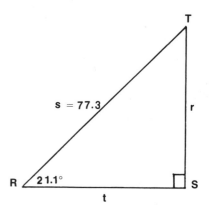

6) Given: $x = 305.0$
           $Z = 41.1°$
           $Y = 90°$

    Find: y, z, X

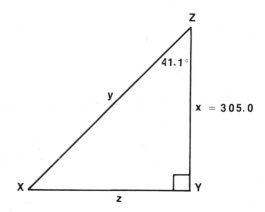

7) In the application at the beginning of this section, what angle does the rafter make with the horizontal?

8) When the shadow of an office building is 25 feet long and the angle of elevation of the sun is 80°, how tall is the building?

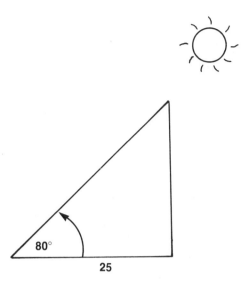

9) Find the length of a guy wire that stretches from the top of a 50-foot pole to the ground if it makes a 63° angle with the ground.

10) A pilot sights a wrecked aircraft at an angle of depression of 35°. How far away is the wreck site if the pilot is flying at 3500 feet? (Hint: the angle of depression is measured from the flight path to the wreck.)

11) The truss of a bridge is shown below. What is the height of the truss?

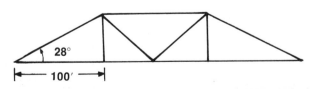

12) A light plane flies 55 miles east from C and then flies 40 miles north. What is the bearing to the nearest degree of the plane from C? (Angle C = ?)

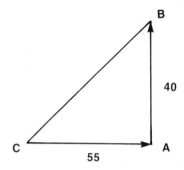

13) A power line tower is known to be 70 feet high and is located on the bank of a river. If the angle from the horizontal, measured from the opposite bank to the top of the tower is 39°, how wide is the river?

12.5   TRIGONOMETRIC RATIOS OF OBTUSE ANGLES

APPLICATION

In order to find the distance across a lake, a surveyor lays out an angle of 130° by sighting the ends of the lake. One calculation in finding the distance is to find the cosine 130°.

OBJECTIVE

137) Find the approximate value of the sine, cosine, and tangent of an obtuse angle using a table or calculator with trigonometric keys.

VOCABULARY

No new vocabulary.

THE HOW AND
WHY OF IT

The trigonometric ratio of an obtuse angle can be approximated by using the tables for acute angles and the following formulas.

> If A is an obtuse angle,
> $\sin A = \sin (180 - A)$
> $\cos A = -\cos (180 - A)$
> $\tan A = -\tan (180 - A)$

These formulas, as well as those for larger angles, are developed in a formal treatment of trigonometry.

EXAMPLES

a)   **Find the sine, cosine, and tangent of $120°$**
$\sin 120° = \sin (180° - 120°) = \sin 60° \approx 0.8660$
$\cos 120° = -\cos (180° - 120°) = -\cos 60° = -0.5000$
$\tan 120° = -\tan (180° - 120°) = -\tan 60° \approx -1.7321$

b)   **Find the sine, cosine, and tangent of $165°$.**
$\sin 165° = \sin 15° \approx 0.2588$
$\cos 165° = -\cos 15° \approx -0.9659$
$\tan 165° = -\tan 15° \approx -0.2679$

Find the value correct to four decimal places of each trigonometric ratio.   **EXERCISES**

1) $\sin 130°$ 2) $\cos 130°$ 3) $\tan 130°$

4) $\sin 127°$ 5) $\cos 135°$ 6) $\tan 98°$

7) $\cos 104°$ 8) $\sin 115°$ 9) $\cos 162°$

10) $\tan 147°$ 11) $\sin 121.7°$ 12) $\cos 135.2°$

13) $\tan 99.3°$ 14) $\tan 108.5°$ 15) $\sin 118.6°$

16) cos 111.1°   17) sin 152.9°   18) tan 140.8°

19) cos 132.4°

20) In the application at the beginning of this section, what is cos 130°?

12.6   OBTUSE ANGLES FROM TRIGONOMETRIC RATIOS

APPLICATION

The major application is in the solution of triangles.

OBJECTIVE

138) Find the measure of an obtuse angle given the value of its sine, cosine, or tangent.

VOCABULARY

No new vocabulary.

THE HOW AND WHY OF IT

To find the obtuse angle given one of its trigonometric ratios, we use the absolute value of the ratio to find an acute angle from the table. This angle is then subtracted from 180°.

> If A is an obtuse angle and
> if $\sin A = w$, then $A = 180° -$ (acute angle whose sin $= |w|$)
> if $\cos A = x$, then $A = 180° -$ (acute angle whose cos $= |x|$)
> if $\tan A = y$, then $A = 180° -$ (acute angle whose tan $= |y|$)

A is identified as being obtuse if the cosine is negative or the tangent is negative.

Find the obtuse angle whose tangent is −0.9753. That is, find A if $\tan A = -0.9753$.

$A = 180° - (\text{acute angle whose tan} = |-0.9753|)$
$A = 180° - (\text{acute angle whose tan} = 0.9753)$
$A \approx 180° - 44.3°$
$A \approx 135.7°$

EXAMPLES

a) Find A if $\cos A = -0.3656$

A is obtuse since the cosine is negative.

$A = 180° - (\text{acute angle whose cos} = |-0.3656|)$
$A \approx 180° - 68.6°$
$A \approx 111.4°$

b) Find A if $\sin A = 0.3567$

A can be either acute or obtuse.
$A = \text{acute angle whose sin} = 0.3567$ or $A = 180° - (\text{acute angle}$
$\qquad\qquad\qquad\qquad\qquad\qquad\qquad\qquad\text{whose sin} = 0.3567)$

$A \approx 20.9°$ or $A \approx 180° - 20.9°$
$\qquad\qquad\qquad\qquad\qquad A \approx 159.1°$

c) Find B if $\tan B = -0.5631$

B is obtuse since the tangent is negative.

$B = 180° - (\text{acute angle whose tan} = |-0.5631|)$
$B \approx 180° - 29.4°$
$B \approx 150.6°$

d) Find C if $\cos C = -0.7695$

The angle is obtuse since the cosine is negative.

$C = 180° - (\text{acute angle whose cos} = |-0.7695|)$
$C \approx 180° - 39.7°$
$C \approx 140.3°$

Find the obtuse angle to the nearest tenth of a degree. **EXERCISES**

1) $\sin A = 0.6123$ 2) $\cos A = -0.2561$

3) $\tan B = -0.8230$

4) $\tan C = -0.5218$

5) $\sin A = 0.9123$

6) $\cos Y = -0.0982$

7) $\tan B = -1.6234$

8) $\cos B = -0.9200$

9) $\sin B = 0.9761$

10) $\cos Y = -0.1111$

11) $\tan A = -1.7111$

12) $\sin A = 0.3456$

13) $\cos A = -0.4682$

14) $\tan B = -2.2461$

15) $\sin X = 0.8901$

16) $\cos X = -0.1098$

17) $\sin C = 0.6534$

18) $\tan A = -0.9820$

19) $\cos A = -0.8902$

20) $\sin B = 0.2089$

12.7   THE LAW OF COSINES

A piece of aluminum 8 inches wide is to be formed into an eaves   APPLICATION
trough with a 5-inch opening at the top. What is the angle formed by the
sides of the trough?

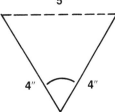

139) Solve an oblique triangle given all three sides and no angles.   OBJECTIVES
140) Solve an oblique triangle given two sides and the angle formed by them.

An *oblique triangle* is any triangle that does not contain a right angle.   VOCABULARY
The *law of cosines* is an expression stating the relationship between
the cosines of the angles of a triangle and the sides. Given the triangle
ABC,

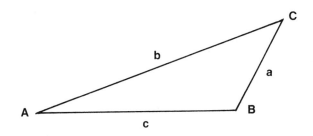

the law of cosines is stated as follows:

$$a^2 = b^2 + c^2 - 2bc \cos A$$
$$b^2 = a^2 + c^2 - 2ac \cos B$$
$$c^2 = a^2 + b^2 - 2ab \cos C$$

As with any formula, if all of the unknowns are given except for one,   THE HOW AND
it can be found by substituting the known ones in the formula. This   WHY OF IT
follows for each of the forms of the law of cosines. Given triangle *ABC*
with $a = 21$, $b = 32$, and $c = 15$, find the size of the angles.

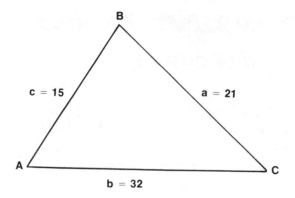

Since all sides are given, any one of the angles can be found by the law of cosines. First find A:

$$a^2 = b^2 + c^2 - 2bc \cos A$$
$$2bc \cos A = b^2 + c^2 - a^2$$
$$\cos A = \frac{b^2 + c^2 - a^2}{2bc}$$
$$\cos A = \frac{(32)^2 + (15)^2 - (21)^2}{2(32)(15)}$$
$$\cos A = \frac{1024 + 225 - 441}{960} = \frac{808}{960}$$
$$\cos A \approx 0.8417$$
$$A \approx 32.7°$$

Now find B:

$$b^2 = a^2 + c^2 - 2ac \cos B$$
$$2ac \cos B = a^2 + c^2 - b^2$$
$$\cos B = \frac{a^2 + c^2 - b^2}{2ac}$$
$$\cos B = \frac{(21)^2 + (15)^2 - (32)^2}{2(21)(15)}$$
$$\cos B = \frac{441 + 225 - 1024}{630} = \frac{-358}{630}$$
$$\cos B \approx -0.5683$$
$$B \approx 124.6°$$

Since the sum of the angles is $180°$,

$$C \approx 180° - 124.6° - 32.7°$$
$$C \approx 22.7°$$

The angles of the triangle are $124.6°$, $32.7°$, and $22.7°$.

EXAMPLES

a) Solve the triangle ABC given $a = 13.7$, $b = 12.5$, and $C = 49.7°$

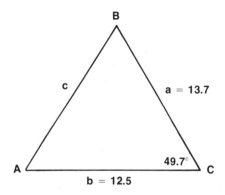

Side c can be found by the law of cosines using

$$c^2 = a^2 + b^2 - 2ab \cos C$$

$$c^2 = (13.7)^2 + (12.5)^2 - 2(13.7)(12.5)(\cos 49.7°)$$

$$c^2 \approx (13.7)^2 + (12.5)^2 - 2(13.7)(12.5)(0.6468)$$

$$c^2 \approx 187.69 + 156.25 - 221.53$$

$$c^2 \approx 122.41$$

$c \approx 11.1$ (nearest tenth)

Angle B can also be found using the law of cosines:

$$\cos B = \frac{a^2 + c^2 - b^2}{2ac}$$

$$\cos B \approx \frac{(13.7)^2 + (11.1)^2 - (12.5)^2}{2(13.7)(11.1)}$$

$$\cos B \approx \frac{187.69 + 123.21 - 156.25}{304.14} = \frac{154.65}{304.14}$$

$$\cos B \approx 0.5085$$

$$B \approx 59.4°$$

The remaining angle must bring the total to $180°$; therefore,

$$A \approx 180° - 59.4° - 49.7°$$

$$A \approx 70.9°$$

The solution of the triangle is $A \approx 70.9°$, $B \approx 59.4°$, $C = 49.7°$, $a = 13.7$, $b = 12.5$, and $c \approx 11.1$.

b) The distance across a lake can be found by measuring the distance from a point to each end of the lake as well as the angle these lines form.

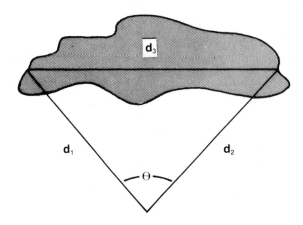

Find the distance across the lake if $d_1 = 2$ mi, $d_2 = 2.6$ mi, and $\theta = 30°$. Using the law of cosines we have

$$(d_3)^2 = (d_1)^2 + (d_2)^2 - 2(d_1)(d_2) \cos \theta$$

$$(d_3)^2 = (2)^2 + (2.6)^2 - 2(2)(2.6) \cos 30°$$

$$(d_3)^2 \approx 4 + 6.76 - 10.4 (0.8660)$$

$$(d_3)^2 \approx 10.76 - 9.01$$

$$(d_3)^2 \approx 1.75$$

$$d_3 \approx 1.3 \text{ (nearest tenth)}$$

Therefore, the lake is approximately 1.3 miles across.

EXERCISES

1) In the application at the beginning of this section, what is the measure of the angle to the nearest degree?

2) Two cars leave point A traveling on different roads that form a 75° angle with each other. How far apart are they at the end of 1 hour if one averages 55 mph and the other one 60 mph? (To the nearest mile.)

3) The distance across a lake is found by sighting both ends from a point away from the lake. If one end is 120 feet and the other 150 feet from the point and the angle between the lines of sight is 70°, what is the distance across the lake? (To the nearest foot.)

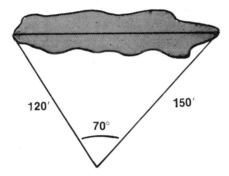

4)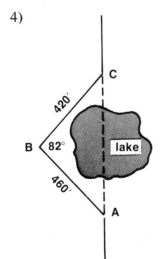

A bridge from point A to point C is to be built to replace the detour around the lake. What is the length of the bridge, using the given information? (To the nearest foot.)

5)

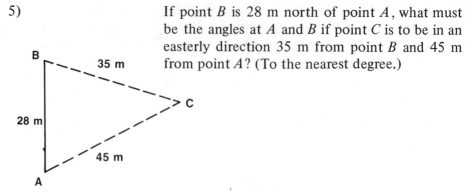

If point B is 28 m north of point A, what must be the angles at A and B if point C is to be in an easterly direction 35 m from point B and 45 m from point A? (To the nearest degree.)

6) A ship travels a course directly west for 4 hours at a speed of 20 knots. For the next four hours the ship travels at 18 knots in a direction that is $25°$ west of north. How far is the ship from its starting point? (To the nearest knot.)

7) A surveyor wants to find the length of a lake. He places three stakes at points A, B, and C. He measures the distance $\overline{AB}$ and $\overline{AC}$ to the nearest ten feet and finds that $\overline{AB}$ = 920 feet and $\overline{AC}$ = 530 feet. The angle at A is found to be $130°$. What is the length of the lake to the nearest 10 feet?

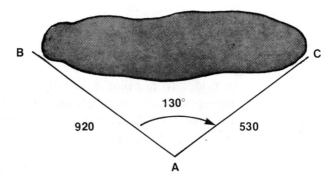

8) In physics when two *forces* act on an object at the same time, the object will not move in the direction of either force, but rather in a direction between the forces along a line called the *resultant*. The resultant is pictured by the diagonal of a parallelogram whose ad-

jacent sides represent the magnitude of the two forces. (See the figure below.)

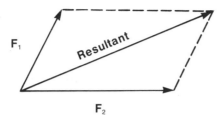

Two forces act on an object so that the angle between them is $57.6°$. If one force has a magnitude of 78 pounds and the other a magnitude of 92 pounds, find the magnitude of the resultant force. (Hint: Solve triangle RST where $\overline{ST} = 92$, $\overline{TR} = 78$, and $T = 180° - 57.6° = 122.4°$. Find $\overline{RS}$ to the nearest pound.)

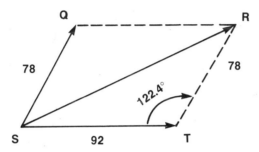

9) When two forces act on an object with magnitudes of 200 pounds and 375 pounds, respectively, the resultant force has a magnitude of 480 pounds. Find the angle between the two forces to the nearest tenth of a degree. (Hint: First find angle B and then subtract from $180°$.)

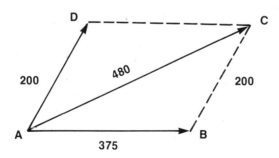

12.8 THE LAW OF SINES

APPLICATION

What will be the width of a house if the rafters are 16′ in length with a 2′ overhang? The angle formed by the rafters at the ridge is 80°. The rafter will form equal angles with the width of the house.

OBJECTIVES

141) Solve an oblique triangle given any two sides and the angle opposite one of them.

142) Solve an oblique triangle given any two angles and the side opposite one of them.

VOCABULARY

The law of sines is a rule giving the relationship between the sides of a triangle and the sines of the angles.

Given triangle ABC,

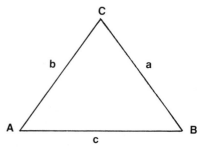

the law of sines is

$$\frac{a}{\sin A} = \frac{b}{\sin B} = \frac{c}{\sin C}$$

THE HOW AND WHY OF IT

The law of sines is a statement of equality among three ratios. If three parts of any pair of the ratios are known, the fourth part can be found. To identify which parts are known, sketch the triangle and label the parts.

Solve the triangle given $A = 75°$, $a = 10$, and $c = 7$. (Solving the triangle means finding B, C, and b.) First sketch the triangle.

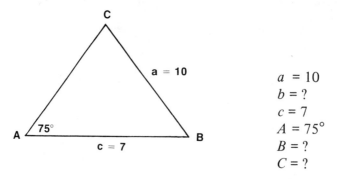

$a = 10$
$b = ?$
$c = 7$
$A = 75°$
$B = ?$
$C = ?$

Since A, a, and c are given, we use this portion of the law of sines:

$$\frac{a}{\sin A} = \frac{c}{\sin C}$$

$$\frac{10}{\sin 75°} = \frac{7}{\sin C}$$

$$\sin C = \frac{7\,(\sin 75°)}{10}$$

$$\sin C \approx \frac{(7)\,(0.9659)}{10}$$

$$\sin C \approx .6761$$

$$C \approx 42.5°$$

Now two angles are known and we can find the third.

$$A + B + C = 180°$$

$$75° + B + 42.5° \approx 180°$$

$$B \approx 62.5°$$

To find b use the following ratios:

$$\frac{a}{\sin A} = \frac{b}{\sin B}$$

$$\frac{10}{\sin 75°} \approx \frac{b}{\sin 62.5°}$$

$$b \approx \frac{10\,(\sin 62.5°)}{\sin 75°}$$

$$b \approx \frac{10\,(0.8870)}{0.9659}$$

$$b \approx 9.2 \text{ (nearest tenth)}$$

The solution of the triangle is $A = 75°$, $B \approx 62.5°$, $C \approx 42.5°$, $a = 10$, $b \approx 9.2$, and $c = 7$.

If side a had been equal to 6.8, there would have been two possible solutions.

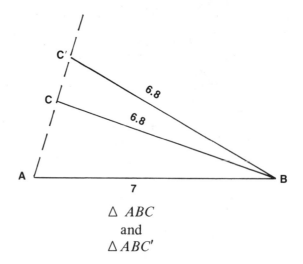

$\triangle\ ABC$
and
$\triangle\ ABC'$

This is known as the "ambiguous case." In this text the problem will dictate which solution is desired. A formal treatment of the ambiguous case is a subject of trigonometry.

EXAMPLES

a)   Solve the triangle given $A = 133.5°$, $B = 36.4°$, and $b = 42$.

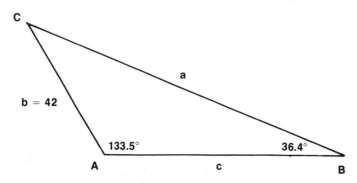

Since two angles are given, the third can be found.

$$A + B + C\ =\ 180°$$

$$133.5° + 36.4° + C\ =\ 180°$$

$$C\ =\ 10.1°$$

Side a can be found using

$$\frac{a}{\sin A}\ =\ \frac{b}{\sin B}$$

$$\frac{a}{\sin 133.5°}\ =\ \frac{42}{\sin 36.4°}$$

$$a = \frac{42\,(\sin 133.5^\circ)}{\sin 36.4^\circ}$$

$$\sin 133.5^\circ = \sin(180^\circ - 133.5^\circ)$$
$$= \sin 46.5^\circ$$

$$a = \frac{42\,(\sin 46.5^\circ)}{\sin 36.4^\circ}$$

$$a \approx \frac{42\,(0.7254)}{0.5934}$$

$$a \approx 51.3 \text{ (nearest tenth)}$$

Side c can be found using

$$\frac{b}{\sin B} = \frac{c}{\sin C}$$

$$\frac{42}{\sin 36.4^\circ} = \frac{c}{\sin 10.1^\circ}$$

$$c = \frac{42\,(\sin 10.1^\circ)}{\sin 36.4^\circ}$$

$$c \approx \frac{42\,(0.1754)}{0.5934}$$

$$c \approx 12.4 \text{ (nearest tenth)}$$

The solution of the triangle is $A = 133.5^\circ$, $B = 36.4^\circ$, $C = 10.1^\circ$, $a \approx 51.3$, $b = 42$, and $c \approx 12.4$.

b) The diagonal of a parallelogram makes angles of 32° and 41.5° with the sides. If the diagonal is 20 inches long, what are the lengths of the sides of the parallelogram?

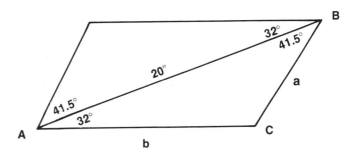

In $\triangle ABC$ we see that $A = 32^\circ$, $B = 41.5^\circ$, and $c = 20''$. Since two angles are known, the third can be found.

$$C = 180^\circ - 32^\circ - 41.5^\circ$$

$$C = 106.5^\circ$$

Side b can be found by using

$$\frac{b}{\sin B} = \frac{c}{\sin C}$$

$$b = \frac{c\,(\sin B)}{\sin C}$$

$$b = \frac{20\,(\sin 41.5°)}{\sin 106.5°}$$

$$b = \frac{20\,(\sin 41.5°)}{\sin 73.5°}$$

$$b \approx \frac{20\,(0.6626)}{0.9588}$$

$$b \approx 13.8 \text{ (to the nearest tenth)}$$

Side a can be found by using

$$\frac{a}{\sin A} = \frac{c}{\sin C}$$

$$a = \frac{c\,(\sin A)}{\sin C}$$

$$a = \frac{20\,(\sin 32°)}{\sin 106.5°}$$

$$a \approx \frac{20\,(0.5299)}{0.9588}$$

$$a \approx 11.1 \text{ (to the nearest tenth)}$$

Therefore, to the nearest tenth the lengths of the sides of the parallelogram are 13.8″ and 11.1″.

EXERCISES

1) In the application at the beginning of this section, what is the width of the house? (To the nearest foot.)

2) An observer at point A is 600 yards directly west of an observer at point B. They are each looking at point C, which is located 55° east of north from A and 65° west of north from B. Which observer is closer to point C and by how many yards? (To the nearest yard.)

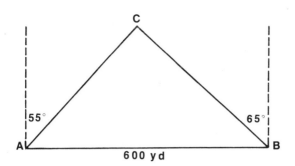

3) Two observers who are 500 m apart observe an airplane from opposite sides at the same point in time. One calculates the angle of elevation to be 46.3°, while the other calculates it to be 58.6°. What is the height of the airplane? (To the nearest ten meters.)

4) Two searchlights spot a plane in the air. If the light beams make angles of 47° and 58° with the ground, and the lights are 5000 feet apart, find the height of the plane to the nearest 100 feet. (Hint: First find $\overline{BC}$.)

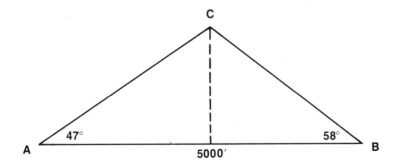

5) Two forces (F_1 and F_2) are to be applied to an object at R to produce a resultant force with magnitude 165 kg. The magnitude of F_1 is 93 kg. If the angle between F_1 and the resultant is 28°, and the angle between the resultant and F_2 is 37°, find the magnitude of F_2 to the nearest kg. (Hint: $U = 180° - 28° - 37°$. Find $\overline{TU}$ in triangle TUR.)

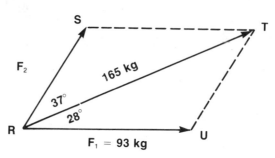

6) An isosceles triangle has equal sides of length 20′ and equal angles of 50.3°. Find the length of the third side. (To the nearest foot.)

7) A 400-foot bridge spans a canyon in central Oregon. In order to calculate the depth of the canyon, engineers measured the angle from the bridge to the deepest point in the canyon from both ends of the bridge. (See the diagram below.) What is the depth (d) of the canyon to the nearest foot?

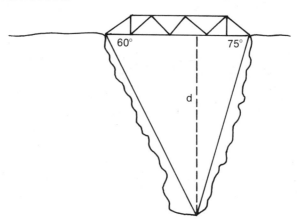

8) Two forces F_1 and F_2 are applied to an object at point P to produce a resultant force. F_1 has magnitude 55 pounds and F_2 has magnitude 40 pounds. If the angle between F_1 and the resultant is to be 70°, find the angle between the two forces to the nearest degree and the magnitude of the resultant to the nearest pound. (Hint: angle $WPT = 70°$. Find angle PTW; then $W = 180° -$ angle $PTW - 70°$, and angle $WPM = 180° - W$)

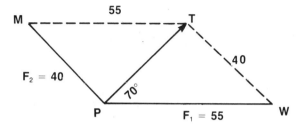

CHAPTER 12 TEST

1. (Obj. 133)   Find the exact value of sin 45°.   _____

2. (Obj. 138)   If sin R = 0.5650, find R to the nearest tenth of a degree (two values).   _____

3. (Obj. 131)   In $\triangle DEF$, E = 16.4° and F = 122°. What is the measure of D?   _____

4. (Obj. 140)   Solve $\triangle ABC$ given a = 5, b = 7, and C = 48°. (Find all measures to the nearest tenth.)   _____

5. (Obj. 136)   In $\triangle DEF$, d = 12, e = 9, and F = 90°. Find f, D, and E. (To the nearest tenth.)   _____

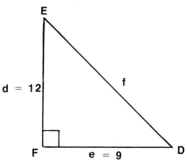

6. (Obj. 141)   Solve $\triangle ABC$ given A = 63°, a = 15, and c = 9. (Find all measures to the nearest tenth.)   _____

7. (Obj. 131)   In $\triangle ABC$, A = 42° and B = 36°. What is the measure of C?   _____

8. (Obj. 138)   If tan Q = −5.2422, find Q. (To the nearest tenth of a degree.)   _____

9. (Obj. 137)   Determine the approximate value of cos 160.4°.   _____

10. (Obj. 132)   In $\triangle ABC$ find sin A, cos A, and tan A.   _____

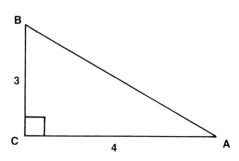

11. (Obj. 132) In $\triangle XYZ$ determine the exact value of $\sin Y$, $\cos Y$, and $\tan Y$. _____

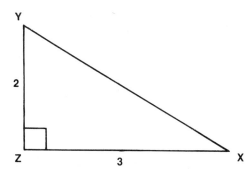

12. (Obj. 134) Determine the approximate value of $\cos 32.7°$. _____

13. (Obj. 136) In $\triangle ABC$, $a = 8$, $C = 90°$, and $A = 40°$. Find c, b, and B. (To the nearest tenth.) _____

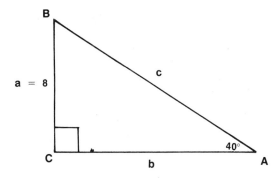

14. (Obj. 137) Determine the approximate value of $\sin 136°$. _____

15. (Obj. 137) Determine the approximate value of $\tan 136°$. _____

16. (Obj. 139) Solve $\triangle ABC$ given $a = 6$, $b = 8$, and $c = 12$. (Find angles to the nearest tenth of a degree.) _____

17. (Obj. 135) If $\cos P = 0.6691$, find P. (To the nearest tenth of a degree.) _____

18. (Obj. 142) Solve $\triangle ABC$ given $A = 28°$, $B = 124°$, and $c = 48$. (Find all measures to the nearest tenth.) _____

19. (Obj. 140) A fishing boat leaves its home port in a due west direction, traveling to a lightship that is 47 miles out to sea. The boat then turns to the right 125° and travels to a buoy that is known to be 22 miles from the lightship. How far is the boat from its home port? (To the nearest mile.)

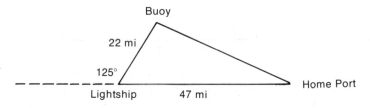

20. (Obj. 142) The base of one face of a pyramid is 350′ and the angles the edges make with the base are both 75°. What is the length of the edge of the pyramid? (To the nearest ten feet.)

Table of Trigonometric Functions

Trigonometric (degrees)

Deg.	Sin	Tan	* Cot	Cos		Deg.	Sin	Tan	Cot	Cos	
0.0	0.00000	0.00000	∞	1.0000	**90.0**	**6.0**	0.10453	0.10510	9.514	0.9945	**84.0**
.1	.00175	.00175	573.0	1.0000	89.9	.1	.10626	.10687	9.357	.9943	83.9
.2	.00349	.00349	286.5	1.0000	.8	.2	.10800	.10863	9.205	.9942	.8
.3	.00524	.00524	191.0	1.0000	.7	.3	.10973	.11040	9.058	.9940	.7
.4	.00698	.00698	143.24	1.0000	.6	.4	.11147	.11217	8.915	.9938	.6
.5	.00873	.00873	114.59	1.0000	.5	.5	.11320	.11394	8.777	.9936	.5
.6	.01047	.01047	95.49	0.9999	.4	.6	.11494	.11570	8.643	.9934	.4
.7	.01222	.01222	81.85	.9999	.3	.7	.11667	.11747	8.513	.9932	.3
.8	.01396	.01396	71.62	.9999	.2	.8	.11840	.11924	8.386	.9930	.2
.9	.01571	.01571	63.66	.9999	89.1	.9	.12014	.12101	8.264	.9928	83.1
1.0	0.01745	0.01746	57.29	0.9998	**89.0**	**7.0**	0.12187	0.12278	8.144	0.9925	**83.0**
.1	.01920	.01920	52.08	.9998	88.9	.1	.12360	.12456	8.028	.9923	82.9
.2	.02094	.02095	47.74	.9998	.8	.2	.12533	.12633	7.916	.9921	.8
.3	.02269	.02269	44.07	.9997	.7	.3	.12706	.12810	7.806	.9919	.7
.4	.02443	.02444	40.92	.9997	.6	.4	.12880	.12988	7.700	.9917	.6
.5	.02618	.02619	38.19	.9997	.5	.5	.13053	.13165	7.596	.9914	.5
.6	.02792	.02793	35.80	.9996	.4	.6	.13226	.13343	7.495	.9912	.4
.7	.02967	.02968	33.69	.9996	.3	.7	.13399	.13521	7.396	.9910	.3
.8	.03141	.03143	31.82	.9995	.2	.8	.13572	.13698	7.300	.9907	.2
.9	.03316	.03317	30.14	.9995	88.1	.9	.13744	.13876	7.207	.9905	82.1
2.0	0.03490	0.03492	28.64	0.9994	**88.0**	**8.0**	0.13917	0.14054	7.115	0.9903	**82.0**
.1	.03664	.03667	27.27	.9993	87.9	.1	.14090	.14232	7.026	.9900	81.9
.2	.03839	.03842	26.03	.9993	.8	.2	.14263	.14410	6.940	.9898	.8
.3	.04013	.04016	24.90	.9992	.7	.3	.14436	.14588	6.855	.9895	.7
.4	.04188	.04191	23.86	.9991	.6	.4	.14608	.14767	6.772	.9893	.6
.5	.04362	.04366	22.90	.9990	.5	.5	.14781	.14945	6.691	.9890	.5
.6	.04536	.04541	22.02	.9990	.4	.6	.14954	.15124	6.612	.9888	.4
.7	.04711	.04716	21.20	.9989	.3	.7	.15126	.15302	6.535	.9885	.3
.8	.04885	.04891	20.45	.9988	.2	.8	.15299	.15481	6.460	.9882	.2
.9	.05059	.05066	19.74	.9987	87.1	.9	.15471	.15660	6.386	.9880	81.1
3.0	0.05234	0.05241	19.081	0.9986	**87.0**	**9.0**	0.15643	0.15838	6.314	0.9877	**81.0**
.1	.05408	.05416	18.464	.9985	86.9	.1	.15816	.16017	6.243	.9874	80.9
.2	.05582	.05591	17.886	.9984	.8	.2	.15988	.16196	6.174	.9871	.8
.3	.05756	.05766	17.343	.9983	.7	.3	.16160	.16376	6.107	.9869	.7
.4	.05931	.05941	16.832	.9982	.6	.4	.16333	.16555	6.041	.9866	.6
.5	.06105	.06116	16.350	.9981	.5	.5	.16505	.16734	5.976	.9863	.5
.6	.06279	.06291	15.895	.9980	.4	.6	.16677	.16914	5.912	.9860	.4
.7	.06453	.06467	15.464	.9979	.3	.7	.16849	.17093	5.850	.9857	.3
.8	.06627	.06642	15.056	.9978	.2	.8	.17021	.17273	5.789	.9854	.2
.9	.06802	.06817	14.669	.9977	86.1	.9	.17193	.17453	5.730	.9851	80.1
4.0	0.06976	0.06993	14.301	0.9976	**86.0**	**10.0**	0.1736	0.1763	5.671	0.9848	**80.0**
.1	.07150	.07168	13.951	.9974	85.9	.1	.1754	.1781	5.614	.9845	79.9
.2	.07324	.07344	13.617	.9973	.8	.2	.1771	.1799	5.558	.9842	.8
.3	.07498	.07519	13.300	.9972	.7	.3	.1788	.1817	5.503	.9839	.7
.4	.07672	.07695	12.996	.9971	.6	.4	.1805	.1835	5.449	.9836	.6
.5	.07846	.07870	12.706	.9969	.5	.5	.1822	.1853	5.396	.9833	.5
.6	.08020	.08046	12.429	.9968	.4	.6	.1840	.1871	5.343	.9829	.4
.7	.08194	.08221	12.163	.9966	.3	.7	.1857	.1890	5.292	.9826	.3
.8	.08368	.08397	11.909	.9965	.2	.8	.1874	.1908	5.242	.9823	.2
.9	.08542	.08573	11.664	.9963	85.1	.9	.1891	.1926	5.193	.9820	79.1
5.0	0.08716	0.08749	11.430	0.9962	**85.0**	**11.0**	0.1908	0.1944	5.145	0.9816	**79.0**
.1	.08889	.08925	11.205	.9960	84.9	.1	.1925	.1962	5.097	.9813	78.9
.2	.09063	.09101	10.988	.9959	.8	.2	.1942	.1980	5.050	.9810	.8
.3	.09237	.09277	10.780	.9957	.7	.3	.1959	.1998	5.005	.9806	.7
.4	.09411	.09453	10.579	.9956	.6	.4	.1977	.2016	4.959	.9803	.6
.5	.09585	.09629	10.385	.9954	.5	.5	.1994	.2035	4.915	.9799	.5
.6	.09758	.09805	10.199	.9952	.4	.6	.2011	.2053	4.872	.9796	.4
.7	.09932	.09981	10.019	.9951	.3	.7	.2028	.2071	4.829	.9792	.3
.8	.10106	.10158	9.845	.9949	.2	.8	.2045	.2089	4.787	.9789	.2
.9	.10279	.10334	9.677	.9947	84.1	.9	.2062	.2107	4.745	.9785	78.1
6.0	0.10453	0.10510	9.514	0.9945	**84.0**	**12.0**	0.2079	0.2126	4.705	0.9781	**78.0**
	Cos	Cot	* Tan	Sin	Deg.		Cos	Cot	Tan	Sin	Deg.

*Interpolation in this section of the table is inaccurate.

(From Flanders, H., and Price, J.J.: Trigonometry. New York, Academic Press, 1975, pp. 222–225.)

Trigonometric (degrees) (continued)

Deg.	Sin	Tan	Cot	Cos	
12.0	0.2079	0.2126	4.705	0.9781	**78.0**
.1	.2096	.2144	4.665	.9778	77.9
.2	.2113	.2162	4.625	.9774	.8
.3	.2130	.2180	4.586	.9770	.7
.4	.2147	.2199	4.548	.9767	.6
.5	.2164	.2217	4.511	.9763	.5
.6	.2181	.2235	4.474	.9759	.4
.7	.2198	.2254	4.437	.9755	.3
.8	.2215	.2272	4.402	.9751	.2
.9	.2233	.2290	4.366	.9748	77.1
13.0	0.2250	0.2309	4.331	0.9744	**77.0**
.1	.2267	.2327	4.297	.9740	76.9
.2	.2284	.2345	4.264	.9736	.8
.3	.2300	.2364	4.230	.9732	.7
.4	.2317	.2382	4.198	.9728	.6
.5	.2334	.2401	4.165	.9724	.5
.6	.2351	.2419	4.134	.9720	.4
.7	.2368	.2438	4.102	.9715	.3
.8	.2385	.2456	4.071	.9711	.2
.9	.2402	.2475	4.041	.9707	76.1
14.0	0.2419	0.2493	4.011	0.9703	**76.0**
.1	.2436	.2512	3.981	.9699	75.9
.2	.2453	.2530	3.952	.9694	.8
.3	.2470	.2549	3.923	.9690	.7
.4	.2487	.2568	3.895	.9686	.6
.5	.2504	.2586	3.867	.9681	.5
.6	.2521	.2605	3.839	.9677	.4
.7	.2538	.2623	3.812	.9673	.3
.8	.2554	.2642	3.785	.9668	.2
.9	.2571	.2661	3.758	.9664	75.1
15.0	0.2588	0.2679	3.732	0.9659	**75.0**
.1	.2605	.2698	3.706	.9655	74.9
.2	.2622	.2717	3.681	.9650	.8
.3	.2639	.2736	3.655	.9646	.7
.4	.2656	.2754	3.630	.9641	.6
.5	.2672	.2773	3.606	.9636	.5
.6	.2689	.2792	3.582	.9632	.4
.7	.2706	.2811	3.558	.9627	.3
.8	.2723	.2830	3.534	.9622	.2
.9	.2740	.2849	3.511	.9617	74.1
16.0	0.2756	0.2867	3.487	0.9613	**74.0**
.1	.2773	.2886	3.465	.9608	73.9
.2	.2790	.2905	3.442	.9603	.8
.3	.2807	.2924	3.420	.9598	.7
.4	.2823	.2943	3.398	.9593	.6
.5	.2840	.2962	3.376	.9588	.5
.6	.2857	.2981	3.354	.9583	.4
.7	.2874	.3000	3.333	.9578	.3
.8	.2890	.3019	3.312	.9573	.2
.9	.2907	.3038	3.291	.9568	73.1
17.0	0.2924	0.3057	3.271	0.9563	**73.0**
.1	.2940	.3076	3.251	.9558	72.9
.2	.2957	.3096	3.230	.9553	.8
.3	.2974	.3115	3.211	.9548	.7
.4	.2990	.3134	3.191	.9542	.6
.5	.3007	.3153	3.172	.9537	.5
.6	.3024	.3172	3.152	.9532	.4
.7	.3040	.3191	3.133	.9527	.3
.8	.3057	.3211	3.115	.9521	.2
.9	.3074	.3230	3.096	.9516	72.1
18.0	0.3090	0.3249	3.078	0.9511	**72.0**
	Cos	**Cot**	**Tan**	**Sin**	**Deg.**

Deg.	Sin	Tan	Cot	Cos	
18.0	0.3090	0.3249	3.078	0.9511	**72.0**
.1	.3107	.3269	3.060	.9505	71.9
.2	.3123	.3288	3.042	.9500	.8
.3	.3140	.3307	3.024	.9494	.7
.4	.3156	.3327	3.006	.9489	.6
.5	.3173	.3346	2.989	.9483	.5
.6	.3190	.3365	2.971	.9478	.4
.7	.3206	.3385	2.954	.9472	.3
.8	.3223	.3404	2.937	.9466	.2
.9	.3239	.3424	2.921	.9461	71.1
19.0	0.3256	0.3443	2.904	0.9455	**71.0**
.1	.3272	.3463	2.888	.9449	70.9
.2	.3289	.3482	2.872	.9444	.8
.3	.3305	.3502	2.856	.9438	.7
.4	.3322	.3522	2.840	.9432	.6
.5	.3338	.3541	2.824	.9426	.5
.6	.3355	.3561	2.808	.9421	.4
.7	.3371	.3581	2.793	.9415	.3
.8	.3387	.3600	2.778	.9409	.2
.9	.3404	.3620	2.762	.9403	70.1
20.0	0.3420	0.3640	2.747	0.9397	**70.0**
.1	.3437	.3659	2.733	.9391	69.9
.2	.3453	.3679	2.718	.9385	.8
.3	.3469	.3699	2.703	.9379	.7
.4	.3486	.3719	2.689	.9373	.6
.5	.3502	.3739	2.675	.9367	.5
.6	.3518	.3759	2.660	.9361	.4
.7	.3535	.3779	2.646	.9354	.3
.8	.3551	.3799	2.633	.9348	.2
.9	.3567	.3819	2.619	.9342	69.1
21.0	0.3584	0.3839	2.605	0.9336	**69.0**
.1	.3600	.3859	2.592	.9330	68.9
.2	.3616	.3879	2.578	.9323	.8
.3	.3633	.3899	2.565	.9317	.7
.4	.3649	.3919	2.552	.9311	.6
.5	.3665	.3939	2.539	.9304	.5
.6	.3681	.3959	2.526	.9298	.4
.7	.3697	.3979	2.513	.9291	.3
.8	.3714	.4000	2.500	.9285	.2
.9	.3730	.4020	2.488	.9278	68.1
22.0	0.3746	0.4040	2.475	0.9272	**68.0**
.1	.3762	.4061	2.463	.9265	67.9
.2	.3778	.4081	2.450	.9259	.8
.3	.3795	.4101	2.438	.9252	.7
.4	.3811	.4122	2.426	.9245	.6
.5	.3827	.4142	2.414	.9239	.5
.6	.3843	.4163	2.402	.9232	.4
.7	.3859	.4183	2.391	.9225	.3
.8	.3875	.4204	2.379	.9219	.2
.9	.3891	.4224	2.367	.9212	67.1
23.0	0.3907	0.4245	2.356	0.9205	**67.0**
.1	.3923	.4265	2.344	.9198	66.9
.2	.3939	.4286	2.333	.9191	.8
.3	.3955	.4307	2.322	.9184	.7
.4	.3971	.4327	2.311	.9178	.6
.5	.3987	.4348	2.300	.9171	.5
.6	.4003	.4369	2.289	.9164	.4
.7	.4019	.4390	2.278	.9157	.3
.8	.4035	.4411	2.267	.9150	.2
.9	.4051	.4431	2.257	.9143	66.1
24.0	0.4067	0.4452	2.246	0.9135	**66.0**
	Cos	**Cot**	**Tan**	**Sin**	**Deg.**

Trigonometric (degrees) (continued)

Deg.	Sin	Tan	Cot	Cos	
24.0	0.4067	0.4452	2.246	0.9135	**66.0**
.1	.4083	.4473	2.236	.9128	65.9
.2	.4099	.4494	2.225	.9121	.8
.3	.4115	.4515	2.215	.9114	.7
.4	.4131	.4536	2.204	.9107	.6
.5	.4147	.4557	2.194	.9100	.5
.6	.4163	.4578	2.184	.9092	.4
.7	.4179	.4599	2.174	.9085	.3
.8	.4195	.4621	2.164	.9078	.2
.9	.4210	.4642	2.154	.9070	65.1
25.0	0.4226	0.4663	2.145	0.9063	**65.0**
.1	.4242	.4684	2.135	.9056	64.9
.2	.4258	.4706	2.125	.9048	.8
.3	.4274	.4727	2.116	.9041	.7
.4	.4289	.4748	2.106	.9033	.6
.5	.4305	.4770	2.097	.9026	.5
.6	.4321	.4791	2.087	.9018	.4
.7	.4337	.4813	2.078	.9011	.3
.8	.4352	.4834	2.069	.9003	.2
.9	.4368	.4856	2.059	.8996	64.1
26.0	0.4384	0.4877	2.050	0.8988	**64.0**
1	.4399	.4899	2.041	.8980	63.9
.2	.4415	.4921	2.032	.8973	.8
.3	.4431	.4942	2.023	.8965	.7
.4	.4446	.4964	2.014	.8957	.6
.5	.4462	.4986	2.006	.8949	.5
.6	.4478	.5008	1.997	.8942	.4
.7	.4493	5029	1.988	.8934	.3
.8	.4509	.5051	1.980	.8926	.2
.9	.4524	.5073	1.971	.8918	63.1
27.0	0.4540	0.5095	1.963	0.8910	**63.0**
.1	.4555	.5117	1.954	.8902	62.9
.2	.4571	.5139	1.946	.8894	.8
.3	.4586	.5161	1.937	.8886	.7
.4	.4602	.5184	1.929	.8878	.6
.5	.4617	.5206	1 021	.8870	.5
.6	.4633	.5228	1.913	.8862	.4
.7	.4648	.5250	1.905	.8854	.3
.8	.4664	.5272	1.897	.8846	.2
.9	.4679	.5295	1.889	.8838	62.1
28.0	0.4695	0.5317	1.881	0.8829	**62.0**
.1	.4710	.5340	1.873	.8821	61.9
.2	.4726	.5362	1.865	.8813	.8
.3	.4741	.5384	1.857	.8805	.7
.4	.4756	.5407	1.849	.8796	.6
.5	.4772	.5430	1.842	.8788	.5
.6	.4787	.5452	1.834	.8780	.4
.7	.4802	.5475	1.827	.8771	.3
.8	.4818	.5498	1.819	.8763	.2
.9	.4833	.5520	1.811	.8755	61.1
29.0	0.4848	0.5543	1.804	0.8746	**61.0**
.1	.4863	.5566	1.797	.8738	60.9
.2	.4879	.5589	1.789	.8729	.8
.3	.4894	.5612	1.782	.8721	.7
.4	.4909	.5635	1.775	.8712	.6
.5	.4924	.5658	1.767	.8704	.5
.6	.4939	.5681	1.760	.8695	.4
.7	.4955	.5704	1.753	.8686	.3
.8	.4970	.5727	1.746	.8678	.2
.9	.4985	.5750	1.739	.8669	60.1
30.0	0.5000	0.5774	1.732	0.8660	**60.0**
	Cos	Cot	Tan	Sin	Deg.

Deg.	Sin	Tan	Cot	Cos	
30.0	0.5000	0.5774	1.7321	0.8660	**60.0**
.1	.5015	.5797	1.7251	.8652	59.9
.2	.5030	.5820	1.7182	.8643	.8
.3	.5045	.5844	1.7113	.8634	.7
.4	.5060	.5867	1.7045	.8625	.6
.5	.5075	.5890	1.6977	.8616	.5
.6	.5090	.5914	1.6909	.8607	.4
.7	.5105	.5938	1.6842	.8599	.3
.8	.5120	.5961	1.6775	.8590	.2
.9	.5135	.5985	1.6709	.8581	59.1
31.0	0.5150	0.6009	1.6643	0.8572	**59.0**
.1	.5165	.6032	1.6577	.8563	58.9
.2	.5180	.6056	1.6512	.8554	.8
.3	.5195	.6080	1.6447	.8545	.7
.4	.5210	.6104	1.6383	.8536	.6
.5	.5225	.6128	1.6319	.8526	.5
.6	.5240	.6152	1.6255	.8517	.4
.7	.5255	.6176	1.6191	.8508	.3
.8	.5270	.6200	1.6128	.8499	.2
.9	.5284	.6224	1.6066	.8490	58.1
32.0	0.5299	0.6249	1.6003	0.8480	**58.0**
.1	.5314	.6273	1.5941	.8471	57.9
.2	.5329	.6297	1.5880	.8462	.8
.3	.5344	.6322	1.5818	.8453	.7
.4	.5358	.6346	1.5757	.8443	.6
.5	.5373	.6371	1.5697	.8434	.5
.6	.5388	.6395	1.5637	.8425	.4
.7	.5402	.6420	1.5577	.8415	.3
.8	.5417	.6445	1.5517	.8406	.2
.9	.5432	.6469	1.5458	.8396	57.1
33.0	0.5446	0.6494	1.5399	0.8387	**57.0**
.1	.5461	.6519	1.5340	.8377	56.9
.2	.5476	.6544	1.5282	.8368	.8
.3	.5490	.6569	1.5224	.8358	.7
.4	.5505	.6594	1.5166	.8348	.6
.5	.5519	.6619	1.5108	.8339	.5
.6	.5534	.6644	1.5051	.8329	.4
.7	.5548	.6669	1.4994	.8320	.3
.8	.5563	.6694	1.4938	.8310	.2
.9	.5577	.6720	1.4882	.8300	56.1
34.0	0.5592	0.6745	1.4826	0.8290	**56.0**
.1	.5606	.6771	1.4770	.8281	55.9
.2	.5621	.6796	1.4715	.8271	.8
.3	.5635	.6822	1.4659	.8261	.7
.4	.5650	.6847	1.4605	.8251	.6
.5	.5664	.6873	1.4550	.8241	.5
.6	.5678	.6899	1.4496	.8231	.4
.7	.5693	.6924	1.4442	.8221	.3
.8	.5707	.6950	1.4388	.8211	.2
.9	.5721	.6976	1.4335	.8202	55.1
35.0	0.5736	0.7002	1.4281	0.8192	**55.0**
.1	.5750	.7028	1.4229	.8181	54.9
.2	.5764	.7054	1.4176	.8171	.8
.3	.5779	.7080	1.4124	.8161	.7
.4	.5793	.7107	1.4071	.8151	.6
.5	.5807	.7133	1.4019	.8141	.5
.6	.5821	.7159	1.3968	.8131	.4
.7	.5835	.7186	1.3916	.8121	.3
.8	.5850	.7212	1.3865	.8111	.2
.9	.5864	.7239	1.3814	.8100	54.1
36.0	0.5878	0.7265	1.3764	0.8090	**54.0**
	Cos	Cot	Tan	Sin	Deg.

Trigonometric (degrees) (continued)

Deg.	Sin	Tan	Cot	Cos	
36.0	0.5878	0.7265	1.3764	0.8090	**54.0**
.1	.5892	.7292	1.3713	.8080	53.9
.2	.5906	.7319	1.3663	.8070	.8
.3	.5920	.7346	1.3613	.8059	.7
.4	.5934	.7373	1.3564	.8049	.6
.5	.5948	.7400	1.3514	.8039	.5
.6	.5962	.7427	1.3465	.8028	.4
.7	.5976	.7454	1.3416	.8018	.3
.8	.5990	.7481	1.3367	.8007	.2
.9	.6004	.7508	1.3319	.7997	53.1
37.0	0.6018	0.7536	1.3270	0.7986	**53.0**
.1	.6032	.7563	1.3222	.7976	52.9
.2	.6046	.7590	1.3175	.7965	.8
.3	.6060	.7618	1.3127	.7955	.7
.4	.6074	.7646	1.3079	.7944	.6
.5	.6088	.7673	1.3032	.7934	.5
.6	.6101	.7701	1.2985	.7923	.4
.7	.6115	.7729	1.2938	.7912	.3
.8	.6129	.7757	1.2892	.7902	.2
.9	.6143	.7785	1.2846	.7891	52.1
38.0	0.6157	0.7813	1.2799	0.7880	**52.0**
.1	.6170	.7841	1.2753	.7869	51.9
.2	.6184	.7869	1.2708	.7859	.8
.3	.6198	.7898	1.2662	.7848	.7
.4	.6211	.7926	1.2617	.7837	.6
.5	.6225	.7954	1.2572	.7826	.5
.6	.6239	.7983	1.2527	.7815	.4
.7	.6252	.8012	1.2482	.7804	.3
.8	.6266	.8040	1.2437	.7793	.2
.9	.6280	.8069	1.2393	.7782	51.1
39.0	0.6293	0.8098	1.2349	0.7771	**51.0**
.1	.6307	.8127	1.2305	.7760	50.9
.2	.6320	.8156	1.2261	.7749	.8
.3	.6334	.8185	1.2218	.7738	.7
.4	.6347	.8214	1.2174	.7727	.6
.5	.6361	.8243	1.2131	.7716	.5
.6	.6374	.8273	1.2088	.7705	.4
.7	.6388	.8302	1.2045	.7694	.3
.8	.6401	.8332	1.2002	.7683	.2
.9	.6414	.8361	1.1960	.7672	50.1
40.0	0.6428	0.8391	1.1918	0.7660	**50.0**
.1	.6441	.8421	1.1875	.7649	49.9
.2	.6455	.8451	1.1833	.7638	.8
.3	.6468	.8481	1.1792	.7627	.7
.4	.6481	.8511	1.1750	.7615	.6
40.5	0.6494	0.8541	1.1708	0.7604	**49.5**
	Cos	Cot	Tan	Sin	Deg.

Deg.	Sin	Tan	Cot	Cos	
40.5	0.6494	0.8541	1.1708	0.7604	**49.5**
.6	.6508	.8571	1.1667	.7593	.4
.7	.6521	.8601	1.1626	.7581	.3
.8	.6534	.8632	1.1585	.7570	.2
.9	.6547	.8662	1.1544	.7559	49.1
41.0	0.6561	0.8693	1.1504	0.7547	**49.0**
.1	.6574	.8724	1.1463	.7536	48.9
.2	.6587	.8754	1.1423	.7524	.8
.3	.6600	.8785	1.1383	.7513	.7
.4	.6613	.8816	1.1343	.7501	.6
.5	.6626	.8847	1.1303	.7490	.5
.6	.6639	.8878	1.1263	.7478	.4
.7	.6652	.8910	1.1224	.7466	.3
.8	.6665	.8941	1.1184	.7455	.2
.9	.6678	.8972	1.1145	.7443	48.1
42.0	0.6691	0.9004	1.1106	0.7431	**48.0**
.1	.6704	.9036	1.1067	.7420	47.9
.2	.6717	.9067	1.1028	.7408	.8
.3	.6730	.9099	1.0990	.7396	.7
.4	.6743	.9131	1.0951	.7385	.6
.5	.6756	.9163	1.0913	.7373	.5
.6	.6769	.9195	1.0875	.7361	.4
.7	.6782	.9228	1.0837	.7349	.3
.8	.6794	.9260	1.0799	.7337	.2
.9	.6807	.9293	1.0761	.7325	47.1
43.0	0.6820	0.9325	1.0724	0.7314	**47.0**
.1	.6833	.9358	1.0686	.7302	46.9
.2	.6845	.9391	1.0649	.7290	.8
.3	.6858	.9424	1.0612	.7278	.7
.4	.6871	.9457	1.0575	.7266	.6
.5	.6884	.9490	1.0538	.7254	.5
.6	.6896	.9523	1.0501	.7242	.4
.7	.6909	.9556	1.0464	.7230	.3
.8	.6921	.9590	1.0428	.7218	.2
.9	.6934	.9623	1.0392	.7206	46.1
44.0	0.6947	0.9657	1.0355	0.7193	**46.0**
.1	.6959	.9691	1.0319	.7181	45.9
.2	.6972	.9725	1.0283	.7169	**.8**
.3	.6984	.9759	1.0247	.7157	**.7**
.4	.6997	.9793	1.0212	.7145	.6
.5	.7009	.9827	1.0176	.7133	.5
.6	.7022	.9861	1.0141	.7120	.4
.7	.7034	.9896	1.0105	.7108	.3
.8	.7046	.9930	1.0070	.7096	.2
.9	.7059	.9965	1.0035	.7083	45.1
45.0	0.7071	1.0000	1.0000	0.7071	**45.0**
	Cos	Cot	Tan	Sin	Deg.

ANSWERS

CHAPTER 1

Page 3, Section 1.1

(1) Two thousand, five hundred two
(2) Two thousand, five hundred twenty
(3) Two thousand, fifty-two
(4) Twenty-five thousand, two hundred
(5) Two hundred fifty-two thousand
(6) 306
(7) 360
(8) 3,006
(9) 3,060
(10) 360,000
(11) Five hundred forty-two
(12) Five hundred forty
(13) Five hundred four
(14) Five thousand, forty-two
(15) Five hundred thousand, forty-two
(16) Fifty million, fifty thousand, five hundred

(17) Five hundred two million, five hundred twenty thousand, fifty-two
(18) Five billion, seven hundred thirteen million, five hundred twenty-two thousand, one hundred seventeen
(19) 243
(20) 243,007
(21) 203,047
(22) 4,002,700
(23) 406,242,713
(24) 315,000,572
(25) 6,000,606
(26) 54,055,056,057
(27) 23,081
(28) Seven hundred eighty-five miles
(29) 4,000,000,000

Page 6, Section 1.2

(1) 43
(2) 33
(3) 24
(4) 131
(5) 159
(6) 32
(7) 28
(8) 86
(9) 56
(10) 59
(11) 206
(12) 309
(13) 424
(14) 1,754
(15) 2,400
(16) 2.777
(17) 13,652
(18) 18,208
(19) 90,106
(20) 2,279
(21) 30,504
(22) 29,118
(23) 1,186
(24) 2,037
(25) 3,348
(26) 1,895
(27) 21,786
(28) 5,979
(29) 16,199
(30) 3,438
(31) 1,815
(32) 4,798
(33) $6,419
(34) 55,412 tickets
(35) 1,391 miles
(36) $3,584

Page 11, Section 1.3

(1) 48
(2) 28
(3) 488
(4) 54
(5) 0
(6) 10
(7) 21
(8) 102
(9) 11 R 1
(10) 51 R 2
(11) 774
(12) 400
(13) 1,833
(14) 2,345
(15) 18,798
(16) 41,013
(17) 37,200
(18) 28,962
(19) 271,404
(20) 3,640,172
(21) 13,683,250
(22) 62
(23) 21
(24) 32
(25) 38 R 19
(26) 38 R 30
(27) 34 R 17
(28) 308
(29) 160
(30) 580
(31) 309 R 3
(32) 191 R 214
(33) 4002 R 10
(34) 89 R 1
(35) $1,638
(36) 2,576 bottles
(37) 11 houses
(38) $5,567
(39) 111 tons
      28 extra bales

Page 14, Section 1.4

(1) 4
(2) 6
(3) 4
(4) 0
(5) 6
(6) 1
(7) 6
(8) 0
(9) 7
(10) 10
(11) 13
(12) 4
(13) 9
(14) 64
(15) 1
(16) 5
(17) 10,000
(18) 1,000,000
(19) 100,000,000
(20) 1
(21) $13,000,000,000
(22) 160,000,000;
      one hundred sixty million
(23) 25,500,000,000,000 miles

471

Page 17, Section 1.5

(1) 32
(2) 17
(3) 3
(4) 18
(5) 6
(6) 4
(7) 4

(8) 36
(9) 14
(10) 2
(11) 40
(12) 33
(13) 6
(14) 13

(15) 43
(16) 16
(17) 70
(18) 8
(19) 24
(20) 16
(21) 12

(22) 6
(23) Time payment costs $675 more
(24) $247
(25) 650 sq ft
(26) 800 ohms

Page 21, Section 1.6

(1) yes
(2) yes
(3) no
(4) yes
(5) yes
(6) no
(7) $1 \cdot 15, 3 \cdot 5$
(8) $1 \cdot 16, 2 \cdot 8, 4 \cdot 4$
(9) $1 \cdot 17$
(10) $1 \cdot 18, 2 \cdot 9, 3 \cdot 6$
(11) 1,2,5,10,25,50
(12) 1,2,3,4,6,9,12,18,36
(13) 1,2,4,5,8,10,20,40
(14) 1,2,4,5,8,10,16,20,40,80
(15) 1,2,3,4,6,8,12,16,24,32,48,96
(16) 1,2,4,5,10,20,25,50,100
(17) 1,101

(18) 1,2,4,5,10,11,20,22,44,55,110,220
(19) 1,3,37,111
(20) 1,5,7,35,49,245
(21) $1 \cdot 100, 2 \cdot 50, 4 \cdot 25, 5 \cdot 20, 10 \cdot 10$
(22) $1 \cdot 80, 2 \cdot 40, 4 \cdot 20, 5 \cdot 16, 8 \cdot 10$
(23) $1 \cdot 96, 2 \cdot 48, 3 \cdot 32, 4 \cdot 24, 6 \cdot 16, 8 \cdot 12$
(24) $1 \cdot 48, 2 \cdot 24, 3 \cdot 16, 4 \cdot 12, 6 \cdot 8$
(25) $1 \cdot 128, 2 \cdot 64, 4 \cdot 32, 8 \cdot 16$
(26) $1 \cdot 131$
(27) $1 \cdot 847, 7 \cdot 121, 11 \cdot 77$
(28) $1 \cdot 500, 2 \cdot 250, 4 \cdot 125, 5 \cdot 100, 10 \cdot 50, 20 \cdot 25$
(29) $1 \cdot 720, 2 \cdot 360, 3 \cdot 240, 4 \cdot 180, 5 \cdot 144, 6 \cdot 120, 8 \cdot 90, 9 \cdot 80, 10 \cdot 72, 12 \cdot 60, 15 \cdot 48, 16 \cdot 45, 18 \cdot 40, 20 \cdot 36, 24 \cdot 30$
(30) $1 \cdot 1311, 3 \cdot 437, 19 \cdot 69, 23 \cdot 57$
(31) 4

Page 25, Section 1.7

(1) composite
(2) prime
(3) composite
(4) composite

(5) composite
(6) prime
(7) composite
(8) prime

(9) prime
(10) composite
(11) prime
(12) prime

(13) composite
(14) prime
(15) composite
(16) composite
(17) composite

(18) prime
(19) composite
(20) prime
(21) prime
(22) composite

Page 28, Section 1.8

(1) 2^3
(2) 3^2
(3) $2 \cdot 5$
(4) 11
(5) $2^2 \cdot 3$
(6) $3 \cdot 5$
(7) $2 \cdot 3^2$
(8) $3 \cdot 7$
(9) $2 \cdot 11$

(10) 3^3
(11) $2 \cdot 3 \cdot 5$
(12) $2^2 \cdot 3^2$
(13) $2^4 \cdot 3$
(14) $2 \cdot 5^2$
(15) 2^6
(16) $3 \cdot 17$
(17) 71
(18) $7 \cdot 13$

(19) 97
(20) $2^2 \cdot 5^2$
(21) $2^3 \cdot 3 \cdot 5$
(22) $2 \cdot 3 \cdot 5^2$
(23) $2 \cdot 3^2 \cdot 11$
(24) $5^2 \cdot 13$
(25) $2 \cdot 3^2 \cdot 23$
(26) $2^8 \cdot 3$
(27) 563

(28) $2^3 \cdot 3 \cdot 37$
(29) 2^9
(30) $3^2 \cdot 73$
(31) $2 \cdot 409$
(32) $3^4 \cdot 11$
(33) $2^2 \cdot 3 \cdot 7 \cdot 17$
(34) $2^4 \cdot 5^3$
(35) $2^3 \cdot 3^3 \cdot 17$

Page 30, Section 1.9

(1) 12
(2) 15
(3) 6
(4) 21
(5) 20

(6) 12
(7) 28
(8) 8
(9) 12
(10) 24

(11) 24
(12) 30
(13) 60
(14) 180
(15) 300

(16) 240
(17) 180
(18) 84
(19) 165
(20) 480

(21) 720
(22) 750
(23) 60
(24) 1440
(25) 630

(26) 720
(27) 204
(28) 6370
(29) 342
(30) 28,800
(31) 36

CHAPTER 2

Page 37, Section 2.1

(1) $\dfrac{4}{7}$ (6) $2\dfrac{1}{8}$ (11) $8\dfrac{13}{15}$ (16) $66\dfrac{2}{3}$ (22) $\dfrac{43}{8}$ (28) 50 ft.

(2) $\dfrac{6}{7}$ (7) $4\dfrac{1}{12}$ (12) 65   (17) $\dfrac{3}{2}$ (23) 5   (29) 195 sections

(3) $\dfrac{6}{8}$ (8) $10\dfrac{3}{4}$ (13) $1\dfrac{9}{23}$ (18) $\dfrac{5}{3}$ (24) $\dfrac{107}{10}$ (30) 3

(4) $\dfrac{6}{8}$ (9) $8\dfrac{7}{10}$ (14) $2\dfrac{5}{35}$ (19) $\dfrac{15}{4}$ (25) $\dfrac{365}{6}$ (31) 9

(5) $\dfrac{3}{6}$ (10) 12   (15) $13\dfrac{1}{6}$ (20) $\dfrac{23}{7}$ (26) $\dfrac{50}{3}$ (32) 37 and 38

(21) $\dfrac{6}{3}$ (27) $\dfrac{317}{100}$ (33) 139 parts

Page 41, Section 2.2

(1) $\dfrac{8}{27}$ (6) $\dfrac{20}{33}$ (11) $\dfrac{15}{88}$ (16) $\dfrac{32}{5}$ (21) $\dfrac{19}{23}$ (26) $\dfrac{28}{33}$

(2) $\dfrac{14}{45}$ (7) $\dfrac{16}{25}$ (12) $\dfrac{135}{8}$ (17) $\dfrac{12}{7}$ (22) $\dfrac{315}{4}$ (27) $\dfrac{3}{14}$

(3) $\dfrac{35}{24}$ (8) $\dfrac{1}{24}$ (13) $\dfrac{55}{36}$ (18) $\dfrac{80}{99}$ (23) $\dfrac{225}{17}$ (28) \$2

(4) $\dfrac{1}{16}$ (9) $\dfrac{4}{45}$ (14) $\dfrac{160}{189}$ (19) $\dfrac{117}{160}$ (24) $\dfrac{255}{304}$ (29) $\dfrac{3}{16}$

(5) $\dfrac{35}{24}$ (10) $\dfrac{15}{32}$ (15) $\dfrac{405}{128}$ (20) $\dfrac{1}{437}$ (25) $\dfrac{21}{40}$

Page 45, Section 2.3

(1) 5   (5) 30   (9) 18   (13) 51   (17) 336   (21) $\dfrac{21}{28}$

(2) 15   (6) 75   (10) 46   (14) 72   (18) 700   (22) $\dfrac{88}{132}$

(3) 35   (7) 28   (11) 180   (15) 66   (19) 875

(4) 63   (8) 8   (12) 140   (16) 57   (20) 200

Page 47, Section 2.4

(1) $\dfrac{1}{4}, \dfrac{3}{4}$ (9) $\dfrac{1}{6}, \dfrac{1}{2}, \dfrac{2}{3}$ (17) $\dfrac{2}{5}, \dfrac{4}{7}, \dfrac{2}{3}$

(2) $\dfrac{13}{17}, \dfrac{18}{17}$ (10) $1\dfrac{3}{10}, 1\dfrac{2}{5}$ (18) $\dfrac{2}{3}, \dfrac{3}{4}, \dfrac{4}{5}$

(3) $\dfrac{1}{2}, \dfrac{5}{6}$ (11) $\dfrac{4}{5}$ (19) $\dfrac{5}{8}, \dfrac{7}{10}, \dfrac{3}{4}$

(4) $\dfrac{2}{3}, \dfrac{5}{6}$ (12) $\dfrac{9}{10}$ (20) $\dfrac{5}{7}, \dfrac{7}{9}, \dfrac{9}{11}$

(5) $\dfrac{1}{2}, \dfrac{3}{5}$ (13) $1\dfrac{3}{4}$ (21) George's

(6) $\dfrac{2}{7}, \dfrac{7}{2}$ (14) $\dfrac{5}{3}$ (22) $5\dfrac{7}{10}$

(7) $\dfrac{2}{5}, \dfrac{2}{3}$ (15) $\dfrac{11}{6}$ (23) $\dfrac{3}{32}, \dfrac{1}{8}, \dfrac{1}{4}, \dfrac{5}{16}, \dfrac{3}{8}, \dfrac{1}{2}, \dfrac{9}{16}$

(8) $\dfrac{3}{2}, \dfrac{5}{2}$ (16) $\dfrac{6}{7}$ (24) $\dfrac{11}{16}, \dfrac{3}{4}, \dfrac{7}{8}, 1\dfrac{1}{16}, 1\dfrac{3}{32}, 1\dfrac{1}{8}$

Page 51, Section 2.5

(1) $\frac{6}{7}$ (7) $\frac{2}{5}$ (13) $\frac{13}{15}$ (19) $\frac{8}{5}$ (25) $\frac{11}{12}$ (32) $\frac{3}{5}$

(2) $\frac{7}{12}$ (8) $\frac{3}{5}$ (14) $\frac{1}{5}$ (20) 4   (26) $\frac{7}{15}$ (33) $\frac{3}{5}$

(3) $\frac{2}{3}$ (9) $\frac{10}{11}$ (15) $\frac{1}{4}$ (21) $\frac{45}{32}$ (27) $\frac{2}{3}$ (34) $\frac{2}{7}$

(4) $\frac{2}{5}$ (10) $\frac{7}{8}$ (16) $\frac{3}{8}$ (22) $\frac{8}{9}$ (28) $\frac{97}{101}$ (35) $\frac{2}{11}$

(5) $\frac{2}{3}$ (11) $\frac{2}{5}$ (17) $\frac{7}{8}$ (23) $\frac{5}{7}$ (29) $\frac{49}{51}$ (36) $\frac{7}{8}$ in

(6) $\frac{2}{3}$ (12) $\frac{3}{5}$ (18) $\frac{7}{12}$ (24) $\frac{9}{25}$ (30) $\frac{3}{5}$ (37) $\frac{25}{3}$ fbm

(31) $\frac{20}{33}$ (38) $\frac{5}{12}$ in^3

Page 55, Section 2.6

(1) $\frac{2}{3}$ (8) $\frac{2}{3}$ (15) $\frac{1}{5}$ (22) 1   (30) $\frac{1}{3}$

(2) $\frac{3}{7}$ (9) $\frac{7}{10}$ (16) $\frac{16}{35}$ (23) $\frac{7}{8}$ (31) 8

(3) $\frac{5}{9}$ (10) $\frac{1}{3}$ (17) $\frac{2}{3}$ (24) 7   (32) 100 miles

(4) $\frac{12}{7}$ (11) $\frac{1}{7}$ (18) $\frac{3}{5}$ (25) 31   (33) 18 drawings

(5) $\frac{10}{11}$ (12) 3   (19) 1   (26) $119\frac{11}{24}$ (34) $15\frac{3}{8}$ lb

(6) $\frac{3}{14}$ (13) $35\frac{5}{8}$ (20) $\frac{7}{8}$ (27) $\frac{63}{125}$ (35) $20,250 less

(7) $\frac{1}{4}$ (14) $3\frac{13}{14}$ (21) $\frac{1}{3}$ (28) $\frac{8}{5}$ (36) 216 in

(29) $\frac{3}{8}$ (37) $40\frac{25}{27}$ lb/in^2

Page 58, Section 2.7

(1) $\frac{6}{7}$ (8) $4\frac{1}{2}$ (15) $\frac{1}{2}$ (22) $\frac{32}{75}$ (29) $6\frac{2}{3}$ turns

(2) $1\frac{1}{2}$ (9) $\frac{6}{7}$ (16) $\frac{2}{3}$ (23) $6\frac{2}{69}$ (30) $3\frac{1}{2}$

(3) $\frac{2}{3}$ (10) $\frac{1}{3}$ (17) $18\frac{1}{3}$ (24) $\frac{22}{27}$ (31) $14\frac{9}{10}$ minutes

(4) $\frac{14}{15}$ (11) 4   (18) $\frac{7}{30}$ (25) $5\frac{13}{28}$ (32) $4\frac{1}{6}$ in

(5) $\frac{14}{25}$ (12) $\frac{1}{3}$ (19) $\frac{9}{14}$ (26) $2\frac{1}{21}$ (33) 6 gerbils

(6) $\frac{1}{30}$ (13) $1\frac{1}{3}$ (20) $1\frac{1}{2}$ (27) $\frac{41}{87}$ (34) $4\frac{1}{2}$ in or $\frac{1}{8}$ yd

(7) 35   (14) $\frac{5}{14}$ (21) $\frac{2}{3}$ (28) $\frac{5}{7}$ (35) 4 boards

Page 62, Section 2.8

(1) 1

(2) $\frac{3}{4}$

(3) $\frac{4}{5}$

(4) $\frac{3}{4}$

(5) $\frac{10}{11}$

(6) $\frac{3}{5}$

(7) $\frac{3}{4}$

(8) $\frac{5}{6}$

(9) 1

(10) $\frac{17}{20}$

(11) $\frac{1}{2}$

(12) $\frac{23}{16}$

(13) $\frac{47}{60}$

(14) $\frac{35}{36}$

(15) $\frac{7}{15}$

(16) $\frac{63}{80}$

(17) $\frac{61}{48}$

(18) $\frac{61}{60}$

(19) $\frac{37}{42}$

(20) $\frac{7}{5}$

(21) $\frac{229}{240}$

(22) $\frac{59}{126}$

(23) $\frac{59}{195}$

(24) $\frac{215}{432}$

(25) $\frac{173}{150}$

(26) $\frac{7}{54}$

(27) $\frac{131}{240}$

(28) $\frac{1381}{2160}$

(29) $\frac{23}{12}$ yd

(30) $\frac{5}{16}$ of a point

(31) $\frac{25}{32}$ in

(32) $\frac{41}{24}$ ohms

Page 66, Section 2.9

(1) $11\frac{5}{8}$

(2) $32\frac{4}{5}$

(3) $14\frac{1}{2}$

(4) $15\frac{1}{4}$

(5) $10\frac{5}{7}$

(6) 14

(7) $6\frac{2}{3}$

(8) $10\frac{1}{3}$

(9) $7\frac{3}{4}$

(10) $9\frac{1}{2}$

(11) $26\frac{3}{10}$

(12) $110\frac{19}{24}$

(13) $719\frac{31}{36}$

(14) $561\frac{2}{15}$

(15) $235\frac{7}{11}$

(16) $86\frac{11}{30}$

(17) $112\frac{19}{24}$

(18) $16\frac{13}{40}$

(19) $180\frac{113}{135}$

(20) $13\frac{14}{15}$

(21) $14\frac{13}{66}$

(22) $448\frac{13}{18}$

(23) $188\frac{23}{36}$

(24) $23\frac{3}{4}$ hr

(25) $80\frac{9}{16}$ lb

(26) $25\frac{15}{24}$

(27) $21\frac{5}{16}$

Page 69, Section 2.10

(1) $\frac{2}{7}$

(2) $\frac{8}{15}$

(3) 3

(4) $\frac{1}{2}$

(5) $\frac{1}{7}$

(6) $\frac{1}{5}$

(7) 1

(8) $\frac{3}{16}$

(9) $\frac{7}{20}$

(10) $\frac{1}{3}$

(11) $\frac{7}{30}$

(12) $\frac{1}{3}$

(13) $\frac{1}{3}$

(14) $\frac{7}{24}$

(15) $\frac{7}{16}$

(16) $\frac{11}{60}$

(17) $\frac{21}{200}$

(18) $\frac{5}{36}$

(19) $\frac{47}{140}$

(20) $\frac{1}{4}$

(21) $\frac{7}{30}$

(22) $\frac{13}{48}$

(23) $\frac{1}{72}$

(24) $\frac{13}{60}$

(25) $\frac{119}{225}$

(26) $\frac{61}{132}$

(27) $\frac{1}{10}$

(28) $\frac{3}{16}$ in

(29) $\frac{5}{12}$ cup

(30) $\frac{5}{24}$ lb

(31) $\frac{13}{60}$ of the loan

Page 73, Section 2.11

(1) $2\frac{1}{5}$

(2) $5\frac{4}{9}$

(3) $11\frac{2}{3}$

(4) $1\frac{1}{3}$

(5) $1\frac{9}{29}$

(6) 7

(7) $8\frac{5}{8}$

(8) $10\frac{1}{2}$

(9) $9\frac{3}{8}$

(10) $7\frac{2}{9}$

(11) $5\frac{2}{7}$

(12) $4\frac{1}{3}$

(13) $103\frac{1}{5}$

(14) $118\frac{1}{6}$

(15) $3\frac{8}{15}$

(16) $3\frac{13}{36}$

(17) $13\frac{3}{5}$

(18) $12\frac{1}{2}$

(19) $155\frac{1}{9}$

(20) $16\frac{3}{4}$

(21) $28\frac{1}{3}$

(22) $64\frac{1}{2}$

(23) $1\frac{17}{36}$

(24) $2\frac{29}{44}$

(25) $9\frac{3}{14}$

(26) $8\frac{11}{14}$

(27) $4\frac{5}{6}$

(28) $\frac{19}{24}$

(29) $\frac{17}{30}$

(30) $2\frac{37}{40}$ yd

(31) $1\frac{5}{6}$ ft

(32) $18\frac{9}{20}$ tons

(33) $10\frac{19}{32}$ in

(34) $10\frac{1}{4}$ oz, $10\frac{3}{4}$ oz

(35) $103\frac{1}{8}$ volts

Page 77, Section 2.12

(1) $\frac{8}{15}$

(2) $\frac{7}{10}$

(3) $\frac{1}{2}$

(4) $\frac{1}{2}$

(5) $\frac{4}{7}$

(6) $\frac{2}{7}$

(7) $\frac{1}{2}$

(8) $\frac{1}{3}$

(9) 1

(10) $1\frac{1}{6}$

(11) $\frac{3}{8}$

(12) $13\frac{1}{2}$

(13) $\frac{2}{27}$

(14) $\frac{3}{5}$

(15) $\frac{9}{10}$

(16) $1\frac{4}{9}$

(17) $1\frac{8}{9}$

(18) $1\frac{4}{9}$

(19) $12\frac{1}{2}$ in

(20) $8\frac{7}{34}$ ohms

CHAPTER 3

Page 85, Section 3.1

(1) Two and five tenths
(2) Sixteen thousandths
(3) Thirty-six and five hundredths
(4) Sixteen and sixteen hundredths
(5) Eighteen ten-thousandths
(6) .4
(7) 2.02
(8) 13.015
(9) .06
(10) 100.01
(11) Four tenths
(12) Two hundred nineteen thousandths
(13) Three and twenty-six hundredths
(14) Forty-eight and four ten-thousandths
(15) Five hundred forty-two thousandths
(16) Five hundred four thousandths
(17) Five and four hundredths
(18) Fifty and four tenths
(19) Five hundred and four tenths
(20) Fifty and four hundredths
(21) Five and four thousandths
(22) Eighteen and two hundred five ten-thousandths
(23) Forty-five
(24) Three hundred eighty-four
(25) Three hundred forty-five thousandths
(26) Three hundred and forty-five thousandths
(27) .15
(28) 500.005
(29) 5000.05
(30) .00005
(31) 5000
(32) 1000.005
(33) 200.031
(34) .231
(35) Seventy-three and seven hundredths
(36) Minimum: 5.25 by 8.30; maximum: 10.06 by 21.75

Page 87, Section 3.2

(1) .8, .801, .81
(2) .05, .06, .07
(3) 1.29, 1.3, 1.31
(4) 16.9, 17.0, 17.5
(5) .0049, .00491, .0051
(6) .09, .11, .13
(7) .112, .113, .115, .119
(8) 99.08, 99.9, 100.6
(9) 2.49, 3.96, 4.35, 4.87
(10) 1.6, 1.65, 1.67, 1.7
(11) .555, .556, .565, .566
(12) .86, .899, .9, .903, .91

(13) .003, .00305, .0031, .00312
(14) .1159, .116, .1163, .117
(15) 17.05, 17.0506, 17.057, 17.16
(16) .072, .0729, .073, .073001, .073015
(17) .30009, .3008, .301, .30101
(18) .88759, .88799, .888, .8881
(19) 1, 1.5, 1.999, 2, 2.006
(20) 8.2975, 8.3401, 8.3599, 8.36
(21) Trupix
(22) .916 cm
(23) 98.35 cents
(24) Too heavy

Page 92, Section 3.3

(1) 2.7, 2.65, 2.653
(2) .8, .84, .836
(3) 12.3, 12.30, 12.302
(4) 1.3, 1.33, 1.335
(5) 10.0, 9.99, 9.989
(6) 10.1, 10.08, 10.075
(7) 53.3, 53.31, 53.313
(8) 9.8, 9.78, 9.777
(9) 2.2, 2.18, 2.179
(10) 3.0, 3.00, 3.001
(11) 0.8, 0.79, 0.793
(12) 14.6, 14.56, 14.555
(13) 0.8, 0.79, 0.789

(14) 1.1, 1.11, 1.114
(15) 1.0, 1.00, 1.000
(16) 6900
(17) 75,700
(18) 4000
(19) 3700
(20) 3000
(21) 26,780
(22) 380
(23) 400
(24) 10
(25) 7450
(26) 24,000

(27) 5000
(28) 1000
(29) 43,000
(30) 1,000,000
(31) 37
(32) 100
(33) 16
(34) 5
(35) 79
(36) 4941.38 cc
(37) 10.1 grams
(38) .50 in

Page 97, Section 3.4

(1) 1.1
(2) .30
(3) 7.09
(4) 9.77
(5) 8.92
(6) .3
(7) .66
(8) 13.06
(9) 3.41
(10) 2.47
(11) 30.1099

(12) 19.447
(13) 872.1691
(14) 471.949
(15) 29.764
(16) 21.1
(17) .023
(18) .948
(19) .125
(20) 2.76
(21) .769
(22) 5.749

(23) 6.418
(24) 3.097
(25) 74.88
(26) 23.56
(27) .3
(28) .472
(29) 198.13 tons, 3.84 tons
(30) 2.7 cc
(31) $3.98
(32) 19.7 miles
(33) 52.4 gal
(34) yes

Page 101, Section 3.5

(1) .18
(2) 9.72
(3) 7.35
(4) 10.8
(5) .0078
(6) .2253
(7) .00865

(8) .0198
(9) 1.15
(10) .654
(11) 2.94
(12) 5.4717
(13) 224.9
(14) .8432

(15) 4.2488
(16) 1.52646
(17) 5.576
(18) 634.744
(19) .1328
(20) 450.375
(21) .3776

(22) 7.742
(23) 1.656
(24) 31.329
(25) .000352
(26) 3.4524
(27) 21.6118
(28) .70824
(29) .16225

(30) 852.066
(31) 78.7626
(32) 29.182
(33) $39.35
(34) $71.69
(35) $8.37
(36) $11.61
(37) $357.30

Page 106, Section 3.6

(1) .04
(2) .4
(3) 40
(4) .4
(5) 400
(6) 23.2
(7) 2.4

(8) .008
(9) .4
(10) 6
(11) .492
(12) .7375
(13) 1.31
(14) .35

(15) .018
(16) .52
(17) .052
(18) 97.06
(19) 6.307
(20) 4.57
(21) .24
(22) 2.39

(23) 1.75
(24) 5.945
(25) 15.555
(26) 2.424
(27) .221
(28) 7.8
(29) 16.5

(30) 6.9
(31) 6.7
(32) 56.3 gal
(33) $ .87
(34) 19 mpg
(35) 48 mph
(36) $ .15
(37) $44.51

Page 111, Section 3.7

(1) .6
(2) .8
(3) .75
(4) .25
(5) .5
(6) .125
(7) .55
(8) .05
(9) .375

(10) .7
(11) .275
(12) .1875
(13) .792
(14) .875
(15) 3.625
(16) .385
(17) .07375
(18) .0112

(19) 3.348
(20) .9825
(21) .3
(22) .8
(23) 1.0
(24) .8
(25) .29
(26) .78
(27) 5.30

(28) .67
(29) .349
(30) .778
(31) .583
(32) .818
(33) .875
(34) $3.71
(35) $6.875
(36) Win .49; Lose .51
(37) .09

Page 114, Section 3.8

(1) 1.0
(2) 5.6
(3) 3.8
(4) .31
(5) .8
(6) .06

(7) .04
(8) .031
(9) .53
(10) 7.6
(11) .9675
(12) 1.83

(13) 7.765
(14) 220.1
(15) .36
(16) .98115
(17) 102.4
(18) .1129

(19) 3.954
(20) 43.4616
(21) 3.59
(22) 1.367
(23) .1742
(24) 30.25

(25) 3.78
(26) 102.786
(27) $158.88
(28) $402.60
(29) $696.65
(30) $650,440

CHAPTER 4

Page 121, Section 4.1

(1) 24
(2) 6
(3) 1
(4) 4
(5) ½
(6) 4
(7) 36

(8) 90
(9) 36
(10) 7000
(11) 60
(12) 3520
(13) 14
(14) 67
(15) 234

(16) 4260
(17) 269
(18) 57
(19) 276
(20) 2400
(21) 8 ft 9 in
(22) 19 ft 7 in
(23) 17 gal 3 qt 1 pt

(24) 17 lb 4 oz
(25) 3 yd 1 ft
(26) 6 min 52 sec
(27) 51 lb 8 oz
(28) 385 miles
(29) 13 qt
(30) $84
(31) 9 cups

Page 125, Section 4.2

(1) 10
(2) 10
(3) .1
(4) 10
(5) 500

(6) 500
(7) 2000
(8) 3000
(9) 6 m 5 cm
(10) 9 kg 140 g

(11) 1300
(12) 1300
(13) 1300
(14) .244
(15) .245
(16) .246

(17) 422
(18) 6500
(19) 700
(20) 16,556
(21) 7150
(22) 173

(23) 18
(24) 11,850
(25) 2.3
(26) 2086
(27) 11.75
(28) $2.86

Page 130, Section 4.3

(1) 3.5
(2) 16
(3) $4\frac{1}{3}$
(4) 1600
(5) 13,100
(6) 10
(7) 309

(8) 45
(9) 7.528
(10) 7.1
(11) 63,360
(12) 7
(13) 30
(14) 1

(15) 1500
(16) 10
(17) 40
(18) .354
(19) 4.25
(20) 2209.09 mph
(21) 10.5 lb
(22) 32,400 pieces

(23) 136.5 lb
(24) $104.52
(25) .012 ℓ
(26) 5 days
(27) $246.40
(28) .6 oz
(29) 32.96 gal
(30) 28.8 pages

Page 134, Section 4.4

(1) 4
(2) 6300
(3) 3
(4) .5 or ½
(5) 1900
(6) 52
(7) 13,500

(8) 43
(9) 8
(10) 10
(11) 49
(12) 50
(13) 94
(14) 82

(15) 40
(16) 210
(17) 40
(18) 210
(19) 112
(20) 176
(21) 40

(22) 104
(23) .2
(24) $\frac{25}{7}$
(25) $\frac{121}{7}$
(26) $\frac{99}{17}$

(27) $A = 464$
(28) $S = 10,100$
(29) $R = .2$
(30) $V = 37.68$
(31) 706.5 sq ft
(32) $19.50
(33) 3825 mi
(34) $1.01

Page 140, Section 4.5

(1) 12 cm
(2) 4.8 yd
(3) 8 ft
(4) 10 m
(5) 11 in
(6) 7.6 cm
(7) 6.28 dm
(8) 3.14 yd
(9) 57 in

(10) 25 ft
(11) 50.24 ft
(12) 49 cm
(13) 11 m
(14) 70 ft
(15) 41.68 in
(16) 33.2 cm
(17) 60.84 in
(18) 113.39 m

(19) 132.48 in
(20) 38.4 cm
(21) $13.30
(22) 360 ft
(23) 12 ft
(24) 1.57 in
(25) $1630
(26) 108 min or 1 hr 48 min
(27) 1261 revolutions

Page 148, Section 4.6

(1) 64 cm^2
(2) 2.25 in^2
(3) 44 ft^2
(4) 72 in^2 or .5 ft^2
(5) 64 cm^2
(6) 32 cm^2
(7) 3.14 dm^2

(8) 4.84 cm^2
(9) 6 in^2
(10) 2.5 in^2
(11) 96 in^2
(12) 432 cm^2
(13) 243 ft^2
(14) 5.29 in^2

(15) 3.36 m^2
(16) 108 in^2
(17) 452.16 in^2
(18) 225 yd^2
(19) $1263.52
(20) 192 in^2
(21) 200.96 cm^2

(22) 326.97 m^2
(23) 297 in^2
(24) 1017.36 in^2
(25) 70 yd^2
(26) 240 oz
(27) 16 A
(28) 128 tiles

Page 154, Section 4.7

(1) 39 cm^2
(2) 13.5 in^2
(3) 9.57 m^2
(4) 2.57 yd^2

(5) 418.08 in^2
(6) 255 ft^2
(7) 150 m^2
(8) 2395 mm^2

(9) 562.5 cm^2
(10) 64.94 ft^2
(11) 659.4 cm^2
(12) 10.9 in^2

(13) 243 yd^2
(14) 49.6 m^2
(15) 6.625 in^2
(16) 64.44 cm^2

(17) $6188.19
(18) 10 lb
(19) 3706.5 yd^2

Page 162, Section 4.8

(1) 6.28 ft^3
(2) 8 m^3
(3) 3.14 dm^3
(4) 20 in^3

(5) 4.19 m^3
(6) 216 in^3
(7) 2461.76 m^3
(8) 267.9 ft^3

(9) 468 cm^3
(10) 389.017 cm^3
(11) 6104.16 in^3
(12) 225,000 ft^3

(13) 14,130 m^3
(14) 35 tons
(15) 69.08 gal
(16) 1008 boxes

(17) 20 loads
(18) a) 1500 ft^3
      b) 11,250 gal

Page 167, Section 4.9

(1) 15,000 cm²   (3) 2813.44 in²   (5) 376 in²   (7) 23.55 ft²   (9) 577.76 ft²
(2) 18.84 ft²   (4) 640 cm²   (6) 50.24 m²   (8) 25.29 ft²   (10) 5.44 gal

CHAPTER 5

Page 175, Section 5.1

(1) 3

(2) −5

(3) $-\frac{1}{2}$

(4) 2.1

(5) $\frac{1}{3}$

(6) 1

(7) 5

(8) $\frac{1}{6}$

(9) 4

(10) 1.2

(11) 31

(12) −13

(13) $\frac{2}{3}$

(14) 2.35

(15) $-3\frac{1}{8}$

(16) $22\frac{2}{3}$

(17) −0.23

(18) 103.6

(19) $4\frac{15}{16}$

(20) 0.625

(21) 6

(22) $\frac{4}{5}$

(23) .035

(24) $3\frac{1}{8}$

(25) 21.75

(26) 2.03

(27) 4.5

(28) 0

(29) 7

(30) .003

(31) $-\frac{3}{8}$

(32) 12° C

(33) 1875 AD

(34) 300

Page 177, Section 5.2

(1) −5

(2) −1

(3) −4

(4) −2

(5) −13

(6) −7
(7) −5
(8) 8
(9) −27
(10) −8
(11) −22

(12) 7

(13) 0

(14) −144

(15) 9

(16) −75

(17) 7
(18) −14
(19) −185
(20) 398
(21) −11.4
(22) 2.6

(23) 2.7

(24) −1.84

(25) $-\frac{1}{6}$

(26) $-\frac{2}{3}$

(27) −54

(28) −63
(29) −27
(30) −4
(31) 4.2
(32) −23.09
(33) −133.73

(34) 1.39

(35) $-2\frac{5}{24}$

(36) $-5\frac{5}{12}$

(37) $1\frac{8}{9}$

(38) $-9\frac{3}{8}$

(39) .605
(40) −95 lb
(41) $6458 profit
(42) 18,041 books
(43) 29,980 ft

Page 181, Section 5.3

(1) 2

(2) −10

(3) 10

(4) −2

(5) −13

(6) 3

(7) −12

(8) −18

(9) 16

(10) −13

(11) $-\frac{3}{4}$

(12) $-\frac{1}{3}$

(13) 41

(14) −21

(15) −9

(16) −92

(17) −90

(18) 0

(19) −.77

(20) 9.81

(21) −13.6

(22) 19.1

(23) −6.3

(24) $-\frac{3}{4}$

(25) $-1\frac{1}{6}$

(26) $-\frac{9}{40}$

(27) $-\frac{3}{70}$

(28) 85°

(29) $378.99

(30) 20,582 ft

Page 184, Section 5.4

(1) −6

(2) 30

(3) −54

(4) 10

(5) −6⁻

(6) 0

(7) −48

(8) −54

(9) −1

(10) 36

(11) −42

(12) −60

(13) −18.12

(14) $\frac{3}{8}$

(15) 30.24

(16) .048

(17) −4

(18) 15

(19) −10.5

(20) $\frac{5}{9}$

(21) 120

(22) 0

(23) .5088

(24) 1.802

(25) −3.07

(26) −25

(27) $-7\frac{7}{9}^{\circ}$ C or −7.8° C

(28) −12.5 lb

(29) −33.96

(30) −$67.20

Page 186, Section 5.5

(1) −2

(2) −5

(3) 2

(4) −4

(5) 4

(6) −5

(7) −18

(8) 4

(9) −5

(10) 7

(11) −3

(12) −1.01

(13) 5.5

(14) −310

(15) −5

(16) −1

(17) 2

(18) −4

(19) 4

(20) −20

(21) 4

(22) $1\frac{1}{4}$

(23) −3

(24) 5

(25) −6

(26) 16.4

(27) −5°

(28) −$69.67

(29) −$30.61

(30) −3.125°

Page 189, Section 5.6

(1) −66

(2) 49

(3) −7

(4) 27

(5) −16

(6) −13

(7) 3

(8) 11

(9) −5

(10) 14

(11) 9

(12) 12

(13) −33

(14) $-3\frac{1}{3}$

(15) −9

(16) 131

(17) −12

(18) −84

(19) yes

(20) 13.8° C

(21) 7° F

(22) yes

(23) yes

(24) no

CHAPTER 6

Page 195, Section 6.1

(1) $x = 6$

(2) $x = 12$

(3) $x = 30$

(4) $x = 3$

(5) $x = 3\frac{1}{3}$

(6) $x = -2$

(7) $x = 1\frac{2}{7}$

(8) $x = 4$

(9) $x = 5$

(10) $a = 1\frac{4}{5}$

(11) $b = -3$

(12) $y = 3$

(13) $z = -2$

(14) $x = 24$

(15) $y = 1$

(16) $z = 1$

(17) $x = 0$

(18) $a = \frac{1}{2}$

(19) $b = 1$

(20) $z = -3$

(21) $x = 3$

(22) $a = -\frac{5}{6}$

(23) $x = \frac{1}{3}$

(24) $c = 35$

(25) 8 cm

(26) 27 ft/sec

Page 198, Section 6.2

(1) 64

(2) 32

(3) 4

(4) $\frac{1}{4}$

(5) 9

(6) x^5

(7) y^{10}

(8) $x^{12}y^{18}$

(9) $\frac{x^4}{y^3}$

(10) x^{11}

(11) 5184

(12) 46,656

(13) $\frac{16}{81}$

(14) x^9

(15) y^{12}

(16) a^6b^4

(17) x^3

(18) $\frac{a^6}{b^3}$

(19) $8a^3b^6$

(20) a^2b

(21) a^{10}

(22) $27a^{12}b^6x^9$

(23) x^4y

(24) a^{14}

(25) x^{5a}

(26) $81a^4b^4x^4y^4$

(27) a^3b^2

(28) x^{25}

(29) x^{6m}

(30) x^{4a-1}

(31) $\frac{1}{4}\pi d^2$ or $\frac{\pi d^2}{4}$

(32) $\frac{4}{3}\pi R^3$

(33) W^6, 64 ft³

Page 203, Section 6.3

(1) $9a$

(2) $-2x$

(3) $16y^2$

(4) $-2a$

(5) $8x^2$

(6) $6a^2b^2$

(7) $-2ab$

(8) $4x^2y^4$

(9) $2a^2b + 2ab$

(10) $16a^5$

(11) $-2y$

(12) $11a^2$

(13) $-7ST$

(14) $-3xy^2$

(15) $5ab + 4a$

(16) $-4xy^2 + 4x^2y^2$

(17) $6x^2$

(18) $2xy^2$

(19) $-48a^2b^2$

(20) b

(21) $-\frac{8}{3}xy$

(22) $-3ax$

(23) $17x^3$

(24) $3a^2b^2c + 12a^2bc^2$

(25) $25z^3$

(26) $x = 3$

(27) $a = 2$

(28) $y = -3$

(29) $a = -2\frac{1}{4}$

(30) $x = \frac{1}{6}$

(31) $x = 20$

(32) $x = 2.4$

(33) $x = 3$

Page 209, Section 6.4

(1) $x - 6$

(2) $2a - 9$

(3) $6x + 14$

(4) $a - 6$

(5) $2a - 2$

(6) $x + 3$

(7) $-2x + 1$

(8) $2a + 4w$

(9) $4x - 3y + 4z$

(10) $-x^2 + 2x - 10$

(11) $-3r + 7s - 9t$

(12) $-10a^2 - b^2 - 8ac$

(13) $-3p - 5q - 6r$

(14) $\frac{17}{12}mn + \frac{1}{3}n$

(15) $-.3bc - 4.3ad$

(16) $-2.91w - 3.08$

(17) $13x - 7y$

(18) $6a + 5b + 5c$

(19) $2x + \frac{7}{4}y$

(20) $.74r - 1.23s - 1.19t$

(21) $2x^2 - 10x + 15$

(22) $x + 15$

(23) $8at + 5a + 2b + 3$

(24) $x - 8y + 2$

(25) $2y - 6$

(26) $b = 6$

(27) $y = 7$

(28) $x = 6.75$

(29) $x = -34$

(30) $x = -2.75$

(31) $x = 3\frac{4}{7}$

Page 214, Section 6.5

(1) $R_1 = 400, R_2 = 250, R_3 = 150$

(2) Father: $36\frac{1}{3}$; son: $15\frac{2}{3}$

(3) $R_1 = 260, R_2 = 410, R_3 = 560, R_4 = 710$

(4) $5.45, $9.30

(5) 10, 12, 5

(6) 187, 64

(7) 23, 24, 25, 26, 27

(8) 53, 55, 57, 59

(9) 400, 200

(10) 15 ft, 15 ft, 18 ft

(11) $L = 310$ ft, $W = 135$ ft

(12) $L = 14$ in, $W = 10$ in

(13) First test 995 seeds, second test 990 seeds, third test 1015 seeds.

(14) 3.15, 5.6

(15) 450 mi, 225 mi, 400 mi

(16) Shoes: $60; suit: $105

(17) 4.75 in, 5.25 in, 5.75 in, 6.25 in, 6.75 in, 7.25 in

(18) Pasture: $162\frac{6}{7}$; timber: $54\frac{2}{7}$; wheat: $132\frac{6}{7}$

(19) $4\frac{21}{32}$ in, $4\frac{15}{32}$ in

(20) Antimony: 1.6 lb; copper: .8 lb; tin: 9.6 lb

Page 221, Section 6.6

(1) $6a - 3$

(2) $4x + 24$

(3) $-6y + 2$

(4) $x^2 + 3x + 2$

(5) $x^2 + 6x + 8$

(6) $x^2 + x - 6$

(7) $-35x - 15$

(8) $-15a^2 + 6ab$

(9) $-28x^2 - 14x^2y$

(10) $-30w^2 + 25wz$

(11) $40x^2y - 70y^2$

(12) $-m^2n^2 + m^3n^2 + m^2n^3$

(13) $6x^2 + 10x - 56$

(14) $-24y^2 - 6wy + 32y + 8w$

(15) $-2r^2 + 11rs - 12s^2$

(16) $x^2 + 4x - 21$

(17) $x^2 + 10x + 25$

(18) $x^2 + x - 2$

(19) $x^2 - 15x + 54$

(20) $6x^2 + x - 1$

(21) $2x^2 - 9x - 18$

(22) $12x^2 + 11x - 15$

(23) $25x^2 - 36$
(24) $x^2 - 36$
(25) $9y^2 - 16$
(26) $15x^2 - 63xy + 54y^2$
(27) $p^2 - q^2$
(28) $y^2 + by + cy + bc$
(29) $12x^2 - 14xy - 10y^2$
(30) $-12x^2 + 26xy + 9wx - 15wy - 10y^2$
(31) $2x^3 + x^2 - 35x - 28$
(32) $8a^2 - 16a^2 b - 2ab + 24ab^2 - 15b^2$
(33) $y = 10$
(34) $b = 6$
(35) $x = -3$

(36) $x = -\dfrac{5}{4}$
(37) $x = -2\dfrac{2}{11}$
(38) $34x + 6$
(39) $12x - 28y + 2$
(40) $-5y - 50$
(41) $123\dfrac{3}{4}$ mi
(42) 3 hr
(43) 6 hr
(44) 7:24 PM
(45) 56 mph

Page 226, Section 6.7

(1) $x + 3$
(2) $y - 3$
(3) $x^2 - 2x - 3$
(4) $x - 1$
(5) $x^2 - x + 1$
(6) $a - b$
(7) $2x - 1$
(8) $x^2 + 2x$
(9) $a^2 - 2$
(10) $3y + 1$
(11) $2bc - 4abc^2$
(12) $-2x^2 + 3$
(13) $6q - 3q^2 - 4q^3$
(14) $x^2 - 2x - 3$
(15) $xy + 2y - 3$
(16) $2 - 4a + b$
(17) $-3rst + 4rt + 6s$
(18) $2c + 3abc^2 + bc^2$
(19) $4x^4 - 5x^3 + 6x^2 - 2x$
(20) $\dfrac{7}{12}y^2 + \dfrac{2}{3}yz + y - \dfrac{5}{4}y^2 z$
(21) $-a^2 bc^3 - ac^2 - bc + a + 1$
(22) $\pi r^2 + \pi rh$

Page 229, Section 6.8

(1) $2(x + 3)$
(2) $4(2x - 3)$
(3) $y(6x - 5)$
(4) $2a(b - 2)$
(5) $x(2x - 3)$
(6) $3b(a - 5)$
(7) $2(a + 4b)$
(8) $x(4x + 5)$
(9) $xz(15y - 7w)$
(10) $4b(3c - 4d)$
(11) Cannot be factored
(12) $\pi h(r^2 - R^2)$
(13) $2(2ab + 4bc - 3cd)$
(14) $5st(5r - 7 + x)$
(15) $8x^2 yz(2y + 3z + 4)$
(16) $\pi r(rh + r + 2)$
(17) $5xy(3x^2 - 5x + 8)$
(18) $5b(ac + 3cd - 2ad - a^2 b)$
(19) Cannot be factored
(20) $3mn(mn - 2 - 4m - 5n)$
(21) $x^2 y^2(1 + xy - x^2 y^2)$
(22) $8a(3c - b)$
(23) $x^2 y^2(8x + 9y - 7)$
(24) $2a(7x + 5y + 3z)$
(25) $12x(2x^3 - x^2 - 3x - 4)$
(26) $2\pi r(h + r)$
(27) $16t(9 - t)$

Page 234, Section 6.9

(1) $(a + 3)(a - 2)$
(2) $(b + 6)(b + 4)$
(3) $(x - 2)(x - 3)$
(4) $(y - 5)(y - 7)$
(5) $(x + 2)(x - 1)$
(6) $(x + 6)(x - 7)$
(7) $(y - 4)(y - 2)$
(8) $(x - 10)(x + 8)$
(9) $(x + 4)(x + 3)$
(10) $(a - 4)(a - 1)$
(11) $(c + 7)(c - 3)$
(12) $b(x - 10)(x + 2)$
(13) $(2y - 5)(y + 1)$
(14) $(3x + 2)(x + 1)$
(15) $(3a - 4)(a - 2)$
(16) Cannot be factored
(17) $(3x - y)(x - 3y)$
(18) $(x + 3)(x + 3)$ or $(x + 3)^2$
(19) $(6x + 7)(x + 1)$
(20) $(2y - 3a)(2y - 3a)$ or $(2y - 3a)^2$
(21) $(x + 10)(x - 9)$
(22) Cannot be factored
(23) $2(y + 1)(2y - 5)$
(24) $(5x - 4y)(5x - 4y)$ or $(5x - 4y)^2$
(25) $(3a - b)(a - 4b)$
(26) Cannot be factored
(27) $(7a + 4)(a - 3)$
(28) $8(2t - 11)(t + 2)$
(29) $(x + 13)(x - 9)$

Page 237, Section 6.10

(1) $(x + 1)(x - 1)$
(2) $(x + 2)(x - 2)$
(3) $(x + 3)(x - 3)$
(4) $(a + 4)(a - 4)$
(5) $(y + 5)(y - 5)$
(6) $(2x + 3)(2x - 3)$
(7) $(3a + 1)(3a - 1)$
(8) $(a + 8)(a - 8)$
(9) $(x + 6)(x - 6)$

(10) $(a + 10)(a - 10)$
(11) $(2x + 3)(2x - 3)$
(12) $(3a + 8b)(3a - 8b)$
(13) $(5x + 6y)(5x - 6y)$
(14) $(\frac{3}{4}b + \frac{4}{5})(\frac{3}{4}b - \frac{4}{5})$
(15) $3(4a + 5)(4a - 5)$
(16) $4(5y + 4z)(5y - 4z)$
(17) $9(x + 2)(x - 2)$

(18) $4a(x + 1)(x - 1)$
(19) $4a(3a + 2)(3a - 2)$
(20) $(5a + 6)(5a - 6)$
(21) $(5b + 4)(5b - 4)$
(22) $2\pi(r + R)(r - R)$
(23) $14a(a + 2b)(a - 2b)$
(24) $(3a + 5b)(3a - 5b)$
(25) $(47)(33)$

Page 241, Section 6.11

(1) $x = -7$ or $x = 8$
(2) $x = -5$ or $x = -6$
(3) $x = 9$ or $x = 4$
(4) $x = -8$ or $x = 11$
(5) $x = 0$ or $x = 1$ or $x = 3$
(6) $x = 9$ or $x = -8$
(7) $x = 8$ or $x = -6$
(8) $x = -3$ or $x = -2$
(9) $x = 4$ or $x = -3$
(10) $a = -7$ or $a = 3$

(11) $y = 12$ or $y = -4$
(12) $x = -7$ or $x = 5$
(13) $x = 9$ or $x = -1$
(14) $a = -12$ or $a = 2$
(15) $b = -5$ or $b = 2$
(16) $y = 7$ or $y = -7$
(17) $x = -\frac{2}{3}$ or $x = -3$
(18) $x = 3$ or $x = -3$
(19) $x = 3$ or $x = -\frac{4}{7}$
(20) $x = 3$ or $x = \frac{1}{3}$
(21) $x = -\frac{1}{2}$ or $x = -\frac{1}{3}$

(22) $x = -\frac{1}{2}$
(23) $s = \frac{3}{4}$ or $s = -\frac{3}{4}$
(24) $e = -10$ or $e = 9$
(25) $x = -3$ or $x = 0$
(26) $5\frac{1}{2}$ sec
(27) Altitude: 8; base: 15
(28) 7, 9
(29) 6, 12
(30) 8 ft by 11 ft
(31) 8 in by 12 in
(32) 9 sec

CHAPTER 7

Page 248, Section 7.1

(1) $\frac{3w}{2z}$
(2) $\frac{6P}{7m}$
(3) $\frac{24}{xy}$
(4) $\frac{72ac}{35bd}$
(5) $-\frac{60x^2 y}{77u^2 v}$

(6) $\frac{2y}{7x^2}$
(7) $\frac{6ax}{25}$
(8) $\frac{18a^2 bd}{35xy}$
(9) $\frac{3x^2 - 2x}{20}$
(10) $\frac{52y + 2yz}{7m^2}$

(11) $-\frac{152ax}{195by}$
(12) $\frac{270r^2 s}{49tv}$
(13) $\frac{45a^2 b + 27ab^2}{3ax + cx}$
(14) $\frac{14x - 7y}{6a^2 - 2ab}$
(15) $\frac{yz + y^2}{z^2 - yz}$

(16) $\frac{x^2 - 8x + 16}{x^2 + 6x + 9}$
(17) $\frac{y^2 - 4y - 21}{6t^2 - t - 2}$
(18) $\frac{7pq - 2q^2}{12r^2 + 14rs}$
(19) $-\frac{77x^2}{150 y^3}$
(20) $\frac{3x^2 + 9x - 12}{2x^3 + 2x^2}$

Page 250, Section 7.2

(1) $\dfrac{a}{3b}$

(2) $\dfrac{y}{2}$

(3) $\dfrac{2b}{3}$

(4) $\dfrac{3}{4}$

(5) $\dfrac{x-y}{2}$

(6) $\dfrac{2x+8}{x+8}$

(7) $8xy$

(8) $14p^2$

(9) axy

(10) 6

(11) $6x$

(12) $5x+10$

(13) $\dfrac{2xy^2}{3}$

(14) $\dfrac{3b}{5d}$

(15) $\dfrac{1}{2}$

(16) 2

(17) $\dfrac{x}{x+4}$

(18) $\dfrac{6}{x+3}$

(19) $10x^2$

(20) $a^2+2ab+b^2$

(21) $4x^2+6xy+2y^2$

(22) x^2+4x+4

(23) $25abc$

(24) $x^3+6x^2+11x+12$

Page 253, Section 7.3

(1) $\dfrac{8}{5y}$

(2) $\dfrac{u^2}{z}$

(3) $\dfrac{7}{4}$

(4) 6

(5) $\dfrac{11}{x-3}$

(6) $\dfrac{x+3}{5}$

(7) $\dfrac{x+4}{x-4}$

(8) $\dfrac{y}{z}$

(9) $\dfrac{6a}{5}$

(10) 1

(11) $\dfrac{xy+2x}{a+y}$

(12) 1

(13) $\dfrac{x+2}{x+5}$

(14) $\dfrac{x+4}{x-4}$

(15) $\dfrac{x^2+2x+3}{x^2+3x+2}$

(16) $\dfrac{a^2x^2y^2}{b}$

(17) $\dfrac{8x+8y}{3}$

(18) $\dfrac{1}{b-4}$

(19) $\dfrac{bc^2}{x+2}$

(20) $\dfrac{12a^3-48a^2-48a+192}{a^2-6a-8}$

(21) $\dfrac{3x^2+3xy}{4xy-4y^2}$

(22) $t=\dfrac{6}{5}$

Page 256, Section 7.4

(1) $6x$
(2) $6x^2$
(3) $15x^2$
(4) $30x^2$
(5) $30\,abc$
(6) $6(x+3)$
(7) $24(3a+7y)$
(8) $20xy(a+1)$

(9) $6(x+y)$
(10) $(x+5)(x-5)$
(11) $8a^2b^2$
(12) $48xyz$
(13) $24a^2b^2$
(14) $(x+4)(x+6)$
(15) $(x+y)(x-y)$
(16) $(a+b)(a-b)$

(17) $7(2a+3)(2a-3)$
(18) $(y+3)(y-3)^2$
(19) $(x+y)^2$
(20) $x(x+y)$
(21) $(b+c)(b-c)$
(22) $12x^2(x+y)$
(23) $(y+5)(y+2)(y+1)$
(24) $2(y+3)(y-4)$
(25) $2x^2$

Page 259, Section 7.5

(1) $\dfrac{2y}{5}$

(2) $\dfrac{z}{3}$

(3) $\dfrac{1}{2w}$

(4) $\dfrac{8}{15a}$

(5) $\dfrac{2x}{3}$

(6) $\dfrac{3x+7}{x+2}$

(7) $\dfrac{7m}{8y}$

(8) $\dfrac{4p+9}{3p^2}$

(9) $\dfrac{6x-4}{7}$

(10) $\dfrac{3x+7}{10}$

(11) $\dfrac{7z+3y}{yz}$

(12) $\dfrac{4x+xy}{3y}$

(13) $\dfrac{3+2b^2+6a^2}{12ab}$

(14) $\dfrac{3xy+5y^2+8x^2}{x^2y^2}$

(15) $\dfrac{19a}{18b}$

(16) $\dfrac{35a+3}{5a}$

(17) $\dfrac{37}{6a}$

(18) $\dfrac{x^2+3x+1}{x}$

(19) $\dfrac{11a+3b}{a^2-b^2}$

(20) $\dfrac{9x+33}{x^2+4x-5}$

(21) $\dfrac{7y+4}{y^2+y}$

(22) $\dfrac{x^2+y^2+y}{x^2-y^2}$

(23) $\dfrac{2a^2-b^2}{a^2+ab}$

(24) $\dfrac{x^2+3x+1}{x}$

(25) $\dfrac{a^3+2a^2b+3ab^2+b^3}{a^2b+ab^2}$

(26) $\dfrac{2x+15}{x^2+15x}$

Page 262, Section 7.6

(1) $\dfrac{a}{3}$

(2) $-\dfrac{3}{a^2}$

(3) $\dfrac{7a - 3}{x - y}$

(4) $\dfrac{7z - 3y}{yz}$

(5) $\dfrac{4b - 3}{b^2}$

(6) $\dfrac{3x^2 - 5}{x^3}$

(7) $\dfrac{2x}{3y}$

(8) $\dfrac{4x - 2}{x}$

(9) $\dfrac{3x - 5y}{x^2 y}$

(10) $\dfrac{3b - 8a}{12ab}$

(11) $\dfrac{9}{10y}$

(12) $\dfrac{5x - 2z}{10y}$

(13) $\dfrac{13a}{18b}$

(14) $\dfrac{3 - 35a}{5a}$

(15) $-\dfrac{29}{6a}$

(16) $\dfrac{x^2 - 3x - 1}{x}$

(17) $\dfrac{-3a - 11b}{a^2 - b^2}$

(18) $\dfrac{-5x - 37}{x^2 + 4x + 5}$

(19) $\dfrac{y + 4}{y^2 + y}$

(20) $\dfrac{x^2 - 2xy - y^2}{x^2 - y^2}$

(21) $\dfrac{-2ab - b^2}{a^2 + ab}$

(22) $\dfrac{b^2}{a^2 + ab}$

(23) $\dfrac{1 - x^2 + 3x}{x}$

(24) $\dfrac{a^3 - 3ab^2 - b^3}{a^2 b + ab^2}$

(25) $T = \dfrac{1000}{R^2 + 10R}$

Page 264, Section 7.7

(1) $\dfrac{bx}{ay}$

(2) $\dfrac{3q}{4p}$

(3) $\dfrac{7r}{4s}$

(4) $2x$

(5) $\dfrac{2}{w}$

(6) $\dfrac{5}{6}$

(7) $\dfrac{x - 3}{2}$

(8) $\dfrac{1}{x + 3}$

(9) $\dfrac{2}{3}$

(10) $\dfrac{a - 1}{3b - 3}$

(11) $\dfrac{8a}{5b}$

(12) $\dfrac{8x}{5y}$

(13) $\dfrac{5b^2}{9a}$

(14) $15ay$

(15) $\dfrac{2a^2 + 4a}{3a - 6}$

(16) $\dfrac{10y}{a}$

(17) $\dfrac{ax + bx - ay - by}{ax + ay - bx - by}$

(18) $\dfrac{1}{x^2 - x - 2}$

(19) $\dfrac{4x}{a}$

(20) $\dfrac{5b^2 + 15b}{3bx + 12x}$

(21) $\dfrac{2a - 1}{x - 4y}$

(22) $\dfrac{5x - 2}{3a - 4b}$

(23) $\dfrac{(b - 4)(b - 6)(b + 1)}{(b + 2)(b + 2)(b - 2)}$

(24) 1

(25) $a = \dfrac{D}{t^2}$

Page 267, Section 7.8

(1) $\dfrac{5}{12}$

(2) $\dfrac{b}{a}$

(3) $\dfrac{2x}{3}$

(4) $\dfrac{x + y}{6xy}$

(5) $\dfrac{17}{6}$

(6) $\dfrac{6x + 3}{4}$

(7) $\dfrac{a^2 c + ab}{c}$

(8) $\dfrac{y + x}{y - x}$

(9) $\dfrac{wx + wy}{xy}$

(10) $\dfrac{axy + bxy}{x + y}$

(11) $\dfrac{ab + 1}{ab - 1}$

(12) $\dfrac{a + b}{ab}$

(13) $-\dfrac{1}{y}$

(14) $\dfrac{b + 1}{b^2 + 1}$

(15) $\dfrac{x^2 z + x^2 y}{y^2 z + xyz^2}$

(16) $\dfrac{b - 2a}{4ab + 3b}$

(17) $\dfrac{10x^2 + 4}{5x}$

(18) $\dfrac{4x}{10x^2 + 5}$

(19) $\dfrac{20x + 8}{x^3}$

(20) $\dfrac{20x}{8 - x^3}$

(21) $\dfrac{x + 1}{x - x^2}$

(22) $I = \dfrac{En}{rn + R}$

Page 270, Section 7.9

(1) $x = 2$

(2) $a = 6$

(3) $b = 38\frac{2}{11}$

(4) $a = -\frac{1}{6}$

(5) $a = \frac{5}{12}$

(6) $x = -10$

(7) $a = -7\frac{2}{3}$

(8) $a = 2$

(9) $x = 7$

(10) $x = 0$

(11) $x = \frac{1}{3}$

(12) $x = 4$

(13) $a = 5\frac{4}{7}$

(14) $b = -\frac{1}{7}$

(15) $x = 4\frac{4}{7}$

(16) $R_t = 3\frac{1}{13}$

(17) $x = 168$

(18) $\frac{7}{19}$

(19) 12 hr

(20) $342\frac{6}{7}$ mph, $267\frac{6}{7}$ mph

(21) $\frac{4}{5}$ mph

(22) $r_1 = 20, r_2 = 5$

(23) 40 mph, 50 mph

(24) $1\frac{1}{5}$ hr

(25) $7\frac{1}{2}$ hr

(26) 12 days

(27) 6 hr, 3 hr

Page 276, Section 7.10

(1) $P = 21$

(2) $h \approx 4$

(3) $g = 16$

(4) $p = 1913.386$

(5) $\ell = 20.5$

(6) $p = 5000$

(7) $h = 8$

(8) $S = 6$

(9) $d \approx 21$

(10) $r = .11$

(11) $w = 1.6$

(12) $t = 0$ or $t = 2$

(13) $d \approx 20$

(14) $R = \frac{E - e}{I}$

(15) $g = \frac{S}{t - \frac{1}{2}t^2}$

(16) $w = \frac{V}{\ell w}$

(17) $a = \frac{S - bh}{b + h}$

(18) $k = \frac{Fr^2}{mM}$

(19) $f = \frac{ts}{t + s}$

(20) $B = \frac{2A}{h} - b$

(21) $N = \frac{k}{T - S}$

(22) $D_2 = V(t_2 - t_1) + D_1$

(23) $m = \frac{y - b}{x}$

(24) $H = \frac{ST}{5252}$

CHAPTER 8

Page 284, Section 8.1

(1) $\frac{8 \text{ people}}{11 \text{ chairs}}$

(2) $\frac{1}{4}$

(3) $\frac{1}{2}$

(4) $\frac{1}{2}$

(5) $\frac{4}{5}$

(6) true

(7) false

(8) true

(9) false

(10) true

(11) $\frac{1 \text{ family}}{3 \text{ children}}$

(12) $\frac{5}{2}$

(13) $\frac{6}{7}$

(14) $\frac{6}{5}$

(15) $\frac{75 \text{ miles}}{1 \text{ hour}}$

(16) $\frac{\$.79}{1 \text{ pound}}$

(17) $\frac{1.3 \text{ television sets}}{1 \text{ house}}$

(18) $\frac{8 \text{ games won}}{5 \text{ games lost}}$

(19) $\frac{27}{2}$

(20) true

(21) true

(22) false

(23) true

(24) false

(25) true

(26) false

(27) true

(28) false

(29) false

(30) true

Page 288, Section 8.2

(1) $x = 16$

(2) $x = 30$

(3) $y = 3\frac{1}{3}$

(4) $y = 15$

(5) $w = 22.5$

(6) $w = 4$

(7) $R = 150$

(8) $R = 125$

(9) $x = 12$

(10) $x = 6$

(11) $y = 7.5$

(12) $y = \frac{3}{8}$

(13) $w = 1.5$

(14) $w = 19$

(15) $y = .01$

(16) $R = 60$

(17) $R = 5\frac{1}{2}$ or 5.5

(18) $x = .1$

(19) $x = 24$

(20) $y = .25$

(21) $y = 2\frac{13}{16}$ or 2.8125

(22) $x = 5\frac{2}{3}$

(23) $A = 2.17$

(24) $w = 7.65$

(25) $x = 234.375$ or $234\frac{3}{8}$

(26) 30

(27) 4.5

(28) 12

(29) 175 rpm

Page 292, Section 8.3

(1) .4 dram or $\frac{2}{5}$ dram

(2) 16 cm

(3) 57 rows

(4) 25 m

(5) $26\frac{2}{3}$ lb

(6) 10 teachers

(7) $1215

(8) 12 cans

(9) $136

(10) 780

(11) 63.4 gal

(12) 5.5 jobs

(13) 487 points

(14) no, $.69

(15) 1200 miles

(16) .4 cc

(17) $900

(18) .0034 inch

(19) 48 boys

(20) 18 lb cashews, 30 lb peanuts

(21) $40,045, $16,018, $16,018

Page 299, Section 8.4

(1) 116.64 dynes

(2) 1812 lb

(3) $4\frac{1}{6}$ hr or 4 hr 10 min

(4) $540

(5) $86

(6) $\frac{15}{22}$ ampere

(7) 237.6 lb

(8) 16 units

(9) $26\frac{2}{3}$ lb

(10) $60,000

(11) 20 hours

Page 304, Section 8.5

(1) 15%

(2) 56%

(3) 63%

(4) 27%

(5) 34%

(6) 160%

(7) 215%

(8) 8, 8%

(9) 48%

(10) 100%

(11) $10\frac{3}{5}$%

(12) $116\frac{1}{4}$%

(13) $77\frac{7}{9}$%

(14) 7%

(15) 10

(16) 62%

(17) 63%

(18) 5%

(19) 6%

(20) 5.25%

(21) 7%

Page 307, Section 8.6

(1) 36%	(6) .7%	(11) 29%	(16) 2700%	(21) 74.5%	(26) $3\frac{1}{3}$%
(2) 595%	(7) 2060%	(12) 74%	(17) 1321%	(22) 2.56%	(27) 575%
(3) 8%	(8) 1100%	(13) 21.4%	(18) $27\frac{2}{3}$%	(23) 40%	(28) $74\frac{1}{6}$%
(4) 833%	(9) 53.1%	(14) 8.3%	(19) .5%	(24) 320%	(29) 1.4%
(5) 160%	(10) 55%	(15) 700%	(20) 127%	(25) 185%	(30) .03%
					(31) 68%

Page 310, Section 8.7

(1) .14	(6) .0081	(11) .0215	(16) .0004	(21) 3.147	(26) $.35\frac{1}{6}$ or $.3516\frac{2}{3}$
(2) .27	(7) .16	(12) 3.12	(17) .00125	(22) 2.613	(27) $.048\frac{1}{3}$ or $.0483\frac{1}{3}$
(3) .923	(8) .059	(13) 5.63	(18) .008	(23) .0012	(28) $.00\frac{1}{4}$ or .0025
(4) .0279	(9) .82	(14) 1.106	(19) 1.35	(24) .005	(29) .01
(5) .03	(10) .36	(15) .537	(20) 1.12	(25) .00007	(30) 1
					(31) .17

Page 312, Section 8.8

(1) 7%	(6) 16%	(11) 115%	(16) $377\frac{7}{9}$%	(21) 33.3%	(26) 83.3%
(2) 50%	(7) 50%	(12) 105%	(17) $5\frac{1}{4}$%	(22) 1.4%	(27) 368.8%
(3) 22%	(8) 60%	(13) $66\frac{2}{3}$%	(18) $93\frac{1}{3}$%	(23) 1.3%	(28) .2%
(4) 30%	(9) 25%	(14) $16\frac{2}{3}$%	(19) $8\frac{1}{2}$%	(24) 255.6%	(29) 62.5%
(5) 85%	(10) 225%	(15) $187\frac{1}{2}$%	(20) $271\frac{3}{7}$%	(25) 38.1%	(30) 11.5%
					(31) 57.1%

Page 315, Section 8.9

(1) $\frac{1}{20}$	(5) $\frac{7}{10}$	(9) $1\frac{3}{25}$	(13) $\frac{23}{250}$	(17) $\frac{1}{300}$	(21) $\frac{1}{40}$
(2) $\frac{7}{20}$	(6) 4	(10) $\frac{13}{200}$	(14) $\frac{163}{500}$	(18) $\frac{1}{62500}$	(22) $\frac{41}{200}$
(3) $1\frac{1}{4}$	(7) $\frac{14}{25}$	(11) $\frac{3}{400}$	(15) $\frac{33}{200}$	(19) $\frac{1}{9}$	(23) $\frac{1}{8}$
(4) $\frac{1}{5}$	(8) $\frac{41}{50}$	(12) $\frac{13}{80}$	(16) $2\frac{17}{20}$	(20) $\frac{1}{400}$	(24) $\frac{4}{25}$
					(25) $\frac{9}{50}$

Page 317, Section 8.10

Percent	Fraction	Decimal
10%		.1
	$\frac{3}{10}$	.3
75%	$\frac{3}{4}$	
90%		.9
	$1\frac{9}{20}$	1.45
37.5% or $37\frac{1}{2}\%$		.375
.1%	$\frac{1}{1000}$	
100%	$\frac{1}{1}$ or 1	
225%		2.25
80%	$\frac{4}{5}$	
	$\frac{11}{200}$	.055
87.5% or $87\frac{1}{2}\%$	$\frac{7}{8}$	
	$\frac{1}{200}$	.005
60%	$\frac{3}{5}$	
	$\frac{5}{8}$	.625
	$\frac{3}{400}$	.0075
$66\frac{2}{3}\%$		$.66\frac{2}{3}$
	$\frac{1}{4}$	.25
20%	$\frac{1}{5}$	
	$\frac{2}{5}$	.4
	$\frac{1}{3}$	$.33\frac{1}{3}$
12.5% or $12\frac{1}{2}\%$	$\frac{1}{8}$	
70%		.7

Page 321, Section 8.11

(1) 36

(2) 80%
(3) 20%

(4) 28.12

(5) 5.58

(6) 2800
(7) 25%

(8) 150%

(9) 52.5

(10) .1%
(11) 10.29

(12) 100

(13) 100%

(14) .0224
(15) .0025

(16) 128.6

(17) 156.7%

(18) 18.3
(19) 219%

(20) $\frac{26}{75}$

(21) 1825

(22) $8.30\frac{1}{3}$

(23) 15.625
(24) 128.7%

(25) 627.27

(26) 1056 lb

Page 326, Section 8.12

(1) 42%
(2) $171.90
(3) 62.5%

(4) $703
(5) 10.3%
(6) $199.96

(7) 142,600
(8) $224
(9) 15%

(10) 16%
(11) $.40
(12) 50%

(13) $8960
(14) $47.92
(15) $.70
(16) 33%

(17) 7.2%
(18) 4761
(19) $13.64
(20) 59%

CHAPTER 9

Page 335, Section 9.1

(1) 1

(2) −1

(3) −1

(4) 3

(5) 3

(6) 9

(7) 8

(8) −5

(9) $\frac{8}{9}$

(10) $-\frac{4}{5}$

(11) −2

(12) 3

(13) 14

(14) $\frac{12}{13}$

(15) $\frac{1}{6}$

(16) 2.83

(17) 14

(18) 31

(19) 4.90

(20) 34.86

(21) 29.87

(22) 24.41

(23) 3.26

(24) 1.87

(25) .20

(26) 8 cm

(27) $5\frac{19}{49}$

(28) 24 ft^2

(29) 10

(30) 21
(31) 15.0 in

Page 338, Section 9.2

(1) 5
(2) 8
(3) 5

(4) 17
(5) 15
(6) 5

(7) 15
(8) 11.40
(9) 36

(10) 2.24
(11) 11.18
(12) yes

(13) no
(14) 13.4 ft
(15) 61.0 ft

(16) 12 ft
(17) 206.2 miles
(18) 127.3 ft

Page 342, Section 9.3

(1) $\sqrt{6}$
(2) $\sqrt{15}$
(3) $\sqrt{ab}$
(4) $\sqrt{10}$
(5) $\sqrt{21}$
(6) $\sqrt{6} + \sqrt{10}$

(7) $\sqrt{14}$
(8) $\sqrt{35}$
(9) $\sqrt{21} + \sqrt{30}$
(10) $\sqrt{70}$
(11) $\sqrt{74}$
(12) 10

(13) 14
(14) $\sqrt{105}$
(15) $\sqrt{xy}$
(16) $\sqrt{abc}$
(17) ab
(18) $\sqrt{15}$

(19) 15
(20) $3\sqrt{2} + \sqrt{6}$
(21) $\sqrt{xywz}$
(22) $6 - \sqrt{30}$
(23) $3\sqrt{6} + \sqrt{30}$
(24) $2\sqrt{3} - 2\sqrt{7}$

(25) $\sqrt{30} + \sqrt{105}$
(26) 14
(27) −4
(28) 2
(29) 0
(30) 2.1

Page 344, Section 9.4

(1) $2\sqrt{2}$
(2) $2\sqrt{3}$
(3) $2\sqrt[3]{2}$
(4) a
(5) $2\sqrt{5}$
(6) $2\sqrt[3]{3}$

(7) $x\sqrt{y}$
(8) $2\sqrt{7}$
(9) $4\sqrt{2}$
(10) $2\sqrt[3]{4}$
(11) $3\sqrt{3}$
(12) $2\sqrt[3]{5}$

(13) $2\sqrt[3]{7}$
(14) $2x\sqrt{3x}$
(15) $5a^2$
(16) $3ab\sqrt{2a}$
(17) $xy^2\sqrt[3]{x^2y^2z}$
(18) $5\sqrt{10}$

(19) $8\sqrt{2}$
(20) $4\sqrt[3]{2}$
(21) $2xy^2\sqrt{3xy}$
(22) $2ac^3\sqrt[3]{2a^2b^2}$
(23) $5\sqrt{3x}$
(24) $12\sqrt{2x}$
(25) $20\sqrt{14}$

Page 347, Section 9.5

(1) $\dfrac{\sqrt{2}}{2}$

(2) $\dfrac{\sqrt{3}}{3}$

(3) $\dfrac{\sqrt{6}}{3}$

(4) $\dfrac{2\sqrt{5}}{5}$

(5) $\dfrac{\sqrt{6}}{6}$

(6) $\dfrac{\sqrt{6}}{3}$

(7) $\dfrac{\sqrt{10}}{5}$

(8) $\dfrac{\sqrt{14}}{7}$

(9) $\dfrac{3-\sqrt{2}}{7}$

(10) $\sqrt{2}$

(11) $\dfrac{\sqrt{42}}{6}$

(12) $\dfrac{\sqrt{65}}{5}$

(13) $\dfrac{3\sqrt{2}}{4}$

(14) $\dfrac{\sqrt[3]{6}}{3}$

(15) $\dfrac{\sqrt{10}}{5}$

(16) $\dfrac{\sqrt{42}}{8}$

(17) $\dfrac{\sqrt{3}}{9}$

(18) $\dfrac{\sqrt[3]{10}}{2}$

(19) $\dfrac{\sqrt{x}}{x}$

(20) $\dfrac{3\sqrt[3]{x}}{x}$

(21) $\dfrac{\sqrt{xy}}{y}$

(22) $\sqrt{3}+\sqrt{2}$

(23) $\dfrac{8+2\sqrt{3}}{13}$

(24) $\dfrac{9+9\sqrt{3}}{-2}$

(25) $\dfrac{\pi\sqrt{15}}{2}$

Page 349, Section 9.6

(1) $2\sqrt{2}$

(2) $3\sqrt{2}$

(3) $2\sqrt[3]{2}$

(4) $3\sqrt{5}$

(5) $3\sqrt{2}$

(6) 7

(7) $\sqrt{7}$

(8) $5\sqrt{2}$

(9) $3\sqrt{2}$

(10) $3\sqrt{6}$

(11) $-3\sqrt[3]{5}$

(12) $5\sqrt{2}$

(13) $8\sqrt{x}$

(14) $5\sqrt{2}$

(15) $\sqrt{7}$

(16) $8\sqrt[3]{2}$

(17) $10a\sqrt{ab}\ -\ 5a\sqrt[3]{a}$

(18) $-3x\sqrt[3]{x}$

(19) $2\sqrt{2}\ +\ 4\sqrt[3]{5}$

(20) $13\sqrt{2}\ -\ 8\sqrt{3}$

(21) $4\sqrt{x}\ -\ 2\sqrt[3]{x}$

(22) $-4\sqrt{6x}$

(23) $2\sqrt{6}\ -\ 4\sqrt{2}$

(24) $8x\sqrt{xy}$

(25) $\dfrac{3\sqrt{2}}{2}$

(26) $\dfrac{7\sqrt{3}}{3}$

(27) $\dfrac{4\sqrt{6}}{9}$

(28) $\dfrac{125\sqrt{2}}{2}$

(29) $\dfrac{1025\sqrt{3}}{4}$ yd² ≈ 443.8 yd²

Page 352, Section 9.7

(1) $2^{\frac{1}{2}}$

(2) $4^{\frac{1}{3}}$

(3) $6^{\frac{1}{2}}$

(4) $a^{\frac{2}{3}}$

(5) $\sqrt[3]{a}$

(6) $\sqrt[3]{9}$

(7) $\sqrt{a}$

(8) $\sqrt{5}$

(9) $\sqrt{ab}$

(10) $7^{\frac{1}{2}}$

(11) $2^{\frac{1}{3}}$

(12) $x^{\frac{1}{2}}$

(13) $y^{\frac{1}{3}}$

(14) $57^{\frac{1}{2}}$

(15) $a^{\frac{1}{2}}b^{\frac{1}{2}}$

(16) $x^{\frac{1}{3}}y^{\frac{1}{3}}$

(17) $x^{\frac{2}{3}}$

(18) $\sqrt{6}$

(19) $\sqrt[3]{12}$

(20) $\sqrt[3]{x}$

(21) $\sqrt{y}$

(22) $\sqrt{7b}$

(23) $x\sqrt[3]{y}$

(24) $\sqrt[3]{rs}$

(25) $a\sqrt{b}$

(26) $\sqrt[3]{a^2}$

(27) $\sqrt[3]{x^2y^2}$

(28) $\sqrt[3]{xy^2}$

(29) $(.32\,A)^{\frac{1}{2}}$

Page 355, Section 9.8

(1) a^{-1}

(2) $2a^{-2}$

(3) $5a^{-3}$

(4) $2x^{-3}$

(5) $\frac{1}{4}$

(6) $\frac{1}{a}$

(7) $\frac{1}{x^2}$

(8) $\frac{4}{a}$

(9) $\frac{1}{4a}$

(10) 4^{-1}

(11) x^{-1}

(12) $7x^{-2}$

(13) $(x+y)^{-1}$

(14) $a^{-1}+b^{-1}$

(15) $2(x+y)^{-1}$

(16) $5(x+y+z)^{-1}$

(17) xy^{-1}

(18) $a^{\frac{1}{2}}bc^{-2}$

(19) $(a+b)^{-\frac{1}{2}}$

(20) $\frac{1}{x}$

(21) $\frac{1}{a^2}$

(22) $\frac{1}{a^3}$

(23) $\frac{1}{(x+y)^{\frac{2}{3}}}$

(24) $\frac{2}{x^2}$

(25) y

(26) $\frac{a^{\frac{1}{2}}}{b^{\frac{1}{2}}}$

(27) xy

(28) $\frac{x+y}{x}$

(29) x^2

(30) $\frac{k}{d^2}$

(31) $\frac{1}{e^{x^2}}$

Page 358, Section 9.9

(1) 2.11×10^5

(2) 3.25×10^6

(3) 1.21×10^{-3}

(4) 3.894×10^7

(5) 6.82×10^{-5}

(6) 210

(7) $.321$

(8) 4310

(9) $.0214$

(10) 2.765×10^3

(11) 1.82×10^5

(12) 4.86×10^6

(13) 2.76×10^{-3}

(14) 1.35×10^{-6}

(15) 1.24×10^{-5}

(16) 3.69×10^{-4}

(17) 7.623×10^{-1}

(18) 1.235×10^9

(19) 3.876×10^8

(20) 6850

(21) $.00276$

(22) $.0000007962$

(23) $382,000$

(24) $.00000009863$

(25) 9.3×10^7 miles

CHAPTER 10

Page 364, Section 10.1

(1) ± 5

(2) ± 9

(3) ± 11

(4) ± 40

(5) $\pm\sqrt{17}$

(6) No real solution

(7) $\pm\sqrt{59}$

(8) $\pm\sqrt{7}$

(9) ± 4

(10) $\pm\sqrt{2}$

(11) $\pm\sqrt{38}$

(12) 0

(13) ± 4

(14) $\frac{\sqrt{35\pi}}{10\pi}$ ft $\approx .33$ ft

(15) 46 ft

(16) 24 ft

(17) 32 ft

Page 368, Section 10.2

(1) $6, -3$

(2) $4, -10$

(3) $-7, -5$

(4) $-2, -9$

(5) $-\frac{2}{3}, \frac{3}{2}$

(6) $7, -\frac{1}{4}$

(7) $-\frac{3}{5}, \frac{7}{4}$

(8) $-6, \frac{1}{2}$

(9) $\frac{1}{4}, \frac{1}{3}$

(10) $\frac{1\pm\sqrt{6}}{2}$

(11) No real roots

(12) $1\pm\sqrt{17}$

(13) No real roots

(14) $\frac{5\pm\sqrt{65}}{2}$

(15) $\frac{1\pm\sqrt{57}}{2}$

(16) $\frac{2\pm\sqrt{10}}{2}$

(17) $\frac{3\pm\sqrt{29}}{10}$

(18) No real roots

(19) $\frac{-1\pm\sqrt{222}}{17}$

(20) $5, -\frac{9}{7}$

(21) $\frac{75\pm\sqrt{921}}{16}$ sec ≈ 2.79 sec or 6.58 sec

(22) $-\frac{3}{5}$ or $\frac{5}{3}$

(23) 20 ft by 32 ft

(24) $\frac{5+\sqrt{1001}}{2}$ in by $\frac{11+\sqrt{1001}}{2}$ in ≈ 18.32 in by 21.32 in

(25) $\frac{8+2\sqrt{17}}{2}$ sec ≈ 8.12 sec

Page 371, Section 10.3

(1) $5, -1$

(2) No real roots

(3) $6, -3\frac{1}{3}$

(4) $\pm 2\sqrt{5}$

(5) $0, -\frac{1}{2}$

(6) $-1\frac{2}{3}, \frac{1}{2}$

(7) $\dfrac{1 \pm \sqrt{109}}{6}$

(8) $\dfrac{-6 \pm 2\sqrt{3}}{3}$

(9) $5, -4$

(10) $\pm\sqrt{29}$

(11) $5 \pm 2\sqrt{7}$

(12) $\pm\sqrt{51}$

(13) No real roots

(14) $1, -\frac{4}{5}$

(15) $1, -1\frac{2}{3}$

(16) $8, -5$

(17) $\dfrac{-7 \pm \sqrt{57}}{4}$

(18) $-5, 6$

(19) $4 \pm 2\sqrt{21}$

(20) $5, 3\frac{2}{3}$

(21) -6

(22) 5

(23) $\dfrac{9 \pm \sqrt{161}}{2}$

(24) $-1, -8$

(25) 6 in by 8 in

(26) 12 cm by 15 cm

(27) $30 \pm 6\sqrt{21} \approx$ 57.5 ft or 2.5 ft

(28) 17 cm by 20 cm

(29) 9 sec

Page 377, Section 10.4

(1) 64

(2) 0

(3) No real solution

(4) 9

(5) $\dfrac{1}{9}$

(6) 16

(7) 36

(8) 12

(9) No real solution

(10) -4

(11) $6, 5$

(12) 8

(13) 18

(14) 18

(15) No real solution

(16) $24\frac{9}{16}$

(17) 9

(18) No real solution

(19) 10

(20) $8, -3$

(21) 9

(22) $-3900 + 30\sqrt{16901}$ mi $\approx$ .115 mi $\approx$ 609.2 ft

(23) 4.68 in

CHAPTER 11

Page 385, Section 11.1

(1) through (10)

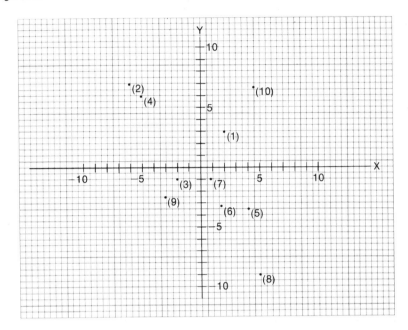

(11) (A) $(1, 1)$ (B) $(-1, 4)$ (C) $(-4, 0)$ (D) $(0, -3)$ (E) $(5, -4)$

(F) $(2\frac{1}{2}, 2)$ (G) $(-2\frac{1}{2}, -\frac{1}{2})$ (H) $(2, -2)$ (I) $(-4, -3\frac{1}{2})$ (J) $(-3, 3)$

(12)   Mazama

Page 388, Section 11.2

(1)

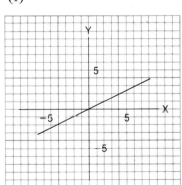

(2)

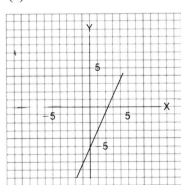

(3)

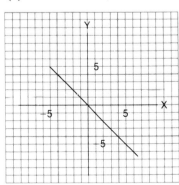

(4)

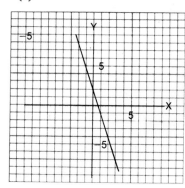

(5)

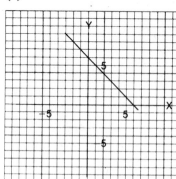

(6)

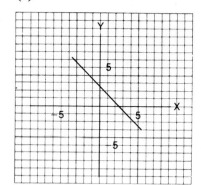

(7)

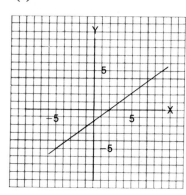

(8)

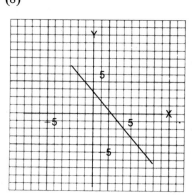

(9)

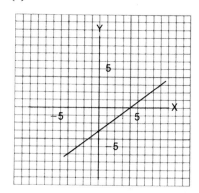

(10)

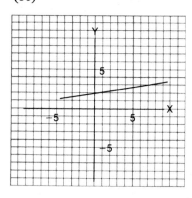

(11)

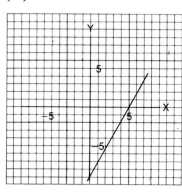

(12)

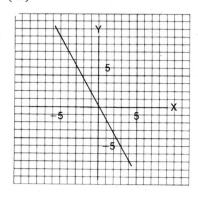

(13)

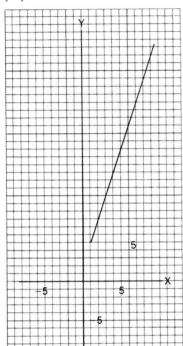

(14)

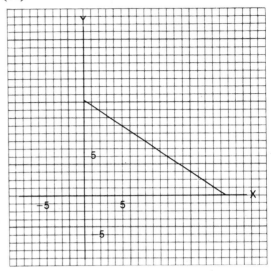

(15)

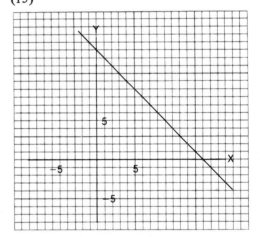

Page 393, Section 11.3

(1) $(2, 0), (0, 5)$

(2) $(3, 0), (0, -14)$

(3) $(2\frac{1}{2}, 0), (0, -3)$

(4) $(0, 0), (0, 0)$

(5) $(\frac{1}{2}, 0), (0, -\frac{1}{9})$

(6) $(-2, 0), (0, 11)$

(7) 1

(8) $-\frac{3}{8}$

(9) 1

(10) $\frac{1}{5}$

(11) $-\frac{5}{8}$

(12) -1

(13) $\frac{3}{8}, (0, -2)$

(14) $-\frac{7}{3}, (0, 5)$

(15) $\frac{1}{8}, (0, -\frac{7}{8})$

(16) $\frac{2}{11}, (0, 0)$

(17) $5, (0, \frac{2}{3})$

(18) $-1, (0, 12)$

(19) $\frac{3}{5}$

Page 396, Section 11.4

(1) 5

(2) $3\sqrt{13}$

(3) $\sqrt{26}$

(4) 5

(5) $\sqrt{36.36} \approx 6.03$

(6) $\sqrt{.5} \approx .71$

(7) $\sqrt{290}$

(8) $2\sqrt{29}$

(9) $10, 4\sqrt{5}, 10$

(10) $4\sqrt{10}$

(11) $2\sqrt{5}$

(12) 40 sq units

Page 399, Section 11.5

(1)

(2)

(3)

(4)

(5)

(6)

(7)

(8)

(9)

(10)

(11)

(12)

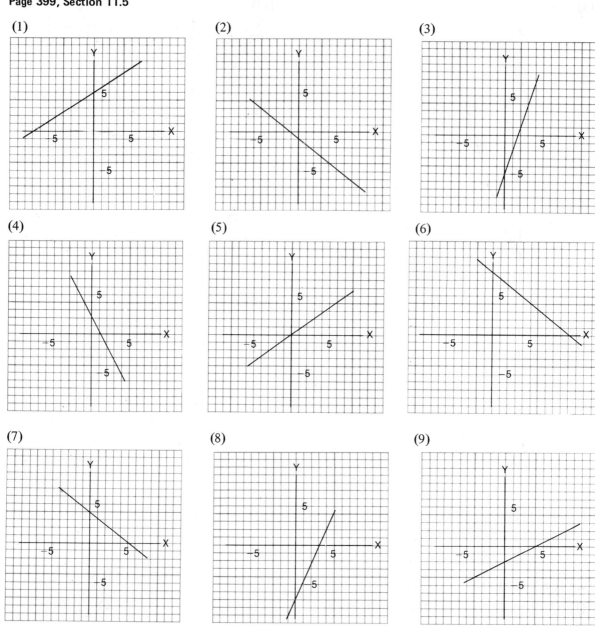

Page 405, Section 11.6

(1)

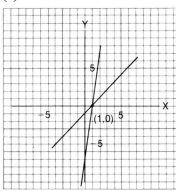

(2)

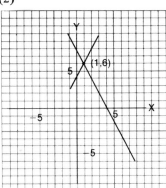

(3)

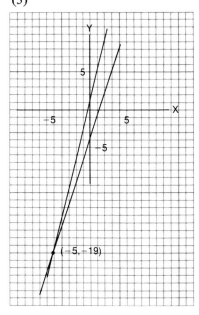

(4)

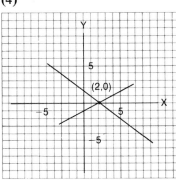

(5)

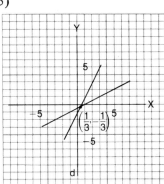

(6)

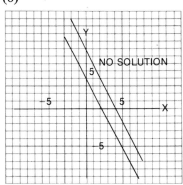

(7)

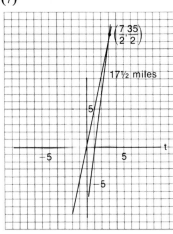

(8)

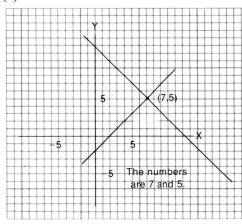

Page 409, Section 11.7

(1) $(4, 7)$

(2) $(-1, 6)$

(3) $(2, 2)$

(4) $(4, -1)$

(5) $(4, 3)$

(6) $(1\frac{7}{13}, -1\frac{1}{13})$

(7) $(-3\frac{2}{3}, -4\frac{1}{3})$

(8) $(0, 4)$

(9) $(32, 3)$

(10) $(\frac{1}{3}, -\frac{2}{3})$

(11) No real solution

(12) $(2, -6)$

(13) $(2, 1)$

(14) $(2, -3)$

(15) 71 g of liver, 16 g of bacon

(16) The numbers are 3 and 4

(17) 25 nickels, 55 dimes

(18) 5 oz of 85%, 10 oz of 70%

(19) 75 gal of 10%, 25 gal of 50%

(20) 8.57 kg of 30% zinc, 3.43 kg of pure zinc

Page 414, Section 11.8

(1) $(-2, 10)$

(2) $(-1, -3)$

(3) $(2, -1)$

(4) $(-5, 5)$

(5) $(-2\frac{15}{32}, -3\frac{3}{32})$

(6) $(25, 19)$

(7) $(-2, 3)$

(8) $(2, 15)$

(9) $(4, 1)$

(10) $(-\frac{1}{2}, \frac{5}{8})$

(11) $(1\frac{3}{10}, \frac{4}{5})$

(12) $(5\frac{1}{5}, -4\frac{4}{5})$

(13) Infinite number of solutions (dependent system)

(14) No solution

(15) 100 lb of 70%, 500 lb of 40%

(16) $(53, 31)$

(17) Larger width 9 cm, smaller width 7 cm

(18) $1200 at 6%, $1800 at 7.5%

(19) 150 mph, 30 mph

(20) 275 g of fruit, 155 g of cereal

(21) 36 quarters, 12 dimes

CHAPTER 12

Page 426, Section 12.1

(1) $90°$

(2) $90°$

(3) $80°$

(4) $10°$

(5) $90°$

(6) $69°$

(7) $98°$

(8) $29°$

(9) $5°$

(10) $107.1°$

(11) $60.2°$

(12) $6.4°$

(13) $28.8°$

(14) $38.1°$

(15) $94.7°$

(16) $60.5°$

(17) $48°$

Page 432, Section 12.2

(1) $\operatorname{Sin} A = \frac{4}{5}$

$\operatorname{Cos} A = \frac{3}{5}$

$\operatorname{Tan} A = \frac{4}{3}$

(2) $\operatorname{Sin} A = \frac{15}{39}$

$\operatorname{Cos} A = \frac{36}{39}$

$\operatorname{Tan} A = \frac{5}{12}$

(3) $\operatorname{Sin} A = \frac{\sqrt{2}}{2}$

$\operatorname{Cos} A = \frac{\sqrt{2}}{2}$

$\operatorname{Tan} A = 1$

(4) $\operatorname{Sin} A = \frac{5}{13}$

$\operatorname{Cos} A = \frac{12}{13}$

$\operatorname{Tan} A = \frac{5}{12}$

(5) $\operatorname{Sin} T = \frac{4}{5}$

$\operatorname{Cos} T = \frac{3}{5}$

$\operatorname{Tan} T = \frac{4}{3}$

(6) $\operatorname{Sin} T = \frac{9}{10}$

$\operatorname{Cos} T = \frac{\sqrt{19}}{10}$

$\operatorname{Tan} T = \frac{9\sqrt{19}}{19}$

(7) $\operatorname{Sin} T = \frac{2\sqrt{13}}{13}$

$\operatorname{Cos} T = \frac{3\sqrt{13}}{13}$

$\operatorname{Tan} T = \frac{2}{3}$

(8) $\operatorname{Sin} T = \frac{\sqrt{2}}{2}$

$\operatorname{Cos} T = \frac{\sqrt{2}}{2}$

$\operatorname{Tan} T = 1$

(9) $\operatorname{Sin} D = \dfrac{2\sqrt{2}}{3}$

 $\operatorname{Cos} D = \dfrac{1}{3}$

 $\operatorname{Tan} D = 2\sqrt{2}$

(10) $\operatorname{Sin} D = \dfrac{\sqrt{10}}{10}$

 $\operatorname{Cos} D = \dfrac{3\sqrt{10}}{10}$

 $\operatorname{Tan} D = \dfrac{1}{3}$

(11) $\operatorname{Sin} D = \dfrac{3\sqrt{10}}{10}$

 $\operatorname{Cos} D = \dfrac{\sqrt{10}}{10}$

 $\operatorname{Tan} D = 3$

(12) $\operatorname{Sin} D = \dfrac{\sqrt{5}}{3}$

 $\operatorname{Cos} D = \dfrac{2}{3}$

 $\operatorname{Tan} D = \dfrac{\sqrt{5}}{2}$

(13) (a)   9%

 (b)   .09

(14) .8250

Page 436, Section 12.3

(1) .3584	(6) .8171	(11) .6921	(17) $75.0°$	(23) $53.1°$
(2) .2756	(7) .8942	(12) .3486	(18) $65.1°$	(24) $59.5°$
(3) 1.0000	(8) .4327	(13) .1477	(19) $66.1°$	(25) $20.1°$
(4) .9205	(9) .9842	(14) $51.6°$	(20) $65.0°$	(26) $80.0°$
(5) .1228	(10) .9917	(15) $32.2°$	(21) $25.0°$	(27) 192.70 ohms
		(16) $19.5°$	(22) $42.2°$	(28) $39.5°$

Page 441, Section 12.4

(1) $f \approx 10.4$
 $d \approx 12.0$
 $F = 60.0°$

(2) $p \approx 14.3$
 $q \approx 10.1$
 $R = 45.0°$

(3) $c \approx 7.1$
 $B \approx 44.4°$
 $C \approx 45.6°$

(4) $c \approx 45.0$
 $A \approx 56.2°$
 $B \approx 33.8°$

(5) $r \approx 27.8$
 $t \approx 72.1$
 $T = 68.9°$

(6) $y \approx 404.7$
 $z \approx 266.1$
 $X = 48.9°$

(7) $18.8°$
(8) 141.8 ft
(9) 56.1 ft
(10) 4998.5 ft
(11) 53.2 ft
(12) $36.0°$
(13) 86.4 ft

Page 445, Section 12.5

(1) .7660	(4) .7986	(7) −.2419	(10) −.6494	(13) −6.1066	(17) .4555
(2) −.6428	(5) −.7071	(8) .9063	(11) .8508	(14) −2.9887	(18) −.8156
(3) −1.1918	(6) −7.1154	(9) −.9511	(12) −.7096	(15) .8780	(19) −.6743
				(16) −.3600	(20) −.6428

Page 447, Section 12.6

(1) $142.2°$	(4) $152.4°$	(7) $121.6°$	(10) $96.4°$	(13) $117.9°$	(17) $139.2°$
(2) $104.8°$	(5) $114.2°$	(8) $156.9°$	(11) $120.3°$	(14) $114.0°$	(18) $135.5°$
(3) $140.5°$	(6) $95.6°$	(9) $102.6°$	(12) $159.8°$	(15) $117.1°$	(19) $152.9°$
				(16) $96.3°$	(20) $167.9°$

Page 452, Section 12.7

(1) $77°$
(2) 70 miles

(3) 157 ft
(4) 578 ft

(5) $A \approx 51°, B \approx 90°$
(6) 128 knots

(7)   1320 ft
(8)   149 lb
(9) $70.6°$

Page 460, Section 12.8

(1) 18 ft
(2) A is closer by 105 yards
(3) 320 m

(4) 3200 ft
(5) 85 kg
(6) 26 ft

(7) 473 ft
(8) $113°, R \approx 54$ lb

INDEX

Measurement, 119–171
    cubic, 158
    of angles, 423–426
Metric denominate numbers, 123
    and equivalents, 124
Minuend, 5
Minute, 424
Mixed numbers, 35–81
    addition of, 64
    changing to decimals, 109
    division of, 57
    multiplication of, 53
    subtraction of, 71
Monomials, 200–205
Multiplication, associative law of, 202
    by power of ten, 357
    commutative law of, 202
    of binomials, 219
    of decimals, 99
    of fractions, 40, 53
    of mixed numbers, 53
    of monomials, 202
    of negative numbers, 183
    of polynomials, 218–224
    of radical expressions, 341
    of rational expressions, 247–252
    of signed numbers, 183
    of square roots, 341
    of whole numbers, 8–10
Multiplication property, of one, 306
Multiplication table, 9

Natural number, 13
Negative exponents, 353
Negative numbers, 173
    addition of, 177
    division of, 185
    multiplication of, 183
Negative square root, 333
Notation, scientific, 356
Numbers, decimal, 83–118. See also *Decimal numbers.*
    denominate, 119–127. See also *Denominate numbers.*
    mixed, 35–81. See also *Mixed numbers.*
    negative, 173. See also *Negative numbers.*
    positive, 173
    signed, 173–192. See also *Signed numbers.*
    whole, 1–34. See also *Whole numbers.*
Numeral form, 1
    of decimals, 83
Numerator, 35

Oblique triangle, 449
Obtuse angles, 424
    and trigonometric ratios of, 444–446
Operations, order of. See *Order of operations.*
Order of operations, 15
    for decimals, 113
    for fractions, 76
    for mixed numbers, 76
    for signed numbers, 188
    for whole numbers, 15
Ordered pairs, 381, 388
Ordinate, 381, 391

Parallelogram, area of, 144, 146
Parentheses, removing from polynomials, 207
Partial quotient, 8
Percent, 301–330
    changing decimals to, 306
    changing fractions to, 311
    changing to decimals, 306
    changing to fractions, 314
Percent comparison, 301
Perfect cube, 333
Perfect square, 333
Perimeter, 137
Place value, of whole numbers, 1
Place value notation, 356
Polynomials, 193–224
    addition of, 206
    combining, 205
    definition of, 205
    division of, 224
    factoring of, 228
        difference of two squares, 236
        trinomials, 231
    multiplication of, 218–224
    removing parentheses from, 207
    subtraction of, 206
Positive numbers, 173
Positive square root, 333
Power, of exponent, 13
Prime factorization, 26
Prime number, 24
Product, 40
    of polynomials, 218–224
    of whole numbers, 8
Proper fraction, 35
Proportion(s), 281
    solving, 286
    true, 281, 282
Pyramid, volume of, 160
Pythagorean theorem, 336–338

Quadrants, 381
Quadratic equations, 239, 363–377
    solved by quadratic formula, 366
    solved by square roots, 363–365
Quadratic formula, 366
Quotient, 8

Radical expressions. See *Square roots.*
Radicals, definition of, 333, 341
    in equations, 374–377
    isolating, 375
Radius, 137, 351
Ratio, 281
    trigonometric, 428–432, 434–436
Rational expressions, 247–278
    addition of, 249, 258
    complex, 266
    division of, 263
    multiplication of, 247–252
    reducing, 249
    renaming, 249
    subtraction of, 249, 261
Rationalizing denominator, 345, 346
Ray, 423
Real roots, 366